TRANSFORMERS

TRANSFORMERS

Bharat Heavy Electricals Limited

McGraw-Hill

New York Chicago San Francisco Lisbon London Madrid
Mexico City Milan New Delhi San Juan Seoul
Singapore Sydney Toronto

The **McGraw·Hill** Companies

Cataloging-in-Publication Data is on file with the Library of Congress.

Copyright © 2005 by the McGraw-Hill Companies, Inc. All rights reserved. Printed in the United States. Except as permitted under the United States Copyright Act of 1976, no part of this publication may be reproduced or distributed in any form or by any means, or stored in a data base or retrieval system, without the prior written permission of the publisher.

1 2 3 4 5 6 7 8 9 0 DOC/DOC 0 1 0 9 8 7 6 5 4

ISBN 0-07-144785-7

First published in India by Tata McGraw-Hill. Copyright © 2003, 1987 by Bharat Heavy Electrical Limited, Piplani, Bhopal, MP.

The sponsoring editor for this book was Stephen S. Chapman and the production supervisor was Sherri Souffrance. The art director for the cover was Anthony Landi.

Printed and bound by RR Donnelley.

 This book was printed on recycled, acid-free paper containing a minimum of 50% recycled, de-inked fiber.

McGraw-Hill books are available at special quantity discounts to use as premiums and sales promotions, or for use in corporate training programs. For more information, please write to the Director of Special Sales, McGraw-Hill Professional, Two Penn Plaza, New York, NY 10121-2298. Or contact your local bookstore.

Information contained in this work has been obtained by the McGraw-Hill Companies Inc. ("McGraw-Hill") from sources believed to be reliable. However, neither McGraw-Hill nor its authors guarantee the accuracy or completeness of any information published herein, and neither McGraw-Hill nor its authors shall be responsible for any errors, omissions, or damages arising out of use of this information. This work is published with the understanding that McGraw-Hill and its authors are supplying information but are not attempting to render engineering or other professional services. If such services are required, the assistance of an appropriate professional should be sought.

To
The Transformer Designers
at
BHEL

Contents

Preface	*xv*
Editorial Committee	*xvii*

1. **Introduction**   1
   - *T.K. Ganguli*
2. **Principles of Transformers**   6
   - *R.K. Singh*
   - 2.1 Working Principle of a Transformer   *6*
   - 2.2 Elementary Theory of an Ideal Transformer   *7*
   - 2.3 EMF Equation of a Transformer   *8*
   - 2.4 Voltage Transformation Ratio (K)   *9*
   - 2.5 Ideal Transformer   *10*
   - 2.6 Transformer Having Winding Resistance but No Magnetic Leakage   *13*
   - 2.7 Magnetic Leakage   *14*
   - 2.8 Transformer with Resistance and Leakage Reactance   *15*
   - 2.9 Equivalent Circuit   *16*
   - 2.10 Regulation   *21*
   - 2.11 Losses in Transformers   *21*
   - 2.12 Efficiency   *22*
   - 2.13 The Auto-Transformer   *23*
   - 2.14 Three-Winding Transformer   *25*
   - 2.15 Parallel Operations of Transformers   *27*
   - *Appendix*   *32*
   - *References*   *33*
3. **Materials Used in Transformers**   34
   - *M.P. Singh, T.K. Ganguli*
   - 3.1 Insulating Oil   *34*
   - 3.2 Insulating Paper   *47*
   - 3.3 Pressboard   *51*

- 3.4 Wood *58*
- 3.5 Insulated Copper Conductor for Windings *59*
- 3.6 Crepe Paper Covered Flexible Copper Cable *66*
- 3.7 Sealing Materials *66*
- 3.8 Cold Rolled Grain-Oriented Electrical Steel Sheet (CRGO) *68*
- 3.9 Structural Steel *81*
- 3.10 Future Trends *84*
    - References *85*

4. Magnetic Circuit 86
- *K.N. Labh, R.C. Agarwal*
- 4.1 Material *86*
- 4.2 Design of Magnetic Circuit *87*
- 4.3 Optimum Design of Core *91*
- 4.4 Manufacturing *99*
    - References *107*

5. Windings and Insulation 108
- *M.V. Prabhakar, S.K. Gupta*
- 5.1 Types of Windings *108*
- 5.2 Surge Voltage Behavior of Windings *116*
- 5.3 Internal Heat Transfer in Windings *125*
- 5.4 Insulation Design *126*
- 5.5 Electric Field Plotting *130*
- 5.6 Finite-Difference Method (FDM) *131*
    - References *136*

6. Voltage Regulation and Tapchanger 138
- *B.L. Rawat, A.K. Ekka*
- 6.1 Off-Circuit Tapchanger *138*
- 6.2 On-Load Tapchanger (OLTC) *141*
- 6.3 Constructional and Operational Features of OLTC *148*
- 6.4 Manual and Electrical Operation of Tapchanger *150*
- 6.5 Automatic Control of Tapchanger *151*
- 6.6 Tapchanger Selection *151*
- 6.7 Latest Trends in Tapchanger Design *152*
    - Reference *153*

7. Electromagnetic Forces in Power Transformers 154
- *M.V. Prabhakar, T.K. Ganguli*
- 7.1 Leakage Flux in a Typical Two Winding Transformer *154*
- 7.2 Nature of Forces *154*

Contents   ix

    7.3 Basic Formula and Methods for Force Evaluation   *158*
    7.4 Radial Forces   *159*
    7.5 Axial Forces   *160*
    7.6 Roth's Method of Force Calculation   *163*
    7.7 Modes of Failure of Windings and Design of Windings   *166*
    7.8 Strengthening of Coils to Withstand Short-Circuit Forces   *168*
    7.9 Design of Clamping Structures   *169*
    7.10 List of Symbols   *174*
        References   *176*

8. Cooling Arrangements   177
- *C.M. Sharma*
    8.1 Various Types of Cooling   *177*
    8.2 Cooling Arrangements   *179*
    8.3 Propeller Type Fans   *185*
    8.4 Transformer Oil Pump   *188*
    8.5 Flow Indicators   *190*
    8.6 Stress and Hydraulic Analysis of Pipework   *191*

9. Design Procedure   198
- *R.C. Agarwal*
    9.1 Specifications of a Transformer   *198*
    9.2 Selection of Core Diameter   *198*
    9.3 Selection of Flux Density   *199*
    9.4 Selection of Type of Core   *200*
    9.5 Selection of Leg Length   *200*
    9.6 Selection of Type of Windings   *201*
    9.7 Selection of Tapchanger   *202*
    9.8 Calculation of Number of Turns   *203*
    9.9 Selection of Conductor and Current Density   *203*
    9.10 Insulation Design   *204*
    9.11 Calculation of Lateral and Axial Dimensions of Coils   *204*
    9.12 Ampere-Turn Balancing   *205*
    9.13 Reactance Calculation   *205*
    9.14 Iron Weight and Losses   *210*
    9.15 Copper Weight and Losses   *210*
    9.16 Stray Losses in Transformer   *211*
    9.17 Stray Loss Control   *214*
    9.18 Impulse Calculation   *218*
    9.19 Mechanical Forces in Windings   *218*
    9.20 Temperature Gradient and Cooling Calculation   *219*

    9.21 Typical Design Calculations for Two Winding and Auto-transformers *219*
           References *226*

10. Structural Design of Transformer Tank   227
• *M.K. Shakya, S.G. Bokade*
  10.1 Types of Tank Constructions *227*
  10.2 Structural Design of Transformer Tanks *230*
  10.3 Testing of Tanks *244*
       References *250*

11. Transformer Auxiliaries and Oil Preservation Systems   251
• *S.C. Verma, J.S. Kuntia*
     Transformer Auxiliaries *251*
  11.1 Gas Operated (Buchholz) Relay *251*
  11.2 Temperature Indicators *253*
  11.3 Pressure Relief Valve *255*
  11.4 Oil Level Indicator *256*
  11.5 Bushing and Cable Sealing Box *257*
     Transformer Oil Preservation Systems *258*
  11.6 Silicagel Breather *260*
  11.7 Gas Sealed Conservators *260*
  11.8 Thermosyphon Filters *263*
  11.9 Bellows and Diaphragm Sealed Conservators *264*
  11.10 Refrigeration Breathers *265*

12. Manufacturing and Assembly   267
• *T.K. Ganguli, M.V. Prabhakar*
  12.1 Core Building *267*
  12.2 Preparation of Windings *268*
  12.3 Winding Assembly *269*
  12.4 Core and Winding Assembly *271*
  12.5 Terminal Gear Assembly *271*
  12.6 Placement of Core and Winding Assembly in Tank *272*
  12.7 Processing *273*
  12.8 Servicing of Transformer *273*
  12.9 Tanking *274*

13. Drying and Impregnation   275
• *M.P. Singh, M.V. Prabhakar*
  13.1 Basic Principles of Drying *276*
     Principles of Drying *277*
  13.2 Conventional Vacuum Drying *278*

13.3 Vapor Phase Drying *281*
   References *290*

14. Testing of Transformers and Reactors 291
- *P.C. Mahajan, M.L. Jain, R.K. Tiwari*
   Section I *292*
   14.1 Testing of Power Transformers *292*
   Section II *305*
   14.2 Impulse Testing *305*
   References *320*
   Section III *321*
   14.3 Partial Discharge Testing *321*
   References *334*
   Section IV *335*
   14.4 Testing of Reactors *335*
   References *349*
   Section V *350*
   14.5 Short Circuit Testing of Power Transformers *350*

15. Standards on Power Transformers 358
- *V.K. Lakhiani, S.K. Mahajan*
   15.1 First Revision of IS: 2026 *359*
   15.2 Other Related Standards *364*
   15.3 New Standards *367*
   15.4 Standard Specification of a Power Transformer *368*
   Appendices *372*

16. Loading and Life of Transformers 377
- *D.P. Gupta*
   16.1 Life of a Transformer *378*
   16.2 Ageing of Insulation *379*
   16.3 Law of Insulation Ageing *381*
   16.4 Significance of Weighted Value of Ambient Temperature *384*
   16.5 Relationship between Weighted Ambient, Winding Rise and Hot-spot Temperatures *386*
   16.6 Determination of Weighted Ambient Temperature *394*
   16.7 Relationship between Weighted Ambient and Load *399*
   16.8 Alternative Approach for the Calculation of Weighted Ambient *400*
   16.9 Transformer Loading Guides *407*
   16.10 Loading by Hot-Spot Temperature Measured by WTI *409*

16.11 Selection and Use of a Transformer *411*
References *413*
Appendices *414*

17. Erection and Commissioning 439
- *C.M. Shrivastava*
17.1 Dispatch *439*
17.2 Inspection upon Arrival at Site *440*
17.3 Handling *440*
17.4 Installation *441*
17.5 Commissioning *447*
17.6 Maintenance *449*
17.7 Dos for Power Transformer *451*
17.8 Don'ts for Power Transformers *453*
17.9 Dos and Don'ts for HV Condenser Bushings *454*

18. Transformer Protection 456
- *B.L. Rawat*
18.1 Protection against External Faults *456*
18.2 Protection against Internal Faults *459*
References *465*

19. Reactors 466
- *C.M. Shrivastava, S.K. Mahajan*
19.1 Series Reactors *466*
19.2 Shunt Reactor *473*
19.3 Neutral Earthing Reactor *482*
19.4 Tuning for Filter Reactors *482*
19.5 Arc Suppression Reactors *483*
19.6 Earthing Transformers (Neutral Couplers) *483*
19.7 Standards on Reactors *484*

20. Traction Transformers 486
- *J.M. Malik*
20.1 Types of Traction Transformers *486*
20.2 Special Considerations *487*
20.3 Design and Constructional Features *489*
20.4 Traction Transformer for Thyristor Controlled Locomotives *493*
References *495*

21. Rectifier Transformers 496
- *J.S. Sastry*
21.1 Comparison between Rectifier Transformer and Power Transformer *497*
21.2 Rectifier Circuits *499*

21.3 Design Features of Rectifier Transformers *503*
21.4 Transductors *517*
21.5 Constructional Features of Rectifier Transformers *518*
21.6 Tank Design *521*
21.7 Testing *523*
References *525*

22. Convertor Transformers for HVDC Systems — 526
- *I.C. Tayal, C.M. Sharma, S.C. Bhageria*
22.1 Insulation Design *526*
22.2 Higher Harmonic Currents *529*
22.3 DC Magnetization *529*
22.4 DC Bushings *530*
22.5 On Load Tapchanger *530*
22.6 Influence of Impedance Variation *530*
22.7 Connections *531*
22.8 Specifications *533*
22.9 Manufacturing Features *533*
22.10 Tests *536*
References *538*

23. Controlled Shunt Reactor — 539
- *S.C. Bhageria, J.S. Kuntia*
23.1 Controlled Shunt Reactor (CSR) Principle *539*
23.2 Controlled Shunt Reactor Transformer (CSRT) *541*
23.3 Special Features of the Controlled Shunt Reactor Transformer *545*
23.4 Controlled Shunt Reactor Much More Than a Shunt Reactor *549*
23.5 Conclusion *550*
References *550*

24. Designing and Manufacturing — A Short-Circuit Proof Transformer — 551
- *T.K. Ganguli, S.K. Gupta*
24.1 Forces During Short-Circuit *552*
24.2 Various Considerations to Design a Transformer Suitable for Short-Circuit Duty *556*
24.3 Manufacturing Aspects *559*
24.4 Quality Aspects *560*
24.5 Conclusion *560*

25. High Voltage Condenser Bushings — 562
- *R.K. Agarwal, Assem Dhamija*
25.1 Introduction *562*

- 25.2 Classification of Bushings *562*
- 25.3 Design of Bushing *565*
- 25.4 Constructional Details and Main Parts of Bushing *567*
- 25.5 Testing of Bushing *571*
- 25.6 Factors Affecting the Performance of Bushing *573*
- 25.7 Condition Monitoring at Site *574*
- 25.8 Dos and Don'ts for HV Condenser Bushings *575*

26. Computerization—A Tool to Enhance Engineering Productivity   **577**
- *R. Mitra, DGM/TRE*

27. Condition Monitoring, Residual Life Assessment and Refurbishment of Transformers   **581**
- *C.M. Shrivastava, T.S.R. Murthy*
  - **Section A** *581*
  - 27.1 Analysis Method *582*
  - 27.2 Physical Inspection *584*
  - 27.3 Residual Life Assessment (RLA) *588*
  - 27.4 Different Techniques for Life Estimation *589*
  - 27.5 Methodology Adopted *591*
  - 27.6 Conventional and Special Tests on Oil Samples *592*
  - 27.7 Refurbishment *593*
  - 27.8 Conclusion *596*
  - Appendix I *597*

28. Transformers: An Overview   **598**
- *S.N. Roy, P.T. Deo*

Solved Examples   **603**

Index   **609**

Preface

Starting from the fundamentals of design, manufacturing technology, materials used, etc. the book provides an in-depth knowledge and understanding of the intricacies of transformers, including the various aspects of latest technological upgradation. Solved examples have also been included in the book to illustrate better the design aspects of a power transformer.

With these enhancements, the book will serve as a useful reference book for practising engineers engaged in design, manu-facturing, planning, testing, erection, operation and maintenance of transformers as well as for students and researchers.

This book is a collective effort of experts in the field of transformers at Bharat Heavy Electricals Limited (BHEL), a leading transformer manufacturer in India. My thanks are due to all the authors who contributed the various chapters to this book. The editorial committee headed by Shri T. K. Ganguli, an eminent transformer designer, went over the complete material to integrate it into a homogeneous book and ensure that all relevant aspects as per the latest practices have been included. To integrate the entire material and present it in a standard format was an uphill task. Our grateful thanks are due to the editorial committee members who put in exemplary efforts.

S.N. Roy
Executive Director
BHEL, Bhopal

Editorial Committee

Shri T. K. Ganguli
Shri M. V. Prabhakar
Shri S. K. Gupta
Shri R. K. Singh
Shri K. Gopal Krishnan

About BHEL

Bharat Heavy Electricals Limited (BHEL) is the largest engineering and manufacturing enterprise of its kind in India and is one of the leading international companies in the field of power equipment manufacture. The first plant of BHEL was set up at Bhopal in 1956, and currently its range of services extend from project feasibility studies to after-sales service, successfully meeting diverse needs through turnkey capability. The company has 14 manufacturing units, 4 power sector regional centres, 8 service centres and 18 regional offices besides project sites spread all over India and abroad.

BHEL manufactures over 180 products under 30 major product groups and caters to core sectors of the Indian economy—Power generation and transmission, industry, transportation, telecommunication, renewable energy, etc. BHEL has acquired certifications of both ISO 9000 and ISO 14000 standards for its operations and has also adopted the concepts of Total Quality Management. BHEL has adopted Occupational Health and Safety Standards as per OHSAS 18001. Two of its divisions have acquired certification to OHSAS 18001 standard and other units, too, are in the process of acquiring the same.

BHEL has

- Installed equipment for over 62,000 MW of power generation—for Utilities, Captive and Industrial users.
- Supplied 2,00,000 MVA transformer capacity and sustained equipment operating in transmission and distribution network up to 400 kV—AC and DC.
- Successfully tested transformers of rating as high as 200 MVA, 1-ph generator transformer and 167 MVA, 1-ph auto transformer at an independent lab in Europe.
- Supplied over 25,000 motors with drive control system to power projects, petrochemicals, refineries, steel, aluminium, fertilizer, cement plants, etc.
- Supplied traction electrics and AC/DC locos to power over 12,000 km railway network.
- Supplied over one million valves to power plants and other industries.

TRANSFORMERS

CHAPTER 1

Introduction

T. K. Ganguli

The history of transformer goes back to the early 1880s. With the sharp increase in demand for electric power, power transformers in 400 kV ratings were produced as early as 1950. In the early 1970s unit ratings as large as 1100 MVA were produced and 800 kV and even higher kV class transformers were manufactured in the early 1980s.

A transformer is a static piece of equipment with a complicated electromagnetic circuit inside. The energy is transferred from one electrical circuit to another through the magnetic field. In its simplest form, a transformer consists of two conducting coils having a mutual inductance. In an ideal case it is assumed that all the flux linked with the primary winding also links the secondary winding. But, in practice it is impossible to realize this condition as magnetic flux cannot be confined. The greater portion of the flux flows in the core while a small portion called the leakage flux links one or the other winding. Depending upon the particular application and type of connection, a transformer may have additional windings apart from the two conventional windings. Chapter 2 deals with the principles of transformers mainly covering the basic electromagnetic force (emf) equation, ideal transformer, equivalent circuit, calculation of regulation, no load and load losses, efficiency and parallel operation of transformers.

Major materials like copper, cold-rolled grain oriented silicon steel, insulating oil, pressboard and paper insulation and certain ferrous and nonferrous items are essential to build a compact and trouble-free transformer. Chapter 3 has been primarily devoted to explain the characteristics of these materials which shall be helpful in selecting the correct material for the equipment.

Designing an insulation system for application in higher voltage class transformer is an art and with the use of the best materials available today, it is possible to economize on size as well as produce a reliable piece of equipment. Keeping in view the transportability, operational limitations and guaranteed technical performance of the transformer, particular type of core construction is adopted. Chapter 4 is devoted to design of magnetic circuit, constructional features, manufacturing, assembly and finally fitting core in the tank.

Windings and insulation arrangement has been covered in Chapter 5. Starting from spiral and layer type of windings, more emphasis is given to helical and disc types which are more commonly used in the latest design practices. The impulse voltage withstand behavior of an ordinary disc winding can be enhanced by interleaving the disc winding. The initial voltage distribution at the line end should be taken care of for high voltage windings. The transient voltage distribution and internal heat transfer in the windings have been described in detail. The insulation design becomes more complex as we move towards higher and higher voltage class of transformer windings. Proper sizing and routing can be further examined by detailed electrostatic field plots. To vary the voltage in a transformer, tap changers are used which have a different type of regulating winding connection, viz. linear, reversing, coarse fine, etc. Chapter 6 describes the tap changer types, constructional features and its control.

Designing a transformer to cater to electromagnetic forces has been covered in Chapter 7. Radial and axial forces occurring during short circuit or line faults can be calculated and winding design can be finalized to withstand these forces. The clamping structures put the coils under a pressure, higher than that produced by short circuit forces. An effort has been made to describe the method for calculating stresses and then dimensioning the clamping structures adequately. The cooling of a transformer becomes more relevant from the point of view of ageing of insulation system and ensuring longer life due to less thermal degradation. Chapter 8 covers various types of cooling, its arrangement and design calculation.

The design procedure described in Chapter 9 gives every detail of a transformer starting from the selection of core-size, winding conductor, reactance calculation and then realization of main guaranteed parameters like percentage impedance, no load loss, load loss, etc. Estimation and control of stray losses, winding gradient has also been described. Typical design calculation for two winding and auto transformers has been covered in brief. Structural design is described in Chapter 10 which gives an idea of stresses that are developed in main tank and other supporting structures.

Proper selection of transformer auxiliaries is essential for ensuring safe operation of the main equipment and provides protection under fault conditions. Chapter 11 covers the major auxiliaries like gas operated buchholz relay, temperature indicators, pressure relief valve, bushings, cable box, oil preservation system, etc. Assembly and manufacturing aspects have been covered in Chapter 12. After the winding and core assembly is completed, drying and impregnation is a vital process that is described in Chapter 13. Cellulose insulation used in power transformers and reactors has approximately 6 to 10% of moisture by weight at ambient temperature, being a hygroscopic material. Vacuum drying and vapor phase drying is an important tool to extract moisture from the insulation items. Uniform heating of the entire active part (core and winding) mass must be ensured so that shrinkage and moisture extraction is optimum.

To ensure quality and conformation to design calculations, testing is an important activity in the transformer. The basic testing requirement and testing codes are set out in national and international standards. Chapter 14 is intended to cover the purpose and methodology of performing the tests. Chapter 15 covers the various standards generally used for transformers. Depending upon the consumer requirement, sometimes the transformers are loaded beyond its nameplate rating for a brief period. Chapter 16 covers the loading and life of a transformer.

Chapter 17 describes some of the main precautions which must be taken during erection and commissioning of a transformer. Maintenance schedule has also been discussed so that continuous trouble free service could be ensured. Inspection upon arrival at site, installation, oil filling, drying of transformer and analysis of gases of power transformers, etc., have been explained for the benefit of the users. The transformer being a vital equipment, its protection is equally important. Some of the basic protection schemes have been dealt with in Chapter 18. Reactors are usually classified according to duty application, viz. current limiting, neutral earthing, shunt, smoothing, etc. These reactors have typical characteristic requirements and call for different constructions, viz. with air core or with gapped iron core for fixed and variable reactance. Chapter 19 deals in detail with the design, and construction, etc., of different types of reactors. Chapters 20 and 21 mainly deal with special types of transformers. For chemical plants, aluminum plants, etc., where electrolytic processes are adopted and the direct current requirement is quite large, rectifier transformers are used. Traction transformers require special considerations for their design due to a limitation of space availability and problems due to vibration. They have to be designed and manufactured to stringent specifications, so as to withstand heavy stresses in this type of application.

The growth of HVDC system for transporting power to long distances calls for discussing transformers for such application, too. Chapter 22 covers the basic aspects of a converter transformer, which is one of the most expensive equipment in HVDC system. Controlled shunt reactors can be employed in place of fixed shunt reactors and Chapter 23 is fully devoted to this new area. The advantages of controlled shunt reactors are mainly fully controllable reactive power, reduced dynamic over voltages, increased power carrying capacity of lines, faster responses, etc. Chapter 24 describes the criteria for designing and manufacturing a short circuit proof transformer. Steady increase in unit ratings of transformers and simultaneous growth of short circuit levels of network have made the short circuit withstand capability of the transformer one of the key aspects of its design. Additionally, aspects like material selection, selection of fittings and manufacturing processes, etc., are to be carefully examined in detail. Chapter 25 covering high voltage condenser bushing has been specially designed to cover application guidelines, use of insulating material, design, construction and testing, etc. The factors affecting the performance of bushing and preventive measures like condition monitoring, etc., have been dealt with in detail. In the present day scenario, it is not possible to function without the application of computer aided design. Accordingly Chapter 26 gives an idea for catering to the high demand for tailormade transformer products and the ways to overcome this problem.

Introduction

The need for a reliable and stable system is being increasingly felt. Thus Chapter 27 is devoted to condition monitoring, residual life assessment and refurbishment of the transformer. The residual life assessment study is carried out to predict the health of the transformer insulation and remaining life of the transformer. Cellulose insulation degrades due to heating or electrical breakdown, resulting in the production of furfural derivatives which dissolve in oil. Hence, chemical analysis of the oil gives evidence of changes that are taking place in the winding insulation during operation. Due to paucity of funds, it is not practical and economical to replace old units with new ones. In such cases, refurbishing/retrofitting is an economical and viable alternative. Finally, Chapter 28 provides an overview of the transformer and describes in general the salient points which contribute an essential part in producing the equipment. Although the transformer is a complex piece of equipment mainly produced manually, efforts on various fronts, viz. design, manufacture, vendor development, process mapping and maintaining history card, etc., contribute as important tools for producing a reliable piece of transformer.

This book is an attempt to include in one volume the various major aspects related to a power transformer. An effort has been made to give a comprehensive and up-to-date coverage of the latest technologies and developments in the field of power transformers. Finally, some solved examples have also been included at the end of the book to aid better understanding especially for students and practicing engineers. The topics covered in the book shall be helpful for design, manufacture, testing, erection, commissioning and maintenance of power transformers in every corner of the world.

Chapter 2

Principles of Transformers

R.K. Singh

2.1 Working Principle of a Transformer

A transformer is a static piece of apparatus used for transferring power from one circuit to another without change in frequency. It can raise or lower the voltage with a corresponding decrease or increase in current. In its simplest form, a transformer consists of two conducting coils having a mutual inductance. The primary is the winding which receives electric power, and the secondary is the one which may deliver it. The coils are wound on a laminated core of magnetic material.

The physical basis of a transformer is mutual inductance between two circuits linked by a common magnetic flux through a path of low reluctance as shown in Fig. 2.1.

The two coils possess high mutual inductance. If one coil is connected to a source of alternating voltage, an alternating flux is set up in the laminated core, most of which is linked up with the other coil in which it produces mutually induced emf (electromotive force) according to Faraday's laws electromagnetic induction, i.e.

$$e = M \frac{di}{dt}$$

where, e = induced emf
M = mutual inductance

If the second circuit is closed, a current flows in it and so electric energy is transferred (entirely magnetically) from the first coil (primary winding) to the second coil (secondary winding).

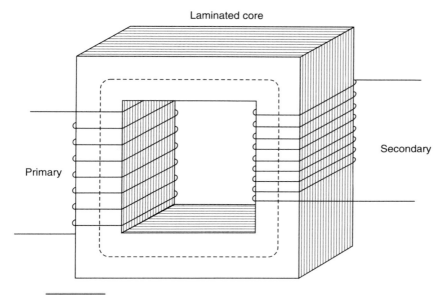

Figure 2.1 *Diagram showing magnetic circuit and windings of a transformer.*

2.2 Elementary Theory of an Ideal Transformer

An ideal transformer is one which has no losses, i.e. its windings have no ohmic resistance and there is no magnetic leakage. In other words, an ideal transformer consists of two coils which are purely inductive and wound on a loss-free core.

It may, however, be noted that it is impossible to realize such a transformer in practice, yet for convenience, we will first analyze such a transformer and then an actual transformer.

Consider an ideal transformer [Fig. 2.2(a)] whose secondary is open and whose primary is connected to a sinusoidal alternating voltage v_1. Under this condition, the primary draws current from the source to build up a counter electromotive force equal and opposite to the applied voltage.

Since the primary coil is purely inductive and there is no output, the primary draws the magnetizing current I_μ only. The function of this current is merely to magnetize the core, it is small in magnitude and lags v_1 by 90°. This alternating current I_μ produces an alternating flux ϕ which is proportional to the current and hence is in phase with it. This changing flux is linked with both the windings. There-

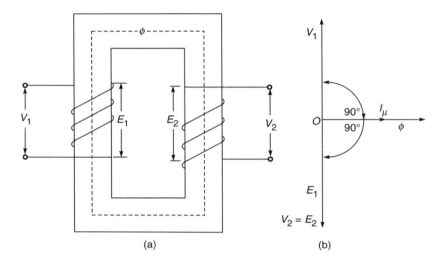

Figure 2.2 *Vectorial representation of induced emf of an ideal transformer under no-load.*

fore, it produces self-induced emf in the primary. This self-induced emf e_1 is, at any instant, equal to and in opposition to v_1. It is also known as counter emf of the primary.

Similarly in the secondary winding, an induced emf e_2 is produced which is known as mutually induced emf. The emf is in phase opposition with v_1 and its magnitude is proportional to the rate of change of flux and the number of secondary turns. Figure 2.2(b) shows the vectorial representations of the above quantities.

2.3 EMF Equation of a Transformer

Let N_1 = Number of turns in primary
N_2 = Number of turns in secondary
ϕ_m = Maximum flux in the core in webers
B_m = Flux density in weber/sq m (Tesla)
A = Net cross-sectional area of core in sq m
f = Frequency of ac input in Hz
v_1 = Instantaneous value of applied voltage in primary winding in volts
v_{1m} = Maximum value of applied voltage in volts.

Principles of Transformers

The instantaneous value of counterelectromotive force e_1 is

$$e_1 = -N_1 \frac{d\phi}{dt} \text{ volt} \qquad (2.1)$$

As discussed above in Sec. (2.2) the counter emf e_1 is equal and opposite to applied voltage v_1, i.e.

$$v_1 = N_1 \frac{d\phi}{dt}$$

If the applied voltage is sinusoidal, that is

$$v_1 = v_{1m} \sin 2\pi f t$$

Then $\phi = \phi_m \sin 2\pi f t$

Hence $e_1 = -N_1 \phi_m \cos 2\pi f t \times 2\pi f$

These equations are expressed as vectors as shown in Fig. 2.2(b), where V_1 and E_1 are the rms values of v_1 and e_1. To obtain the rms value of counter emf e_1, divide its maximum value given above by $\sqrt{2}$.

Then

$$E_1 = \frac{2\pi}{\sqrt{2}} f N_1 \phi_m \qquad (2.1.1)$$

The cosine term has no significance except to derive the instantaneous values.

i.e. $E_1 = 4.44 f N_1 \phi_m$

or $E_1 = 4.44 f N_1 B_m A \qquad (2.2)$

Similarly rms value of emf induced in secondary is

$$E_2 = 4.44 f N_2 B_m A \qquad (2.3)$$

In an ideal transformer

$$V_1 = E_1$$

and $V_2 = E_2$

where V_2 is the secondary terminal voltage [Fig. 2.2(b)].

2.4 Voltage Transformation Ratio (K)

From Eqs. (2.2) and (2.3), we get

$$\frac{E_2}{E_1} = \frac{N_2}{N_1} = K \qquad (2.4)$$

This constant is known as voltage transformation ratio.

(a) If $N_2 > N_1$, i.e., $K > 1$, then the transformation is called a step-up transformer.

(b) If $N_2 < N_1$, i.e., $K < 1$, then the transformer is called a step-down transformer.

Again for an ideal transformer

$$\text{Input} = \text{Output}$$
$$V_1 I_1 = V_2 I_2 \text{ (neglecting } I_\mu\text{)}$$

or
$$\frac{I_2}{I_1} = \frac{V_1}{V_2} = \frac{1}{K} \qquad (2.5)$$

where I_1 and I_2 are primary and secondary currents.

Hence the currents are in the inverse ratio of the transformation ratio.

2.5 Ideal Transformer

We will consider two cases,
(a) when such a transformer is on no-load and
(b) when it is loaded.

(a) Transformer on No-load

The primary input current under no-load condition has to supply (i) iron-loss in the core, i.e., hysteresis loss and eddy current loss and (ii) a very small amount of copper-loss in primary. Hence the no-load primary input current I_0 is not at 90° behind v_1 but lags it by an angle θ_0 which is less than 90°. No-load primary input power $W_0 = V_1 I_0 \cos \theta_0$. No-load condition of an actual transformer is shown vectorially in Fig. 2.3.

As seen from Fig. 2.3, primary current I_0 has two components.

(i) One in phase with V_1. This is known as active or working or iron-loss component I_w, because it supplies the iron-loss plus a small quantity of primary Cu-loss.
$$I_w = I_0 \cos \theta_0 \qquad (2.6)$$

(ii) The other component is in quadrature with V_1 and is known as magnetizing component because its function is to sustain the alternating flux in the core. It is wattless.
$$I_\mu = I_0 \sin \theta_0 \qquad (2.7)$$

Obviously I_0 is the vector sum of I_w and I_μ, hence
$$I_0 = \sqrt{(I_\mu^2 + I_w^2)} \qquad (2.8)$$

The no-load primary current I_0 is very small as compared to full-load primary current. As I_0 is very small, hence no-load primary copper-

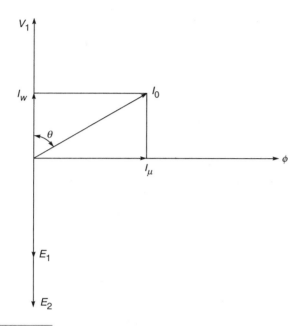

Figure 2.3 *No-load current of an ideal transformer (having core losses only).*

loss is negligibly small which means that no-load primary input is practically equal to the iron-loss in a transformer.

(b) *Transformer on-load*

When the secondary is loaded, secondary current I_2 is set up. The magnitude of I_2 is determined by the characteristic of the load. The secondary current sets up its own mmf ($= N_2 I_2$) and hence its own flux ϕ_2 which is in opposition to the main primary ϕ, which is due to I_0. The opposing secondary flux ϕ_2 weakens the primary flux momentarily and primary back emf E_1 tends to reduce. For a moment V_1 gains the upper hand over E_1 and hence causes more current (I'_2) to flow in primary.

The current I'_2 is known as load component of primary current. This current is in phase opposition to current I_2. The additional primary mmf $N_2 I'_2$ sets up a flux ϕ'_2 which opposes ϕ_2 (but is in the same direction as ϕ) and is equal to it in magnitude. Thus, the magnetic effects of secondary current I_2 get neutralized immediately by additional primary current I'_2. The whole process is illustrated in Fig. 2.4. Hence, whatever may be the load conditions, the net flux

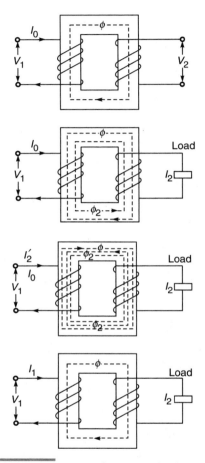

Figure 2.4 *Transformer on-load.*

passing through the core is approximately the same as at no-load. Due to this reason the core-loss is also practically the same under all load conditions.

As
$$\phi_2 = \phi_2'$$
$$N_2 I_2 = N_1 I_2'$$
$$I_2' = \frac{N_2}{N_1} \times I_2 = K I_2$$

Hence, when transformer is on-load, the primary winding has two currents I_0 and I_2' (which is antiphase with I_2 and K times its magnitude). The total primary current is the vector sum of I_0 and I_2'. In

Fig. 2.5 are shown the vector diagrams for a loaded transformer. In Fig. 2.5(a), current I'_2 is in phase with E_2 (for non-inductive loads). In Fig. 2.5(b), it is lagging behind E_2 (for inductive loads).

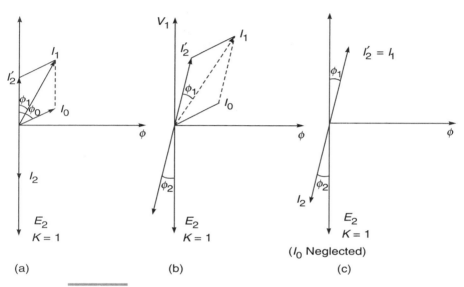

Figure 2.5 *Vector diagram of an ideal transformer.*

If we neglect I_0 as compared to I'_2 as shown in Fig. 2.5(c), then $\phi_1 = \phi_2$ and thus

$$N_1 I'_2 = N_1 I_1 = N_2 I_2$$
$$\frac{I_1}{I_2} = \frac{N_2}{N_1} = K$$

It shows that under load conditions, the ratio of primary and secondary currents is constant.

2.6 Transformer Having Winding Resistance but No Magnetic Leakage

An ideal transformer was supposed to possess no resistance but an actual transformer has primary and secondary windings with some resistances. Due to these resistances, there is some voltage drop in the two windings. The result is that:

(a) The secondary terminal voltage V_2 is equal to the vector difference of the secondary induced emf E_2 and $I_2 R_2$ where R_2 is the resistance of the secondary winding.
$$V_2 = E_2 - I_2 R_2 \qquad (2.9)$$
(b) Similarly primary induced emf E_1 is equal to the vector difference of V_1 and $I_1 R_1$ where R_1 is the resistance of the primary winding.
$$E_1 = V_1 - I_1 R_1 \qquad (2.10)$$

The vector diagrams for non-inductive, inductive and capacitive loads are shown in Fig. 2.6(a), (b) and (c) respectively.

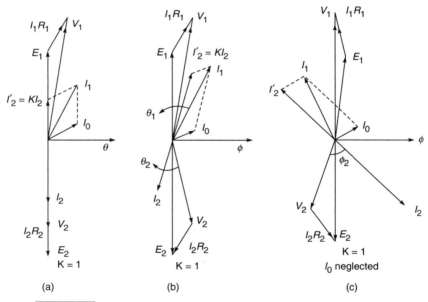

Figure 2.6 *Vector diagram of an ideal transformer (having core losses and winding resistance).*

2.7 Magnetic Leakage

In an ideal case it is assumed that all the flux linked with the primary winding also links the secondary winding. But, in practice it is impossible to realize this condition as magnetic flux cannot be confined. The greater portion of the flux (i.e., the mutual flux) flows in the core while a small proportion (Fig. 2.7) called the leakage flux links one or the other winding, but not both. On account of the leakage flux, both the primary and secondary windings have leakage reactance, that is, each will become the seat of an emf of self-induction, of a magnitude equal to a small fraction of the emf due to main flux. The terminal voltage V_1 applied to the primary must, therefore, have a component $I_1 X_1$ (where X_1 is leakage reactance of primary) to balance the primary leakage emf. In the secondary, similarly, an emf of self-induction $I_2 X_2$ (where X_2 is leakage reactance of secondary) is developed. The primary and secondary coils in Fig. 2.7 are shown on

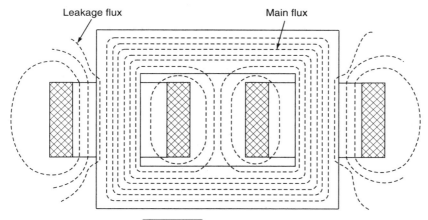

Figure 2.7 *Leakage flux.*

separate limbs, an arrangement that would result in an exceptionally large leakage. Leakage between primary and secondary could be eliminated if the windings could be made to occupy the same space. This, of course, is physically impossible, but an approximation to it is achieved if the coils of primary and secondary are placed concentrically. Such an arrangement leads to a marked reduction of the leakage reactance. If on the other hand, the primary secondary are kept separate and widely spaced, there will be much room for leakage flux and the leakage reactance will be greater.

2.8 Transformer with Resistance and Leakage Reactance

Figure 2.8 shows the primary and secondary windings of a transformer with resistance and leakage reactances taken out of the

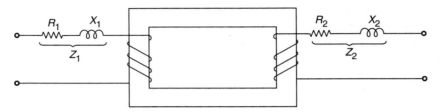

Figure 2.8 *Resistances and leakage reactances.*

windings. The primary impedance is given by

$$Z_1 = \sqrt{(R_1^2 + X_1^2)}$$

And the secondary impedance is given by

$$Z_2 = \sqrt{(R_2^2 + X_2^2)}$$

$$V_1 = E_1 + I_1(R_1 + jX_1) = E_1 + I_1 Z_1 \quad (2.11)$$
and
$$E_2 = V_2 + I_2(R_2 + jX_2) = V_2 + I_2 Z_2 \quad (2.12)$$

The vector diagram of such a transformer for different kinds of loads is shown in Fig. 2.9. In these diagrams vectors for resistive drops are drawn parallel to current vectors, whereas reactive drops are perpendicular to the current vectors. The angle θ_1 between V_1 and I_1 gives the power-factor angle of the transformer.

2.9 Equivalent Circuit

The transformer shown in Fig. 2.10(a) can be represented by an equivalent circuit in which the resistance and leakage reactance of the transformer are imagined to be external to the winding. The no-load current I_0 is simulated by a pure inductance X_0 taking the magnetizing component I_μ and a noninductive resistance R_0 taking the active component I_w connected in parallel across the primary circuit as shown in Fig. 2.10(b).

To make the transformer calculations simpler, it is preferable to transfer voltage, current and impedance to the primary side.

The primary equivalent of the secondary induced voltage is

$$E_2' = E_2/K = E_1$$

(where K is the transformation ratio)

Principles of Transformers

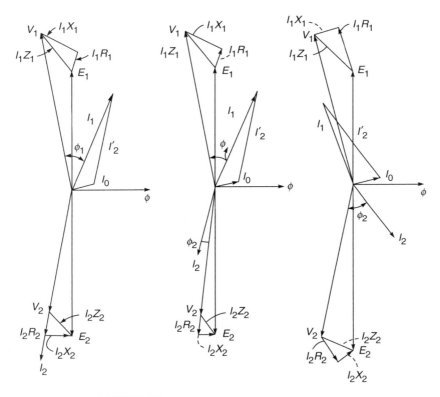

Figure 2.9 *Vector diagram of transformer.*

Similarly, primary equivalent of secondary terminal or output voltage is
$$V_2' = V_2/K$$
Primary equivalent of secondary current is
$$I_2' = I_2 K$$
and
$$R_2' = R_2/K^2,$$
$$X_2' = X_2/K^2,$$
$$Z_2' = Z_2/K^2$$

The same relationship is used, for shifting an external load impedance to primary. The secondary circuit is shown separately in Fig. 2.11(a) and its equivalent primary values in Fig. 2.11(b).

The equivalent circuit with secondary parameters transferred on primary side is given in Fig. 2.12. This is known as exact equivalent

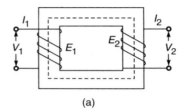

(a)

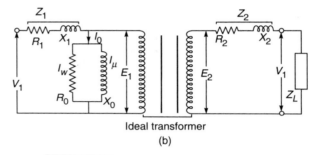

Ideal transformer
(b)

Figure 2.10 *Parameters of transformer.*

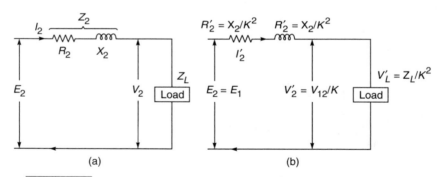

Figure 2.11 *Secondary resistance and reactance referred to primary.*

circuit. A simplification of the equivalent circuit can be further made by transferring the exciting circuit across the input terminals as shown in Fig. 2.13 or in Fig. 2.14.

It should be noted that in this case $R_0 = V_1/I_w$ and $X_0 = V_1/I_\mu$. Further simplification may be achieved by omitting I_0 altogether as shown in Fig. 2.15(a).

Equivalent circuit for a three-winding transformer described in cl. 2.14 is given in Fig. 2.15(b) as a star network.

Principles of Transformers

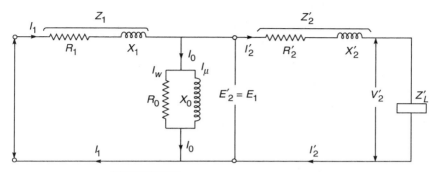

Figure 2.12 *Exact equivalent circuit.*

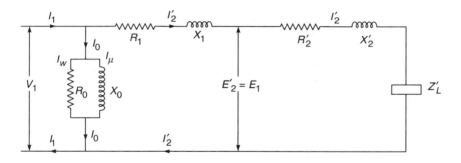

Figure 2.13 *Simplified equivalent circuit.*

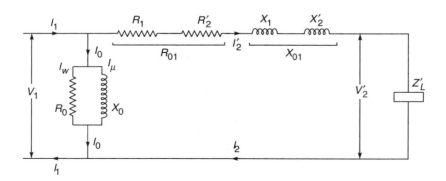

Figure 2.14 *Simplified equivalent circuit.*

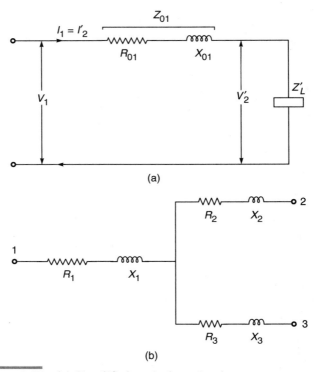

Figure 2.15 *(a) Simplified equivalent circuit.*
(b) Equivalent circuit for three-winding transformer.

From Fig. 2.12 it is found that total impedance between the input terminals is

$$Z = \left[z_1 + \cfrac{1}{\cfrac{1}{z_2' + z_1'} + \cfrac{1}{z_m}} \right] \quad (2.13)$$

where $Z_2' = R_2' + jX_2'$
Z_m = impedance of the exciting circuit
    = $R_0 + jX_0$

$$V_1 = I_1 \left[Z_1 + \cfrac{1}{\cfrac{1}{Z_2' + Z_L'} + \cfrac{1}{Z_m}} \right] \quad (2.14)$$

2.10 Regulation

When the transformer is loaded, with a constant primary voltage, then the secondary terminal voltage drops (assuming lagging power-factor); it will increase if power factor is leading, because of its internal resistance and leakage reactance.

Let V_2 = secondary terminal voltage at no-load
V_2' = secondary terminal voltage at load

$$\% \text{ Regulation} = \left[\frac{V_2 - V_2'}{V_2}\right] \times 100$$

Then, % regulation of a loaded transformer at any power factor is given as

$$= (R \cos \theta + X \sin \theta) + \frac{(X \cos \theta - R \sin \theta)^2}{200} \quad (2.15)$$

where R = percentage resistive drop
X = percentage reactive drop
$\cos \theta$ = lagging power-factor

It is to be noted that in case of leading power-factor, θ will change to $-\theta$.

Typical calculations for computation of percentage regulation at different load and power factor are given in Appendix 2.1.

2.11 Losses in Transformers

Losses in transformers are

(a) No-load Losses

It includes both hysteresis loss and eddy current loss. As the core flux in a transformer remains practically constant at all loads, the core-loss is also constant at all loads.

Hysteresis loss $W_h = K_h f B_m^{1.6}$ watts (2.16)
Eddy current loss $W_e = K_e f^2 K_f^2 B_{m2}$ watts (2.17)
where K_h = the hysteresis constant
K_e = the eddy current constant
K_f = the form factor

These losses are minimized by using steel of high silicon content for the core and by using very thin laminations. The input power of a transformer, when on no-load, measures the core-loss.

(b) Load Losses

This loss is mainly due to the ohmic resistance of the transformer winding. Copper-loss also includes the stray loss occurring in the mechanical structure and winding conductor due to the stray fluxes. Copper-loss (I^2R-loss plus stray loss) is measured by the short-circuit test.

2.12 Efficiency

$$\% \text{ Efficiency} = \frac{\text{Output}}{\text{Input}} \times 100$$

$$= \frac{(\text{Input} - \text{Losses})}{\text{Input}} \times 100$$

$$= \left(1 - \frac{\text{Losses}}{\text{Input}}\right) \times 100$$

2.12.1 Condition of Maximum Efficiency

Copper-Loss $W_c = I_1^2 R_{01} = I_2^2 R_{02}$
where R_{01} and R_{02} are equivalent resistance referred to primary and secondary sides respectively.

$$\text{Iron-Loss } W_i = W_h + W_e$$
$$\text{Primary input} = V_1 I_1 \cos \theta_1$$

$$\text{Efficiency } \eta = \left[1 - \frac{(I_1^2 R_{01} - W_i)}{V_i I_1 \cos \theta_1}\right]$$

$$= \left[1 - \frac{I_1 R_{01}}{V_1 \cos \theta_1} - \frac{W_i}{V_1 I_1 \cos \theta_1}\right] \quad (2.18)$$

Differentiating both sides with respect to I_1, we get

$$\frac{d\eta}{dI_1} = 0 - \left(\frac{R_{01}}{V_1 \cos \theta_1}\right) + \left(\frac{W_i}{V_1 I_1^2 \cos \theta_1}\right)$$

for η to be maximum,

$$\frac{d\eta}{dI_1} = 0$$

hence the above equation becomes

$$\frac{R_{01}}{V_1 \cos\theta_1} = \frac{W_i}{V_1 I_1^2 \cos\theta_1}$$

or
$$W_i = I_1^2 R_{01}$$
or
$$= I_2^2 R_{02}$$
or
$$\text{Iron-loss} = \text{Copper-loss}$$

i.e., maximum efficiency occurs when iron-loss is equal to copper-loss.

The load corresponding to maximum efficiency is given by

$$= \sqrt{\left(\frac{\text{Iron-loss}}{\text{Full-load copper-loss}}\right)} \times \text{Full-load}$$

Typical calculations for computation of percentage efficiency at different loads and power factors are given in Appendix 2.1.

2.13 The Auto-Transformer

The auto-transformer has a single continuous winding which is used for the input and output voltages, as shown in the Fig. 2.16. A portion BC of the primary winding AB is used as secondary winding. It is used where transformation ratio differs a little from unity. Its theory and operation principles are similar to that of a two-winding transformer. If N_1 is the primary winding turns and N_2 (a portion of N_1) is secondary winding turns, the transformation ratio K can be represented in the same way as a two-winding transformer

$$\frac{V_2}{V_1} = \frac{N_2}{N_1} = K$$

The current in the secondary winding (may be called common winding) is the vector difference of I_2 and I_1. But as the two currents are practically in phase opposition, the common winding current may be taken as arithmetical difference of I_2 and I_1 i.e. $(I_2 - I_1)$ where I_2 is greater than I_1.

2.13.1 Savings of Copper in Auto-transformers

With reference to Fig. 2.16.
Weight of copper in AC section (series winding) is proportional to
$$(N_1 - N_2)I_1$$

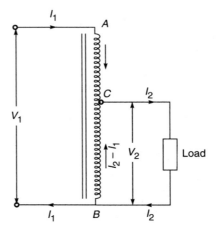

Figure 2.16 *Auto-transformer.*

Weight of copper in BC section (common winding) is proportional to
$$N_2(I_2 - I_1)$$
So, total weight of copper in auto-transformer is proportional to
$$(N_1 - N_2)I_1 + N_2(I_2 - I_1)$$
In a two-winding transformer,
total weight of copper is proportional to $(N_1 I_1 + N_2 I_2)$

$$= \frac{\text{Weight of copper in auto-transformer}}{\text{Weight of copper in two-winding transformer}}$$

$$= \frac{(N_1 - N_2)I_1 + N_2(I_2 - I_1)}{N_1 I_1 + N_2 I_2}$$

$$= 1 - \frac{2 \times \dfrac{N_2}{N_1}}{\left(1 + \dfrac{N_2}{N_1} \times \dfrac{I_2}{I_1}\right)}$$

$$= 1 - \left(\frac{2K}{2}\right) = (1 - K)$$

$\left(\text{where } \dfrac{N_2}{N_1} = \dfrac{I_1}{I_2} = K\right)$

or weight of copper in auto-transformer (W_a)

$$= (1 - K) \times \text{[weight of copper in ordinary two-winding transformer } (W_0)\text{]}$$

Hence, saving $= W_0 - W_a = W_0 - (1 - K)W_0 = KW_0$

or   Saving = Transformation ratio $(K) \times$ weight of copper in ordinary two-winding transformer

Thus, it can be visualized that saving will increase as K approaches unity.

2.14 Three-Winding Transformer

2.14.1 Purpose

A transformer may have additional windings apart from the two conventional main windings depending upon the particular application and type of connection (of the main windings). In three-winding transformers, the third winding is normally called as tertiary winding and it is provided to meet one more of the following requirements:

(a) For an additional load which for some reason must be kept isolated from that of secondary.
(b) To supply phase-compensating devices, such as condensers, operated at some voltage not equal to primary or secondary or with some different connection (e.g., mesh).
(c) In star/star-connected transformers, to allow sufficient earth fault current (zero-sequence component current) to flow for operation of protective gear, to suppress harmonic voltages and to limit voltage unbalance when the main load is asymmetrical, the tertiary winding is delta-connected.
(d) As a voltage coil in a testing transformer.
(e) To load charge split winding generators.
(f) To interconnect three supply systems operating at different voltages.

Tertiary windings are mostly delta-connected.

Consequently, when faults occur on the primary or secondary sides (particularly between lines and earth), considerable unbalance of phase voltage, may be produced which is compensated by large circulating currents.

In case of single line-to-ground fault, either on primary or secondary sides, the zero sequence current flowing through the delta-connected tertiary winding is given as

$$I(Z_0 \text{ Tert.}) = \frac{I_r \times 100}{2Z_{H-L} + Z_{0L-T} \text{ (or } Z_{0\,H-T})} \qquad (2.19)$$

(Z_{0L-T} to be taken if the fault is on LV i.e., secondary side and $Z_{0\,H-T}$, to be taken if fault is on HV i.e., primary side.)

where
- I_r = Rated current in tertiary winding at rated capacity of main (primary/secondary) winding.
- Z_{H-L} = Positive sequence impedance (in percentage) between primary and secondary windings at rated capacity of main windings.
- Z_{0L-T} = Zero sequence impedance (in percentage) between secondary and tertiary windings at rated capacity of main windings.
- Z_{0H-T} = Zero sequence impedance (in percentage) between primary and tertiary windings at rated capacity of main windings.

Thus the reactance of the winding must be such as to limit the circulating current to that value, which can be carried by the copper, otherwise the tertiary windings may overheat and mechanically collapse under fault conditions.

2.14.2 Tertiary Windings in Star/Star Transformers

Star/star transformers comprising single-phase units, or three-phase five-limb core-type units, suffer from the following disadvantages:

(a) that they cannot readily supply unbalanced loads between line and neutral, and
(b) that their phase voltage may be distorted by third harmonic emfs.

By use of delta-connected tertiary windings, induced currents are caused to circulate in it, apportioning the load more evenly over the three phases. The delta-connected tertiary provides a path for the third harmonic currents.

2.14.3 Rating of Tertiary Winding

Rating of tertiary winding depends upon its use. If it has to supply additional loads, its winding cross-section and design philosophy is decided as per load and three-phase dead short-circuit on its terminal with power flow from both sides of HV and MV.

In case it is to be provided for stabilization purposes only, its cross-section and design has to be decided from the thermal and mechanical considerations for the short duration fault currents during various fault conditions single, line-to-ground fault being the most onerous.

2.15 Parallel Operations of Transformers

When operating two or more transformers in parallel, their satisfactory performance requires that they have

(a) The same voltage-ratio;
(b) The same per unit (or percentage) impedance;
(c) The same polarity;
(d) The same phase sequence and zero relative phase displacement.

Of these, (c) and (d) are absolutely essential, and (a) must be satisfied to a close degree. There is more latitude with (b), but the more nearly it is true, the better will be the load division between the several transformers.

(a) *Voltage Ratio*

If voltage readings on the open secondaries of various transformers, to be run in parallel do not show identical values, there will be circulating currents between the secondaries (and therefore between primaries also) when the secondary terminals are connected in parallel. The impedance of transformers is small, so that a small percentage voltage difference may be sufficient to circulate a considerable current and cause additional I^2R-loss. When the secondaries are loaded, the circulating current will tend to produce unequal loading conditions and it may be impossible to take the combined full-load output from the parallel connected group without one of the transformers becoming excessively hot.

Case I: Equal Voltage Ratios

The equivalent circuit of two transformers running in parallel are shown in Fig. 2.17. (Magnetizing current of both the transformers have been assumed more or less equal, resulting in voltages coincident in phases also.)

Let Z_1, Z_2 = the impedances of the two transformers
I_1, I_2 = their currents

28 *Transformers*

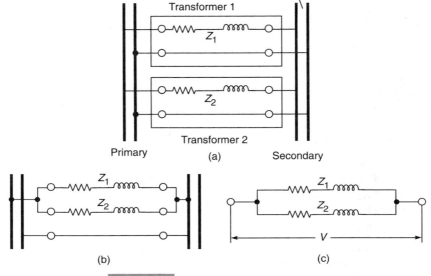

Figure 2.17 *Transformers in parallel.*

V = common terminal voltage
I = total combined current
v = voltage drop across the impedances in parallel

From Fig. 2.17, $v = I_1 Z_1 = I_2 Z_2 = I Z_{12}$, where Z_{12} is the impedance of Z_1 and Z_2 in parallel.
Therefore,

$$Z_{12} = \frac{Z_1 Z_2}{(Z_1 + Z_2)}$$

So
$$I_1 = \frac{v}{Z_1} = \frac{I Z_{12}}{Z_1} = I \left[\frac{Z_1 Z_2}{Z_1 (Z_1 + Z_2)} \right]$$

$$= \frac{I Z_2}{(Z_1 + Z_2)}$$

Similarly,

$$I_2 = \left[\frac{I Z_1}{(Z_1 + Z_2)} \right]$$

Multiplying both sides by common terminal voltage V

Principles of Transformers

$$VI_1 = \frac{IZ_2 \times V_1}{(Z_1 + Z_2)}$$

and
$$VI_2 = VI\left[\frac{Z_2}{(Z_1 + Z_2)}\right]$$

Writing $VI \times 10^{-3} = S$, the combined load kVA, the kVA carried by each transformer is

$$S_1 = S\left[\frac{Z_2}{(Z_1 + Z_2)}\right]$$

and
$$S_2 = S\left[\frac{Z_1}{(Z_1 + Z_2)}\right] \quad (2.20)$$

These expressions are complex, so that S_1 and S_2 are obtained in magnitude and phase angle.

Case II: Unequal Voltage Ratio

Let = E_1, E_2 = no-load secondary emf's
and Z = load impedance at secondary terminals
Other notations are taken as in case I above.
Then $v = IZ_{12} = (I_1 + I_2) Z_{12}$

Also the emf's of the transformers will be equal to the total drops in their respective circuits.

$$E_1 = I_1 Z_1 + (I_1 + I_2)Z$$
and
$$E_2 = I_2 Z_2 + (I_1 + I_2)Z \quad (2.21)$$

Whence
$$E_1 - E_2 = (I_1 Z_1 - I_2 Z_2)$$

as might be expected from this expression

$$I_1 = \frac{(E_1 - E_2) + I_2 Z_2}{Z_1}$$

giving I_1 in terms of I_2. Substituting this Eq. (2.21) above for E_2 gives

$$E_2 = I_2 Z_2 + \left[\frac{(E_1 - E_2) + I_2 Z_2}{Z_1} + I_2\right] \times Z$$

Whence

$$I_2 = \frac{E_2 Z_1 - (E_1 - E_2) Z}{Z_1 Z_2 + Z(Z_1 + Z_2)}$$

and by symmetry

$$I_1 = \frac{E_1 Z_2 - (E_1 - E_2) Z}{Z_1 Z_2 + Z(Z_1 + Z_2)}$$

(2.22)

It is clear from Eq. (2.21) that, on no-load there will be a vector, circulating current between the two transformers, of amount
$$I_1 = -I_2 = (E_1 - E_2)/(Z_1 + Z_2)$$
a result obtainable from Eq. (2.22) by writing
$$Z = \infty \text{ (infinity)}$$
On short circuit the expected result from Eq. (2.22) is given by
$$I_1 = E_1/Z_1$$
and $\quad\quad\quad\quad I_2 = E_2/Z_2$

(b) Impedance

The currents carried by the two transformers are proportional to their ratings if their numerical or ohmic impedances are inversely proportional to those ratings and their per-unit impedances are identical.

A difference in the quality factor (i.e., ratio of reactance to resistance) of the per-unit impedance results in a divergence of the phase angle of the two currents, so that one transformer will be working with a higher, and the other with a lower power factor than that of the combined output.

(c) Polarity

The primary and secondary windings of any individual transformer may, under certain conditions of coil winding, internal connections and connections to terminals, have the same or opposite polarity. When, respective induced terminal voltages for primary and secondary windings are in the same direction, the polarity of the two windings is the same. This polarity is generally spoken as *subtractive*. When, on the other hand, the induced terminal voltages are in opposite direction, the windings are of opposite polarity which is usually referred to as *additive*.

Figure 2.18 shows the test connections at (a) and (b) respectively for single-phase transformers having subtractive and additive polarity. When single-phase voltage is applied to terminals A_1 and a_2, the measured voltage between terminals A_1 and A_2 shall be less than applied voltage in case of (a) and greater than applied voltage in case of (b).

For three-phase transformer the testing procedure is similar, except that the windings must be excited from three-phase supply and more voltage measurements have to be taken for determination of exact polarity and phase sequence. Figure 2.18(c) shows the test connections and results for a star/star-connected transformer with subtractive polarity.

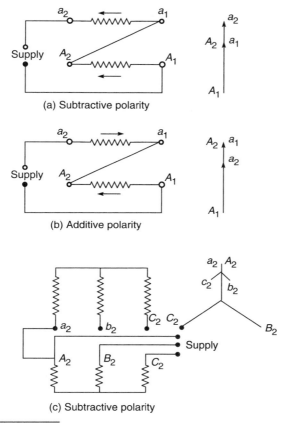

Figure 2.18 *Test connection for determining polarity.*

For parallel operation, the polarity should be the same. A wrong polarity results in a dead short-circuit.

(d) Phase Sequence

The phase sequence, or the order in which the phases reach their maximum positive voltages, must be identical for two paralleled transformers. Otherwise, during the cycle, each pair of phases will be short-circuited.

Any mixture of internal connections can be worked out of it is remembered that the primary and secondary coils on any one limb have induced emf's that are in time-phase. The several connections produce various magnitudes and phases of secondary voltage; the magnitude can be adjusted for parallel operation by suitable choice of turn-ratio, but the phase divergence cannot be compensated. Thus two sets of connections giving secondary voltages with a phase displacement cannot be used for transformers intended for parallel operation.

The following are typical of the connections for which, from the view point of phase sequence and angular divergence, transformers can be operated in parallel:

| Transformer | 1 | Yy | Yd | Yd |
| Transformer | 2 | Dd | Dy | Yz |

Internal connections of different windings can be done in a variety of ways, giving different vector groups. Various vector group arrangements and their connection symbols are given in Indian Standard IS : 2026 (Part IV).

Appendix
Calculation of % Regulation and Efficiency

Transformer Specification   250 MVA, 15.75/240 kV, three-phase, 50 Hz generator transformer having no-load loss 120 kW, load-loss 500 kW at full-load at 75°C and reactance voltage drop 14% at rated load.

(a) Regulation at full-load, unity power-factor
$$R = 0.2\%, X = 14\%$$
$$\cos \theta = 1.0, \sin \theta = 0$$
Substitute values in Eq. (2.15)
Percentage regulation
$$= 0.2 \times 1.0 + \frac{(14 \times 1)^2}{200} = 1.18\%$$

(b) Regulation at 3/4-load, 0.8 lagging power-factor
$$R = 0.15\%, X = 10.5\%$$
$$\cos \theta = 0.8, \sin \theta = 0.6$$
Substitute values in Eq. (2.15)
Percentage regulation
$$= 0.15 \times 0.8 + 10.5 \times 0.6 + \frac{[(10.5 \times 0.8) - (0.15 \times 0.6)]^2}{200}$$
$$= 6.77\%$$

(c) Efficiency at full-load, unity power-factor
Percentage efficiency
$$= \left(1 - \frac{\text{Total loss}}{\text{Input power}}\right) \times 100$$

$$= \left(1 - \frac{500 + 120}{250 \times 10^3}\right) \times 100$$
$$= 99.75\%$$
(d) Efficiency at 3/4 load, 0.8 lagging power-factor
No-load loss = 120 kW

$$\text{Load loss} = 500 \times \left(\frac{3}{4}\right)^2 = 281.25 \text{ kW}$$

Input power $= 250 \times 10^3 \times \frac{3}{4} \times 0.8 = 150 \times 10^3$ kW

Percentage efficiency
$$= \left(1 - \frac{281.25 + 120}{150 \times 10^3}\right) \times 100$$
$$= 99.73\%$$

REFERENCES

1. Say, M.G., *The Performance and Design of AC Machines* (Book), The English Language Book Society and Sir Isaac Pitman & Sons, London.
2. Puchstein, A.F., T.C. Lloyd, and A.G. Conrad, *Alternating Current Machines,* Asia Publishing House.
3. Hellman, Charles I, *Elements of Radio* (Book), D. Van Nostrand Company, INC, New York.
4. Theraja, B.L., *A Text Book of Electrical Technology,* Sultan Chand & Co., New Delhi.

CHAPTER 3

Materials Used in Transformers

M. P. Singh
T. K. Ganguli

Apart from active materials like copper and cold rolled grain-oriented silicon steel, a number of ferrous, nonferrous and insulating materials are employed for building up a transformer. Optimum utilization of all materials in consonance with their electrical, mechanical, physical, chemical and thermal characteristics is necessary for obtaining a compact size transformer. Strict quality control measures like testing of raw materials, therefore, become imperative. One basic requirement for all materials used in an oil-filled transformer is that they should be compatible with insulating oil and should not react with or deteriorate oil.

Table 3.1 gives application, applicable national and international standards of various materials used in a transformer.

3.1 Insulating Oil

Insulating oil forms a very significant part of the transformer insulation system and has the important functions of acting as an electrical insulation as well as a coolant to dissipate heat losses. The basic raw material for the production of transformer oil is a low-viscosity lube termed as transformer oil base stock (TOBS), which is normally obtained by fractional distillation and subsequent treatment of crude petroleum. Important characteristics of TOBS given in Table 3.2 must be kept within permissible limits in order to produce good insulating oils. TOBS is further refined by acid treatment process to yield transformer oil.

Table 3.1 Materials Used in Power Transformer and Reactor

Sl. No	Material	Applicable Standards and Grade	Application
1	2	3	4
A.	**Insulating Materials**		
1.	Transformer oil	IS : 335, BS : 148 IEC 296	Liquid dielectric and coolant
2.	Electrical grade paper		
	(i) Kraft insulating paper of medium air permeability	IEC 60554–3–1	Layer winding insulation, condenser core of oil impregnated bushing
	(ii) Kraft insulating paper of high air permeability	IEC 60554–3–1	Covering over rectangular copper conductor Covering over continuously transposed copper conductor
	(iii) Crepe kraft paper	BS 5626–3–3 Base paper as per (i) above	Covering over stranded copper cable Covering over flexible copper cable Insulation of winding lead
	(iv) Press paper	IS : 8570, BS : 3255	Insulation over shield
	(v) Kraft paper with aluminum bands	Base paper as per (i) above	Backing paper for axial cooling duct Line and common shield in winding
	(vi) Crepe kraft paper with aluminum foil	Base paper as per (i) above	Metallization of high-voltage lead and shield

Contd.

Table 3.1 *(Contd.)*

Sl. No	Material	Applicable Standards and Grade	Application
1	2	3	4
3.	Pressboard		
	(i) Pressboard moulding from wet sheet or wet wood pulp	IEC 60641-3-1	Angle ring, cap, sector, snout, square tube, lead out and moulded piece of intricate profile for insulating ends of windings, insulation between windings, and numerous other applications
	(ii) Soft calendered pressboard—solid	Type C of IS : 1576 IEC : 60641-3-1	Cylinder, barrier, wrap, spacer, angle washer, crimped washer, and yoke insulation, etc.
	(iii) Soft pressboard—laminated	BS EN 60763-1.2	Block, block washer, terminal-gear cleat and support, spacer, etc.
	(iv) Precompressed pressboard—solid	IEC 60641-3.2	Dovetail block and strip, clack-band, cylinder, warp, barrier, spacer, block, block washer, corrugated sheet, yoke bolt, washer, etc.

Contd.

Table 3.1 (Contd.)

Sl. No	Material	Applicable Standards and Grade	Application
1	2	3	4
	(v) Precompressed pressboard—laminated	IEC 60763–3.1	Top and bottom coil clamping ring, block, block washer, dovetail strip, spacer, etc.
4.	Wood and laminated wood		
	(i) Unimpregnated densified laminated wood—low density	IEC : 61061	Cleat and support, core/yoke clamp, wedge block, winding support block, sector, core-to-coil packing, etc.
	(ii) Unimpregnated densified laminated wood—high density	IEC : 61061	Coil clamping ring, block, cleat support, etc.
5.	Insulated copper conductor and cable		
	(i) Paper covered rectangular copper conductor	IEC : 60317 IS : 13730	For making different types of windings
	(ii) Paper covered continuously transposed copper conductor	IEC : 60317	For making different types of windings
	(iii) Paper covered stranded copper cable	IS : 8572 conductor to IS : 8130, IEC : 60228	For making lead and terminal
	(iv) Crepe paper covered flexible copper cable	Conductor to IS : 8130, IEC : 60228	For making lead and terminal required to be bent to a small radius

Contd.

▶ **Table 3.1** (*Contd.*)

Sl. No	Material	Applicable Standards and Grade	Application
1	2	3	4
	(v) PVC insulated copper cable—single and multicore	IS : 1554, BS : 6346 IEC : 60502	Control wiring in marshalling box, nitrogen sealing system
6.	Insulating Tape		
	(i) Cotton tape	IS : 1923	For various taping purposes
	(ii) Cotton newar tape	—	For taping and banding
	(iii) Glass woven tape	IS : 5352, IEC : 61067–1	Used in core bolt insulation
	(iv) Woven terylene tape	IS : 5351, IEC : 61068–1	For taping purposes at places requiring higher strength
	(v) Polyester resin impregnated weftless glass tape	—	Banding of transformer cores
7.	Phenolic laminated paper base sheet	IS : 2036, BS : 2572	Terminal-gear support and cleat, gap filler in reactor, tap changer components
8.	Phenolic laminated cotton fabric sheet	IS : 2036, BS : 2572	Terminal board, for making core duct, support and cleat
B.	**Sealing Materials**		
9.	Synthetic rubber bonded cork	IS: 4253 (Part II)	As gasket in different places to prevent oil leakage from joints viz. tank rim, turret opening, inspection cover and

Contd.

Table 3.1 *(Contd.)*

Sl. No	Material	Applicable Standards and Grade	Application
1	2	3	4
10.	Nitrile rubber sheet and moulding	BS : 2751	with mounting flange of various fittings, etc. As gasket in different places to prevent oil leakage from joints, 'O' ring in bushings, moulded component in fittings
C.	**Ferrous Materials**		
11.	Cold rolled grain oriented silicon steel (CRGO)	BS : 6404/ASTM A876M/ DIN 46400	For making transformer core
12.	Cold rolled carbon steel sheet	IS : 513 ASTM A 620 M BS : 1449–1.1	For making radiator
13.	High tensile strength structural steel plate	IS : 8500	Core clamp plate, anchoring and clamping core to bottom tank
14.	1.5% nickel-chromium-molybdenum steel bar and sections hardened and tempered	IS : 5517	Lifting pin, roller shaft
15.	Austenitic chromium nickel steel titanium stabilized plate (stainless steel)	IS : 6911, BS : 1449	Turret opening, non-magnetic insert, etc., to neutralize the effect of eddy currents

Contd.

Table 3.1 (Contd.)

Sl. No	Material	Applicable Standards and Grade	Application
1	2	3	4
16.	Stainless steel sections (austenitic)	IS : 6603, BS : 970	Non-magnetic bar for high current applications
17.	Structural steel—standard quality (plate, section, flat, bar, channel, angle, etc.)	IS : 2062	Tank, end frame, clamp plate, 'A' frame for radiator, conservator, turret, cable box and for other structural purposes
18.	Bright steel bar and sections—cold drawn	IS : 7270	Threaded and machined components
D.	**Non-Ferrous Materials**		
19.	High conductivity copper		
	(i) Sheet, strip, foil—hard and soft	IS : 1897	For various current-carrying applications, e.g., bushing conductor, terminal lead, divertor and selector contacts of on-load tap changers, winding shield, cable box components, off-circuit switch items, etc.
	(ii) Rod	IS : 613	
	(iii) Tube	BS : 1977	
	(iv) Casting and forging	BS : EN 1982	
	(v) Tinned foil	IS : 3331	
	(vi) Flexible cable	IS : 8130, IEC : 60228	
	(vii) Flat flexible braid	—	
20.	Copper alloys		
	(i) Free machining brass rod, square and hexagon	IS : 319	Tie rod and for making different components

Contd.

Table 3.1 (Contd.)

Sl. No	Material	Applicable Standards and Grade	Application
1	2	3	4
	(ii) Phosphor bronze rod	IS : 7811	Tap-changer components
	(iii) Nickel silver strip	IS : 2283	For making winding shield
21.	Aluminum		
	(i) Aluminum alloy plate	Alloy 54300M (NP 8—M) of IS : 736	Flange in bushing, cable box, other non-magnetic applications
	(ii) Aluminum plate (99.0 percent)	Alloy PIC of IS : 736	Shielding of reactor tank,
	(iii) Aluminum foil	—	Condenser layer in bushings

➤ **Table 3.2** Characteristics of TOBS

Sl. No	Characteristic	Requirement
1.	Viscosity at 40°C	9–14 cSt.
2.	Pour, point, Max.	– 9°C
3.	Flash point (Pensky—Marten closed-cup method) Min.	145°C

3.1.1 Chemical Composition

Transformer oil consists of four major generic classes of organic compounds, namely, paraffins, naphthenes, aromatics and olefins. All these are hydrocarbons and hence insulation oil is called a pure hydrocarbon mineral oil. For good fresh insulating oil, it is desirable to have more of saturated paraffins, less of aromatic and/or naphthenes and none of olefins. However, for better stability of properties, it is necessary to have optimum aromatic and/or naphthenic hydrocarbons. Such an optimum balance is struck by a carefully controlled refining process. Depending upon the predominance, oil is usually termed as of paraffinic base or naphthenic base.

3.1.2 Characteristics of Oil and Their Significance

Table 3.3 gives characteristics of oil as per IS : 335. The typical approximate values (all relating to 60°C) of some other physical properties of oil considered in design calculations are given in Table 3.4. Figures in parenthesis indicate the approximate temperature coefficient (per °C rise) for the property concerned. The significance of various tests conducted on oil to evaluate its quality is given below.

A. Physical Properties

(i) *Density*. This test has special significance when transformer is operated in a very low temperature zone. The maximum value to density fixed at 29.5°C ensures that water in the form of ice present in oil remains at the bottom and does not tend to float on the oil up to a temperature of about –10°C.

(ii) *Interfacial tension (IFT)*. This is a measure of the molecular attractive force between oil and water molecules at their interface. This test provides a means of detecting soluble polar

Materials Used in Transformers 43

> **Table 3.3** Schedule of Characteristics of Insulating Oil

Sl. No.	Characteristic	Requirement as per IS : 335
1.	Density at 29.5°C, max	0.89 g/cm^3
2.	Kinematic viscosity at 27°C, max.	27 cSt
3.	Interfacial tension at 27°C min.	0.04 N/m
4.	Flash point, Pensky—Marten (closed), min.	140°C
5.	Pour point, max.	–6°C
6.	Neutralization value (total acidity), max.	0.03 mg KOH/g
7.	Corrosive sulphur	Non-corrosive
8.	Electric strength (breakdown voltage) min.	
	(a) As received	30 kV (rms)
	(b) After filtration	60 kV (rms)
9.	Dielectic dissipation factor (tan-delta) at 90°C, max.	0.002
10.	Specific resistance (resistivity), min.	
	(a) At 90°C	30×10^{12} Ω cm
	(b) At 27°C	1500×10^{12} Ω cm
11.	Oxidation stability	
	(a) Neutralization value after oxidation, max.	0.40 mg KOH/g
	(b) Total sludge after oxidation, max.	0.10% by weight
12.	Presence of oxidation inhibitor	The oil shall not contain antioxidant additives
13.	Water content (as received), max.	50 ppm by weight

> **Table 3.4** Physical Constants of Oil

Sl. No.	Property	Value
1.	Permittivity	2.2 (–0.001)
2.	Specific heat	2.06 kJ/kg°C (0.0038)
3.	Thermal conductivity	0.12 W/m°C
4.	Coefficient of expansion	0.00078/°C
5.	Mean density correction factor over the normal range of operating temperature	0.00065/°C

contaminants and products of deterioration, which decrease molecular attractive force between oil and water. It is considered that IFT gives an indication of degree of sludging of oil (Fig. 3.1).

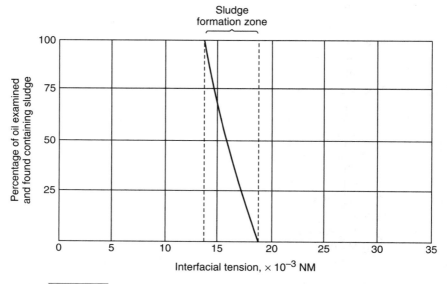

Figure 3.1 *Presence of sludge in oils related to the interfacial tension values at an ambient temperature of 27°C.*

(iii) *Moisture content.* The amount of free and dissolved water present in the oil is its moisture content and is expressed in ppm (parts per million by weight, i.e., mg/kg). Presence of moisture is harmful since it adversely affects the electrical characteristics of oil and accelerates deterioration of insulating paper.

(iv) *Flash point.* It is the temperature at which oil gives so much vapor that this vapor, when mixed with air, forms an ignitable mixture and gives a momentary flash on application of flame under prescribed conditions. A minimum flash point is specified in order to prevent the risk of fire that might result by accidental ignition.

(v) *Viscosity.* It is a measure of oil resistance to continuous flow without the effect of external forces. The oil must be mobile, as heat transfer in transformers occurs mainly by convection currents. Since viscosity increases with decrease in temperature, it is necessary that viscosity be as low as possible at low temperatures.

(vi) *Pour point.* The temperature at which oil will just flow under the prescribed conditions is known as the pour point. If the oil becomes too viscous or solidifies, it will hinder the formation of convection currents and thus cooling of equipment will be severely affected.

B. Electrical Properties

(i) *Electric strength (breakdown voltage).* BDV is the voltage at which breakdown occurs between two electrodes when oil is subjected to an electric field under prescribed conditions. Electric strength is the basic parameter for insulation system design of a transformer. It serves to indicate the presence of contaminating agents like moisture, fibrous materials, carbon particles, precipitable sludge and sediment.

(ii) *Resistivity (specific resistance).* This is the most sensitive property of oil requiring utmost care for its proper determination. Resistivity in Ω cm is numerically equivalent to the resistance between opposite faces of a centimeter cube of the liquid. Insulation resistance of windings of a transformer is also dependent upon the resistivity of oil. A low value indicates the presence of moisture and conductive contaminants.

(iii) *Dielectric dissipation factor (DDF).* DDF is numerically equal to sine of the loss angle (approximately equal to tangent of loss angle for dielectrics) and is a good tool to indicate the quality of an insulation. A high value of DDF is an indication of the presence of contaminants or deterioration products such as water, oxidation products, metal soaps, soluble varnishes and resins.

C. Chemical Properties

(i) *Neutralization value (total acidity).* It is a measure of free organic and inorganic acids present in the oil and is expressed in terms of milligrams of KOH required to neutralize the total free acids in one gram of oil. Oxidation of oil in service is a consequence of reaction between hydrocarbons present in the oil and oxygen. The oxygen may be atmospheric since oil comes into contact with the atmospheric air during breathing of transformer, or may have been dissolved in oil if oil is not degassed properly, or may be liberated due to the effect of heat on cellulose insulation. Oxidation of oil is a chain reaction by which organic acids and sludge are formed. Copper present in a large quantity in transformer, acts as a strong catalyst in oxidation. Hence, as far as possible, no bare copper is allowed to be used in power transformers. Copper is covered with paper or enamel coated or electrotinned. The products of oxidation are injurious to the insulation system of transformer. Acids formed give rise to formation of sludge which precipitates out and deposits on windings and other parts of transformer. This causes hindrance to proper oil circulation and heat dissipation. The acids also encourage deterioration of cellulose insulation, i.e., paper, pressboard and wood. Water is produced during oxidation which reduces electric strength of oil and also accelerates corrosion of metals and deterioration of insulating materials. Hence the measurement of total acidity is the most convenient and direct method of assessing the capability of oil for nonformation of acids during service.

(ii) *Oxidation stability.* This test is the measure of neutralization value and sludge after oil is aged by simulating the actual service conditions of a transformer. The oxidation stability test is very important for new oil but not for oil in service and shows the presence of natural inhibitors which impart anti-oxidation characteristics to oil.

(iii) *Sediment and precipitable sludge.* These are oil deterioration products or contaminants which are insoluble after dilution of the oil with n-heptane under prescribed conditions. However, precipitable sludge is soluble in the solvent mixture of equal parts of toluene, acetone and alcohol but sediment is insoluble in this solvent mixture. These contaminants are determined for oils in service. Oil is considered unsatisfactory for use if sediment or precipitable sludge is detected.

(iv) *Corrosive sulphur.* Crude petroleum usually contains sulphur compounds, most of which are removed during the refining processes. This test is designed to detect any traces of free corrosive sulphur that may be present in oil. Presence of corrosive sulphur in oil will result in pitting and black deposit on the surface of bare copper used in transformer, which will adversely affect the dissipation of heat and consequently performance of the equipment.

3.1.3 Stability of Oil

Paraffinic base insulating oils available indigenously have shown deterioration of electrical properties during use, at a rate which is a little bit faster than of naphthenic base oils produced in the USA, the UK and other countries. Efforts are afoot to produce paraffinic base oil in the country having stability comparable to naphthenic base oil.

3.2 Insulating Paper

Paper is a fabric made from vegetable fibers which are felted to form a web or sheet. The fibrous raw materials are obtained from plants including cotton, hemp, manila, straw and coniferous/deciduous trees. The cell or fibers of such plants consist mainly of cellulose. The molecular formula for cellulose is $(C_6H_{10}O_5)_n$. The degree of polymerization of the molecular unit indicated by the letter n, varies widely in various plants. The value of n can go up to 2500 or more for cotton fibers and up to 1200 or more for wood pulp. Various other materials, e.g., lignin, hemicellulose, mineral matter and resins are associated with the cellulose in the fibers. These contaminants are removed by sulphate process treatment of wood pulp and careful water washing. The different types of paper used and their application are given in Table 3.1. Terminal-gear and high voltage leads are normally insulated using crepe kraft paper to obtain higher flexibility and smaller radii at bends. These papers have high stretchability varying between 50 to 100%.

3.2.1 Characteristics of Insulating Papers and Their Significance

Characteristics of different papers are given in Table 3.5 and their significance are described as follows:

> **Table 3.5** Schedule of Characteristics of Electrical Grade Papers

Sl. No.	Characteristic	Requirement		
		Kraft paper of medium air permeability	Kraft paper of high air permeability	Press paper
1	2	3	4	5

I. Physical Properties

1.	Substance, g/m^2	100 ± 5 for 100 μm thk.	100 ± 5 for 125 μm thk.	—
2.	Density g/cm^3	0.95 to 1.05	0.75 to 0.85	1.0 to 1.3
3.	Moisture content, % (max.)	8.0	8.0	8.0
4.	Oil absorption, % (min.)	—	—	3 to 21% depending upon density
5.	Water absorption, % (min.)	5	10	—
6.	Air permeability μm/Pa S	0.2–0.5	0.5–1.0	—

II. Mechanical Properties

7. Tensile strength (expressed as tensile index), N m/g (min.)

Machine direction	83	78	69
Cross direction	30	28	24

8. Elongation at break, % (min.)

Machine direction	2	2	—
Cross direction	4	4	—

9. Internal tearing resistance (expressed as tear index), mN m^2/g, (min.)

Substance (g/m^2)	Tear resistance			
	MD	XD	MD	XD
≥40–80	5	6	5	6	—
>80–120	6	7	6	7	—
>120	8	9	—	—	—

10. Heat stability (decrease in bursting strength), % (max.) 20   20   60

Contd.

Materials Used in Transformers **49**

➢ **Table 3.5** *(Contd.)*

Sl. No.	Characteristic	Requirement		
		Kraft paper of medium air permeability	Kraft paper of high air permeability	Press paper
1	2	3	4	5
III.	**Electrical Properties**			
11.	Electric strength (BDV), kV/mm (min.)			
	(i) In air at 90°C	7.5	7.0	10
	(ii) In oil at 90°C	—	—	65
12.	Dissipation factor of unimpregnated paper at 105°C (max.)	0.003	0.003	—
13.	Conducting paths	should be free from any conducting path		
IV.	**Chemical Properties**			
14.	Mineral ash, % (max.)	1.0	1.0	2.0
15.	Conductivity of 5% aqueous extract mS/m (max.)	10	10	13.5
16.	pH value	6–8	6–8	5.0–8.5
17.	Chloride content of aqueous extract, mg/kg (max.)	50	50	—
18.	Conductivity of organic extract, nS/m (max.)	10	10	—
V.	**Methods of Test**	IEC 60554–3 IS : 9335 (Part II)	IEC 60554–3 IS : 9335 (Part II)	IS : 8570

A. *Physical Properties*

(i) *Substance (grammage)*. It is the ratio of mass to the area and is a fundamental parameter which influences most of the mechanical and electric properties. This is a basic property for papermaking process.

(ii) *Density*. It is a function of thickness and grammage of paper. Density is also a basic property for setting process parameters for the manufacture of paper. Papers in density range of 0.6 to 1.3 g/cm^3 are employed in transformer manufacture.

(iii) *Moisture content.* Cellulosic fibers are hygroscopic. Water has the effect of plasticizing the cellulose fibers and of relaxing and weakening the interfiber bonding. Hence moisture content has a significant effect on many properties of paper.

(iv) *Oil and water absorption.* Paper attains a high value of electric strength when impregnated in oil under vacuum, since place of moisture liberated during vacuum drying is taken up by oil in the molecular structure of the paper. Oil absorption is dependent upon density and air permeability of paper. Sometimes as a substitute to oil absorption, water absorption is monitored.

(v) *Air permeability.* It is the measure of the rate at which paper allows air to penetrate through it. It is influenced by both the internal structure and the surface finish of the paper. Electric strength of the paper is inversely proportional to its air permeability.

B. Mechanical Properties

(i) *Tensile strength and elongation.* Paper should be able to withstand tension exerted during its wrapping over layer winding coils. Paper tape is wound over conductors and cables under specified tension at a high speed by paper-lapping machines. If specified tensile strength is met, breakages of paper will not occur in paper-lapping operation. It has been proved theoretically that the displacement of windings during short circuit of a transformer depend very much on the Young's modulus of insulating materials used in windings which shows the importance of tensile strength and elongation tests.

(ii) *Internal tearing resistance.* This gives the load under which paper will tear off under specified conditions and shows its capability against tearing.

(iii) *Bursting strength.* This is the pressure required to burst a disc of paper which is gripped firmly around its periphery, and to one side of which pressure is applied at a uniform rate, using liquid as a medium. This test also gives an idea about the mechanical strength of paper.

(iv) *Heat stability*. It is the ability of paper to withstand thermal stresses during service life of a transformer and is determined by measuring decrease in internal tearing resistance, bursting strength and degree of polymerization after subjecting it to accelerated temperature.

C. Electrical Properties

(i) *Electric strength*. This is a basic parameter in deciding the insulation system design of a transformer. Electric strength depends on density and air permeability of paper.

(ii) *Dissipation factor*. It is a good tool to indicate the quality of a dielectric. A high value of dissipation factor shows presence of contaminants or depolymerization of paper.

(iii) *Freedom from conducting paths*. Electric strength of paper is adversely affected by the presence of conducting paths and hence these are undesirable impurities.

D. Chemical Properties

The conductivity of aqueous extract is a measure of the electrolytes present in the paper which is extracted by gently boiling paper in distilled water. The electrolytic impurities are present as ionizable acids, bases, salts or a mixture of these. This property indicates chemical purity of paper and is a basic parameter for papermaking process. Insulation resistance of paper is dependent on this property. Conductivity of organic extract, pH value of aqueous extract, ash content and chloride content of aqueous extract are also measured to check purity. Mineral ash, solid residue remaining after complete combustion of paper are injurious to paper and should be within the permissible limits.

3.3 Pressboard

Pressboard is a widely used insulating material for making a variety of components used in electrical, mechanical and thermal design of transformers. Types of pressboard used and their applications are given in Table 3.1. Like paper, pressboard is also made entirely from vegetable fibers, whose cells contain mainly cellulose. Pressboard is manufactured using the following raw materials:

(a) Sulphate wood pulp
(b) Cotton
(c) Mixture of sulphate wood pulp and cotton or jute hemp
(d) Mixture of cotton and jute hemp

3.3.1 Characteristics of Pressboard

The characteristics of different qualities of pressboard commonly used are given in Table 3.6. Solid pressboard up to 6 mm thick calendered and up to 8 mm thick precompressed is made. Precompressed pressboard is manufactured by pressing wet sheet at an elevated temperature. For higher thickness, laminated calendered and precompressed pressboard is manufactured by building up laminae of solid pressboard using an adhesive (e.g., polyester resin, polyvinyl alcohol, casein glue) for bonding under pressure and high temperature.

3.3.2 Moulding

The most difficult practical insulation problems in high voltage transformers occur at the ends of the windings and at the leadouts from the windings. Pressboard moulded components can be made to any required shape to follow the contour of the equipotentials determined from the field plot. Angle rings and caps are the widely used mouldings (Fig. 3.2 and Fig. 3.3). Other types of mouldings used in EHV transformers are snouts, angle sectors and leadouts, etc.

Evaluation of moulding is another difficult task due to intricate profiles. These cannot be tested without destroying them. Normally, bend/neck of the component is the weakest point. Hence, these are tested on regular, irregular and neck portions to ensure the overall quality.

3.3.3 Advantage of Pressboard Over Other Solid Insulating Materials

In EHV transformers care is taken to use such insulating materials in stress zones which will have the least partial discharge. Synthetic resin bonded paper based laminates and cylinder and laminated wood are prone to give more P.D. in such zones compared to pressboard, since air voids are trapped due to extensive use of resin

Table 3.6 Schedule of Characteristics of Different Types of Pressboard

Sl. No.	Characteristic	Requirement				
		Precompressed solid	Soft calendered solid	Precompressed laminated	Soft calendered laminated	
1	2	3	4	5	6	
	Composition	100% wood pulp	Cotton, jute and wood pulp	100% wood pulp	Cotton, jute and wood pulp	

I. Physical Properties

1. Density, g/cc	1.1–1.3	0.9–1.1	1.2–1.3	0.95–1.15
2. Moisture content, % (max.)	6	8	5	7.5
3. Oil absorption, % (min.)	9	20	5	12
4. Shrinkage, % (max.)				
(i) *In Air*				
Machine direction	0.5	1.0	0.4	1.0
Cross direction	0.7	1.5	0.6	1.5
Perpendicular	5.0	6.0	4.0	5.0
(ii) *In Oil*				
Machine direction	—	0.3	—	0.5
Cross direction	—	0.3	—	0.5
Perpendicular	—	1.5	—	0.5
5. Cohesion between plies	The specimen shall not readily split by delamination, i.e., between two adjacent plies and the exposed torn surfaces shall have a distinctly hairy or ragged appearance.				

Contd.

Table 3.6 (Contd.)

Sl. No.	Characteristic	Requirement			
		Precompressed solid	Soft calendered solid	Precompressed laminated	Soft calendered laminated
1	2	3	4	5	6
	Composition	100% wood pulp	Cotton, jute and wood pulp	100% wood pulp	Cotton, jute and wood pulp

II. Mechanical Properties

6.	Tensile strength, MPa (min.)					
	(i) Machine direction	100	47	—	—
	(ii) Cross direction	75	22	—	—
7.	Cross breaking strength, MPa (min.)					
	(i) Machine direction	NA	NA	100	32.5
	(ii) Cross direction	NA	NA	85	30
8.	Compressibility, % (max.)					
	(i) In air	4–10 (dependent on thickness)	13	3	9
	(ii) In oil	—	20	—	13

III. Electrical Properties

9.	Electric strength in oil 90°C kV/mm (min.)		
	(i) 1.5 mm thick	35 BDV	15.3 for one min.
	(ii) 3 mm thick	30 BDV	11.3 for one min.
	(iii) 6 mm thick	30 BDV	8.3 for one min.
	(iv) 8 mm thick	25 BDV	—

70 kV for one min. on 25 mm width (edgewise)

Contd.

Table 3.6 (Contd.)

Sl. No.	Characteristic	Requirement				
		Precompressed solid	Soft calendered solid	Precompressed laminated	Soft calendered laminated	
1	2	3	4	5	6	
	Composition	100% wood pulp	Cotton, jute and wood pulp	100% wood pulp	Cotton, jute and wood pulp	
IV. Chemical Properties						
10.	Conductivity of 5% aqueous extract, mS/m (max.)	10	11	10	15	
11.	pH value of aqueous extract	5.5–9	7–9.5	5–8	5–9	
12.	Mineral ash, % (max.)	1	2	2	2	
13.	Effect on insulating oil					
	(i) Increase in acidity, mg KOHg (max.)	N.A.	N.A.	0.1	0.1	
	(ii) Increase in sludge content, % (max.)	N.A.	N.A.	0.05	0.05	
V. Method of Test		IEC 60641-3-1	IS : 1576	IEC : 60763-3.1	BS : 5354	

N.A. = Not Applicable.

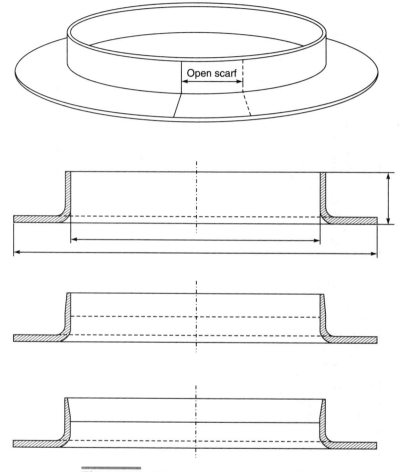

Figure 3.2 *Typical pressboard angle rings.*

as the bonding material. Oil absorption in resin bonded laminates and cylinders is negligible and hence these are electrically inferior to pressboard. Efforts, are made to use cylinders, winding cleats and supports, etc., made of pressboard as far as possible.

3.3.4 Significance of Pressboard Characteristics

Properties like density, tensile strength, elongation, ash content, pH of aqueous extract, conductivity of aqueous extract, oil absorption,

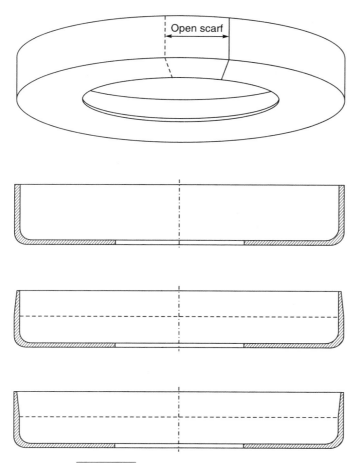

Figure 3.3 *Typical pressboard caps.*

moisture content, freedom from conducting particles and electric strength, described in Sec. 3.2 are also determined on pressboard. Apart from the above, the details of additional tests conducted for the evaluation of pressboard are given below:

(i) *Compressibility*. It is a very important property and is a measure of reduction in thickness of dried specimen when compressed under load, under prescribed conditions. Transformer windings are designed based on expected compressed thickness of dovetail blocks used for providing radial ducts between turns and sections. If compressibility of pressboard is more than the value taken in the design, the coil height will become less than the design value after vacuum drying and oil impregnation of transformer and this reduction can lead to serious problems of short circuit forces if not rectified.

(ii) *Shrinkage.* It is a measure of reduction in length, width and thickness of specimen after it is dried under specified conditions. Effect of shrinkage is also taken into consideration in the design.

(iii) *Cross-breaking strength.* This is determined to check the proper adhesion of laminae in a laminated pressboard and ensures against delamination.

(iv) *Electric strength edgewise.* It measures the creep strength across the glue line of laminated pressboard.

3.4 Wood

Wood based laminates are manufactured from selected veneers (mostly 2 mm thick) obtained from various timbers. The veneers are dried and partially or fully impregnated with neutral phenol formaldehyde resin depending upon type of laminate required and then densified under heat and pressure. The placement of veneers one over the other is such as to obtain the desired grain orientation to achieve specific mechanical strength in the required direction. The following types of wood based laminates are used in transformers whose applications are given in Table 3.1 and properties in Table 3.7.

Type	*Direction of Veneers*
(a) Unimpregnated densified laminated wood—low density and high density.	Alternate veneers with grain directions at right angles to each other.
(b) Resin impregnated densified laminated wood type V.	All veneers have grain in the same direction.
(c) Resin impregnated densified laminated wood type VI.	Same as (a).

The areas which require higher mechanical and lower electric strength, densified laminated wood is used for making a variety of insulation components like coil clamping ring, cleat, support, core and yoke clamp, wedge block, bolt and nut, terminal board and core step filler, etc. Haldu and teak wood seasoned planks are also used as a filler material between core limbs and enveloping coils and as yoke-step fillers.

3.4.1 Evaluation of Wood

Details and significance of tests conducted on densified laminated wood as per Table 3.7 have been described is Secs. 3.2 and 3.3. The quality of laminated wood may vary widely from batch to batch and sometimes even sheet to sheet. The reason has been attributed to many parameters, e.g., species of wood selected for veneers, their location and age, number of joints in veneers to make up the required width, gap in joints and defects in timber, etc. Hence, utmost care is taken in their use.

3.5 Insulated Copper Conductor for Windings

The following types of conductors are used for making spiral, helical, continuous disc, layer and interleaved disc windings of transformers described in Chapter 5.

(a) Paper covered rectangular copper conductor.
(b) Twin paper covered rectangular copper conductors bunched together.
(c) Paper covered continuously transposed copper conductor (CTC).
(d) Twin transposed copper conductors bunched together.
(e) Twin rectangular copper conductors bunched together and provided with a common paper covering and a glued (epoxy coated) paper strip between the two conductors.
(f) Epoxy coated continuously transposed conductor.

Conductors at (b) and (d) above are used to improve the winding space factor. Individual conductor is covered with only three or four layers of paper to electrically separate them instead of giving full insulation as shown in Fig. 3.4. The two conductors/cables are kept either in axial height or in radial depth of coils, thereby reducing total coil axial height and radial depth respectively.

Conductors at (e) and (f) further improve the mechanical strength of the winding, besides improving the space factor of the winding.

> **Table 3.7** Characteristics of Densified Laminated Wood

Sl. No.	Characteristic	Unimpregnated laminated wood —low density	Unimpregnated laminated wood —high density
I.	**Physical Properties**		
1.	Density, g/cm^3	0.90 to 1.09	1.10 to 1.20
2.	Moisture content, % (max.)	7	7
3.	Oil absorption, % (min.)	9	6
4.	Cohesion between laminae	The specimen shall not readily split by delamination and the exposed torn surfaces shall have a distinctly hairy or ragged appearance.	
II.	**Mechanical Properties**		
5.	Tensile strength, MPa (min.)	70	95
6.	Cross-breaking strength, MPa (min.)		
	(i) Along the grain	90	112
	(ii) Across the grain	67	85
7.	Compressive strength, MPa (min.)		
	Perpendicular to laminae	160	185
8.	Shear strength, MPa (min.) perpendicular to laminae	35	50
III.	**Electrical Properties**		
9.	Electric strength (one minute proof) in oil at 90°C		
	(i) Flatwise, kV/mm (min.)	4	4
	(ii) Edgewise, kV (min.)	60	60
IV.	**Chemical Properties**		
10.	Effect of wood on oil		
	(i) Increase in acidity, mg KOH/g, (max.)	0.1	0.1
	(ii) Increase in sludge content, % (max.)	0.05	0.05
V.	**Methods of Test**	IS : 1998	IS : 1998

These conductors are used to obtain better mechanical withstand properties without sacrificing the electrical properties which are difficult to meet in case single conductors are used.

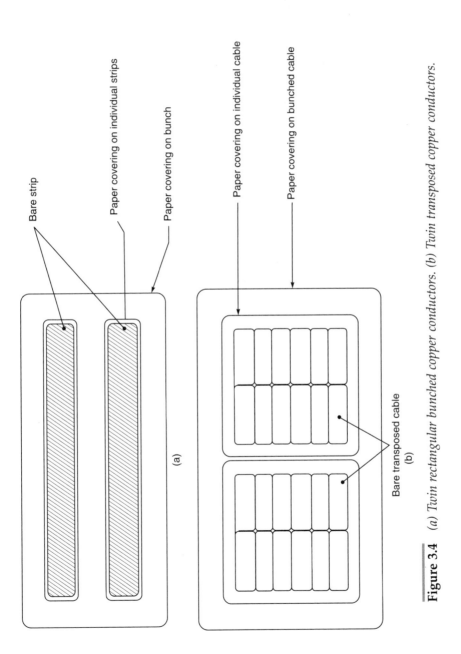

Figure 3.4 (a) Twin rectangular bunched copper conductors. (b) Twin transposed copper conductors.

However, due care is required to be taken for manufacturing of windings with these conductors, as these conductors especially conductors at (e) have limited shelf life and the glued (epoxy) strip is susceptible to lose its mechanical properties at an exponential rate when exposed to higher temperatures. The shelf life of these conductors is expected to reduce by half for every 10°C increase in temperature.

The bare strips are made from electrolytic tough pitch (ETP) grade copper wire bars to IS : 191 with high conductivity (99.14% IACS or 57.5 Sm/mm^2) and annealed complying with requirements of IS : 6160. The manufacturing operations involved are hot rolling, pickling, shaving, drawing, flattening, annealing and finally covering with requisite number of kraft paper layers. Specified radius is given on the four corners of a strip to avoid sharp edges. Since transformer windings are to withstand power frequency, impulse and switching surge voltages during factory tests and operation, it is required that a strip is perfectly smooth. The surface should be free from various defects like spills, cracks, slivers, scratches, pits, black spots and copper dust. Micro projection on the strip surface may puncture a few layers of insulating paper. Surface defects are likely to produce partial discharges due to nonuniformity of field, which may ultimately lead to conductor insulation breakdown. This is the reason for giving importance to electric strength test on these conductors. Table 3.8 gives test voltage for electric strength test on rectangular and transposed conductors as well as on paper covered stranded and flexible cables.

High air permeability kraft paper with properties in accordance with Table 3.5 is used as covering. All layers except the outermost are butt wound, the outermost layer being overlap wound. Maximum thickness of paper for butt wound layers is normally 75 μm. Increase in dimension due to paper covering is measured by difference method as well as by adding up the thickness of paper used in each layer to ensure reasonable tightness of covering.

3.5.1 Continuously Transposed Copper Conductor (CTC)

CTC comprises an odd number of high conductivity annealed copper strips insulated with a polyvinyl-acetal based enamel, arranged in two side-by-side stacks with the individual enamelled conductors

Table 3.8 Test Voltage for Electric Strength Test on Rectangular/Transposed Conductor and Flexible/Stranded Cable

Radial insulation thickness (mm)	Test voltage breakdown voltage in oil at 90°C (kV rms)
0.5	20
1.0	35
1.5	48
2.0	61
2.5	72
3.0	83
3.5	93
4.0	104
4.5	113
5.0	123
7.0	160
10.0	199

continuously transposed throughout their length. Additional insulation is provided by covering the transposed stack by a number of paper layers, as for rectangular single conductors. Figures 3.5(a) and (b) show the general view and constructional details and Fig. 3.6 depicts the transposition of individual strips in a CTC.

Pitch of transposition is the distance between adjacent crossovers measured on one side of the conductor and is normally 15 times the individual strip width. At least one full transposition should be made in a full turn of the winding. Hence, the minimum winding diameter is decided by the relationship.

$$D_{min} = \frac{\text{pitch of transposition} \times \text{no. of conductors}}{\pi}$$

CTC offers the following technical advantages over single rectangular conductor (economic aspects are also taken into account for optimum design):

(a) Winding time is reduced, since the need for hand transpositions is partially or completely eliminated.
(b) The winding space factor improves noticeably compared with single rectangular conductors.

64 Transformers

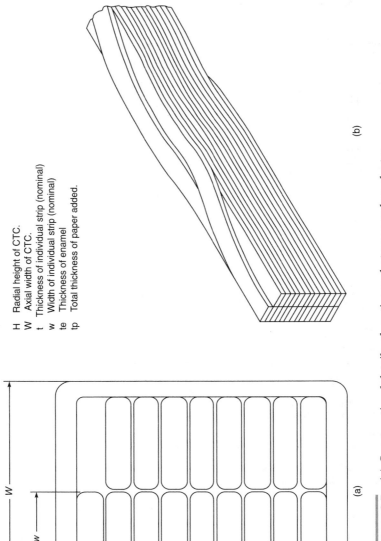

H Radial height of CTC.
W Axial width of CTC.
t Thickness of individual strip (nominal)
w Width of individual strip (nominal)
te Thickness of enamel
tp Total thickness of paper added.

Figure 3.5 *(a) Constructional details of continuously transposed conductor. (b) General view of transposed-strip conductor (27 strips in parallel).*

Materials Used in Transformers

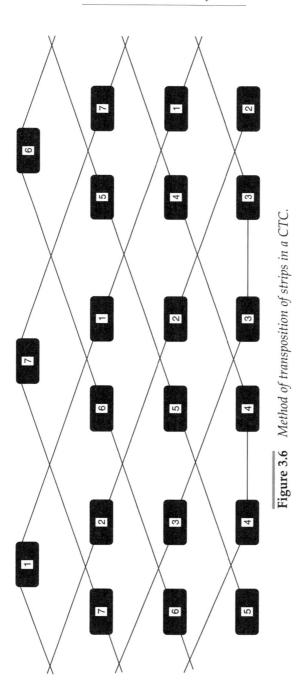

Figure 3.6 *Method of transposition of strips in a CTC.*

(c) By employing a number of small conductors which are continuously transposed, eddy losses, due to circulating currents between adjacent turns are reduced and hence total copper-losses of transformer due to windings are also reduced.

(d) Transposed conductor offers an improvement in the mechanical strength of the winding.

3.6 Crepe Paper Covered Flexible Copper Cable

Tapping leads and high voltage leads are required to be given sharp bends up to a radius of four to five times the overall diameter of cable. The cables used have high flexibility due to use of a fairly large number of thin wires. Clubbed with the benefits of crepe paper mentioned in Sec. 3.2, these cables provide extra-high flexibility and damage to paper insulation at sharp bends is eliminated.

3.7 Sealing Materials

For all sealing purposes, to avoid oil leakage at joints, gasket and moulded components made from nitrile rubber or nitrile rubber bonded cork are used. Characteristics of these materials are given in Table 3.9. PTFE and polyamide washer, tape, etc., are used for plugs and valves.

3.7.1 Nitrile Rubber Bonded Cork

This is the most widely used general purpose gasket. The cork gasket is adaptable even to rough unmachined surface by virtue of its compressibility under light seating loads. This gasket is made by the compounding of the granular cork with medium nitrile butadiene rubber. The cork used is clean, soft-grade type, uniformly granulated and free from hard board, wood flour, dust and other foreign materials. Nitrile rubber contributes to the "sealability" factor that enables the bonded cork to effect a tight seal under very light seating loads and also permits considerable distortion of gasket without crack and damage. The presence of cork allows compression without excessive spreading associated with solid rubbers.

Table 3.9 Schedule of Characteristics of Gasket Materials

Sl. No.	Characteristic	Requirement	
		Nitrile rubber bonded cork	Nitrile rubber
I.	**Mechanical Properties**		
1.	Hardness, IRHD	70 ± 10	60 ± 5
2.	Tensile strength, MPa (min.)	1.75	8.4
3.	Elongation at break % (min.)	—	400
4.	Compressibility at 2.8 MPa, % (max.)	25–35	—
5.	Recovery % (min.)	80	—
6.	Compression set, % (max.)	85	30
7.	Flexibility	No breakage through the granules of cork or separation of granules should occur.	—
II.	**Effect of Ageing in Air**		
8.	Increase in hardness, IRHD (max.)	N.A.	10
III.	**Compatibility with Insulating Oil**		
9.	Increase in acidity, mg KOH/g (max.)	0.1	—
10.	Increase in sludge content, % (max.)	0.05	—
IV.	**Methods of Test**	IS : 4253	IS : 3400

3.7.2 Nitrile Rubber

This is used as gasket normally with a metal limiter to avoid excessive compression and as sealing rings and mouldings, etc. Nitrile rubber is a combination of butadiene and acrylonitrile rubber conforming to compound BA 60 of BS 2751. The resilience of nitrile rubber makes it possible to reuse the gasket after opening a gasketed joint. However, this requires machined flange surfaces. For smaller flanged openings, a groove is cut into the flange for seating of nitrile rubber gasket. In such a case, limiter is normally not required.

3.8 Cold Rolled Grain-Oriented Electrical Steel Sheet (CRGO)

CRGO electrical steel with an approximate silicon content of 3% is used for magnetic circuits of a transformer. The following features influence selection of the type of steel sheet.

(a) Maximum magnetic induction to obtain a high induction amplitude in an alternating field.
(b) Minimum specific core-loss for low no-load loss.
(c) Low apparent power input for low no-load current.
(d) Low magnetostriction for low noise level.
(e) High grade surface insulation.
(f) Good mechanical processing properties.

CRGO made from a ferrous base present maximum magnetizability, i.e., permitting a high induction. Iron crystallizes into a body-centered cubic lattice, with the cube edges of the lattice pointing in the direction of easiest magnetizability and lowest, core-loss. Grain-oriented electrical sheet consists of a silicon-iron alloy, with the crystallites being predominantly oriented by means of a specific manufacturing process, in such a way as to have four cube edges pointing in the rolling direction and the diagonal plane being parallel to the sheet surface. In this way the rolling direction becomes the direction of maximum magnetic properties and approaching the ideal properties of the individual crystallite. The more pronounced the texture, the nearer the properties become to those of an individual crystallite.

3.8.1 Grades of CRGO

CRGO steel for transformers is available in different sheet thicknesses, ranging from 0.18 thickness, to 0.35 thickness, and accordingly the maximum watt loss/kg of the material differs. Further, based on the grain orientation they are also categorized as conventional grain-oriented (CGO) and HI-B steels. HI-B steels have low watt loss/kg when compared to CGO steel. The schedule of characteristics of some typical grades commonly used in transformers is given in Table 3.10.

> **Table 3.10** Schedule of Characteristics of CRGO Electrical Steel (Typical)

Sl. No.	Characteristic	Grade CGO	Grade HI-B
1	2	3	4
I.	**Physical Checks**		
1.	Thickness, mm	(0.18, 0.23, 0.27, 0.30, 0.35)	
2.	Edge camber, mm (max.)	2 for 2000 mm sample length	
3.	Edge burr, mm (max.)	0.05	0.05
4.	Waviness (max.)	5 mm for 2000 mm long sample, height of any wave not to exceed 1/80 of the length of a wave, or 1 mm, whichever is the greater.	
5.	Stacking factor, % (min.)	93.5 to 96.5 depending on thickness	
II.	**Magnetic Properties**		
6.	Specific core-loss, W/kg (max.) at 50 Hz.		
	(i) At 1.5 T	0.69–1.11	—
	(ii) At 1.7 T	1.17 to 1.57	0.95–1.17
7.	Ageing test (increase in core-loss after ageing at 150°C for 14 days), % (max.)	3	3
8.	Magnetic induction at magnetizing force of 800 A/m (B_8), T (min.)	1.75	1.85
9.	Specific apparent power at induction of 1.5 T, VA/kg (max.)	1.25	1.0
10.	Saturation induction, T	2.03	2.03

Contd.

> **Table 3.10** (Contd.)

Sl. No.	Characteristic	Grade CGO	Grade HI–B
1	2	3	4
III.	**Electrical Properties**		
11.	Resistance of surface coating, Ω (min.)	2 for 80% of the readings 5 for 50% of the readings	
IV.	**Mechanical Properties (Typical)**		
12.	Ductility (Bend test)	Min. one 180° bend without fracture	
13.	Tensile strength, MPa		
	(i) Longitudinal	350	330
	(ii) Transverse	420	390
14.	Yield point, MPa		
	(i) Longitudinal	330	315
	(ii) Transverse	360	325
15.	Elongation, %		
	(i) Longitudinal	6	8
	(ii) Transverse	24	30
16.	Hardness, HV (load 50 N)	175	170
V.	**Physical Properties (Typical)**		
17.	Density, g/cm^3	7.65	7.65
18.	Specific electric resistance, $\mu\Omega$ cm	48	45
19.	Thermal conductivity at 25°C, J/m sK	26	28
VI.	**Methods of Test**	BS 6404—Part 5	
VII.	**Applicable standards**	BS 6404—Part 2 DIN 46400 JIS C 2553 ASTM 876 M	

Specific core-loss is made up of hysteresis and eddy current losses. Hysteresis loss is usually reduced by making an improvement in grain orientation. As the (100) pole figures in Fig. 3.7 show, the mean deviation angle of the (100) axis from the rolling direction is about 3° for HI-B, which is far lower than the average 7° for CGO. Thus in terms of hysteresis loss, HI-B is one or two grades superior

Materials Used in Transformers 71

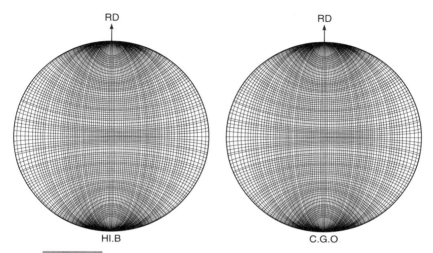

Figure 3.7 *(100) pole figures showing the grain orientation of HI-B and CGO.*

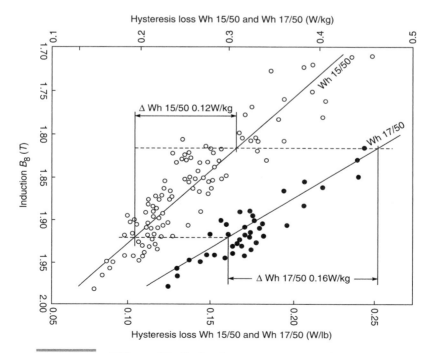

Figure 3.8 *Effects of B_8 (induction at magnetizing force of 800 A/m) value on hysteresis loss.*

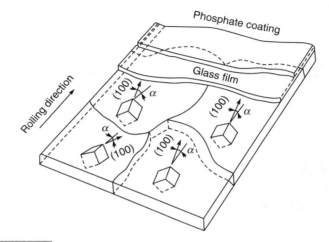

Figure 3.9 *Grain arrangement of GO—Heavy arrows show the (100) direction.*

than conventional material (Fig. 3.8). Eddy current loss depends to a large extent on the sheet thickness, the frequency of the alternating field and to a minor degree on the electrical conductivity. Eddy current loss is usually improved by applying a surface coating which generally consists of a glass film and a phosphate coating (Fig. 3.9). This surface insulation, uniformly coated on both sides withstands stress relief annealing, without deterioration in its adhesion or electrical insulation value, at temperatures up to 840°C. The mean value of the insulation resistance of the coating is at least 10 Ω cm^2 on each side.

Figures 3.10 and 3.11 show the improvements in core-loss of HI-B steel over CGO when the induction is over 1.5 T. Therefore, HI-B permits the manufacture of transformers with the same no-load loss as before, but a higher induction amplitude with the resulting economic advantages such as smaller size, less weight and reduced amount of material as shown in Fig. 3.12.

Reduction in core-loss in CRGO and HI-B grade steel has been possible with reduced thickness, and by change/modifications in the metallurgical process. Now, with present day techniques, it is further possible to reduce the core-losses by refining the magnetic domain structure of the steel. This technology involves in further subdivision of the width of the magnetic domains by physical means, or by electrolytic etching, or by laser irradiation.

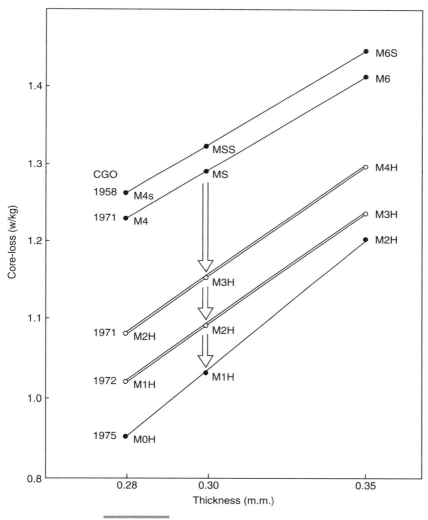

Figure 3.10 *Improvement in core-loss.*

Table 3.11 lays down the possible methods of reducing core-loss:

However, when domain refined steel is used for manufacturing of the transformer core, it is very important to keep in mind that this material should not be subjected to annealing, unlike CGO and HI-B grade steels, as this process would totally disturb the domain refinement achieved, and would lose the purpose of its use, i.e., low core-loss.

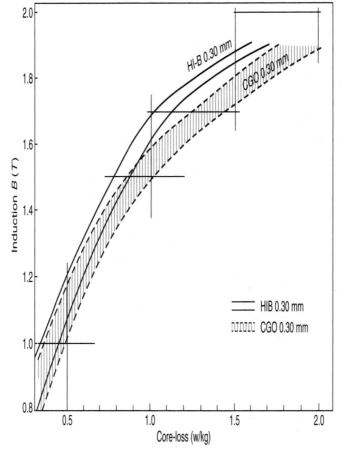

Figure 3.11 Comparison of B.W. curves between HI-B and CGO.

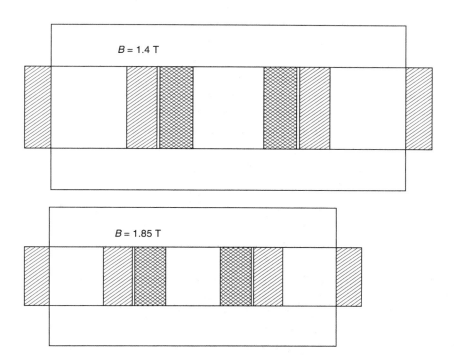

Figure 3.12 *Schematic comparison between a three-phase transformer with a mean induction amplitude and one with a high induction amplitude.*

➢ **Table 3.11** Methods of Reducing Core-loss

Core-loss		Core-loss reducing method
Reduction in hysteresis loss		1. Improvement in orientation.
		2. Reduction in inclusions and impurity.
		3. Reduction in internal strain.
Reduction in Eddy current	Classical eddy current losses	1. Reduction in thickness.
		2. Increase in silicon content to increase resistivity.
	Anomalous eddy current loss	1. Grain size.
		2. Refinement of magnetic domain.

Alternatively, transformers can be manufactured with 5 to 20% reduction in no-load loss depending upon the grade and induction amplitude using HI-B steel.

3.8.2 Core-loss in Cross-Grain Direction

If magnetization is applied in directions other than the rolling direction, the core-loss increases substantially. Core-loss is more than three-folds at 90° to the rolling direction and more than four-folds at 60°. This aspect is taken care of while designing the magnetic circuit of transformers. Core-loss of HI-B steel in 90° direction is inferior to those of conventional steel. The B-H curves in Fig. 3.13 indicate that HI-B has a high permeability. However, saturation induction of both HI-B and CGO is comparable.

Due to lower specific apparent power of HI-B steel, 10 to 50% reduction in transformer exciting volt-amperage is obtained.

3.8.3 Magnetostriction

For environmental protection, the noise level produced by the transformers has recently been a significant factor for the user. Such transformer noise is generated by the iron core of the transformer, as well as from that made by the fans. In an ideal core, the laminations of which are pressed together in such a way as to prevent any fluttering or shifting, the noise is generated by the magnetostriction of the laminations in the alternating field, causing the core to vibrate and thus act as an acoustic source. If the above mechanical conditions of the core manufacture are not observed, other causes (such as magnetic vibrations) may also be the reason for noise.

Magnetostriction, i.e., the change in configuration of a magnetizable body in a magnetic field, leads to periodic changes in the length of the body in an alternating magnetic field. The frequency of this magnetostriction is twice as large as that of the existing alternating field, i.e. when magnetizing an electrical sheet with 50 Hz the fundamental frequency of magnetostriction is 100 Hz. The process of change in length as a function of induction is not strictly a linear one, so that the magnetostriction of an alternating field gives both fundamental and harmonic frequencies. The fundamental frequency is predominant in an alternating magnetic field up to an induction

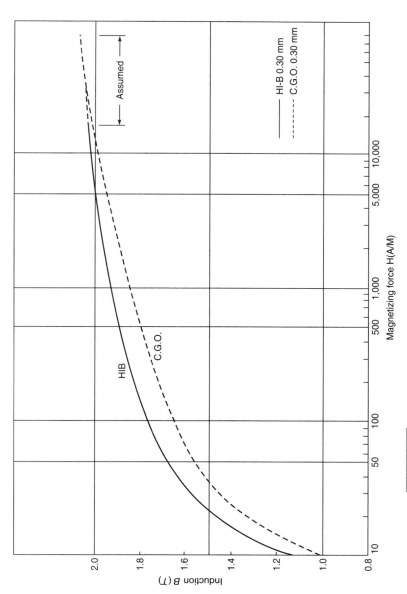

Figure 3.13 Comparison of D-C B-H curves between HI-B and CGO.

1.7 T, while as a result of the sophisticated process of alternating contractions and expansions, the harmonic frequency prevails for induction above 1.7 T. Magnetostricton is minimum in the rolling direction, while in the 90°-direction it is maximum. Except for the 90°-direction, all other directions that present major deviations from the rolling direction have a high percentage of harmonic frequencies in the spectrum, which are particularly undesirable because they generate harsh noise in frequency ranges, where the human ear is especially sensitive. Therefore, for sheets used in transformers, all transverse and oblique effects are kept to a minimum and very high induction amplitudes are prevented in the rolling direction.

Magnetostriction of HI-B is lower than that of CGO (Fig. 3.14) and consequently 2 to 7 dB reduction in the noise level of transformer is achieved.

3.8.4 Stress Sensitivity

Mechanical stresses exert a significant effect on the magnetic properties of CRGO electrical sheet. Such stresses are introduced in a sheet by way of external forces and by way of plastic deformation (internal stresses). External stresses are developed due to the following:

(a) When the sheets of a core are exposed to tensile or compressive stresses exerted by the fixings with the longitudinal forces in the limbs and the yokes.
(b) Bending stresses, where the sheets are forced into waves (for instance, as a result of heavily tightened clamp plates or end frames).
(c) Bending stresses due to deflection (for instance, in the case of non-uniform corner jointing).
(d) When use is made of wavy or bent sheets, which are necessarily flattened in the core.
(e) When sheets cannot adopt their natural condition as a result of friction existing between them.

Internal stresses occur along the cut edges during each cutting operation and as a result of bending the sheet or subjecting it to tension beyond the yield point (in slitting lines, continuous furnaces, etc.). Bending stresses can be divided into components of

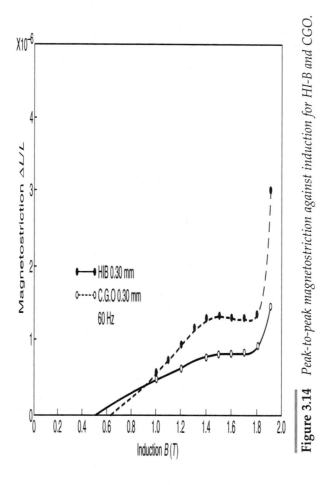

Figure 3.14 Peak-to-peak magnetostriction against induction for HI-B and CGO.

tension and compression. Similarly in simple cases, internal stresses can be resolved into pairs of tensile and compressive components. A few examples are given in Fig. 3.15.

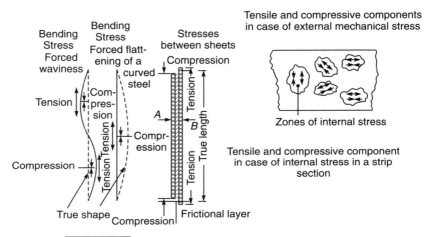

Figure 3.15 *Examples of mechanical stress in a sheet.*

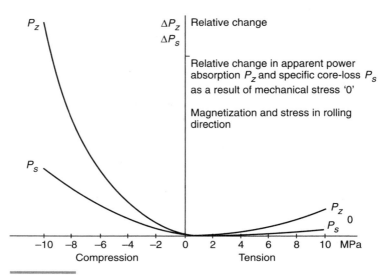

Figure 3.16 *Core-loss and apparent power absorption of CRGO steel as a function of external mechanical stress.*

Figure 3.16 displays the specific core-loss P_s and the apparent power P_z for an induction of 1.7 T and various levels of mechanical stresses. Both P_s and P_z are highly dependent on the compressive stress but only slightly dependent on the tensile stress. Thus, the apparent power is more sensitive than the core-loss.

Figure 3.17 shows the dependence of magnetostriction upon the tensile and compressive stresses for an induction of 1.7 T. Here too, effect of compressive stress is much bigger than that of tensile stress. Hence compression beyond a certain level occurring in the plane of the sheet is harmful and therefore, stresses must be removed by stress relief annealing. Sheets affected by waviness or coil-set are maintained to a flat shape, since they are subject to elastic stresses when being nested into the final shape of the transformer cores due to pressure. Buckled sheets must be discarded. Stress relief annealing of laminations is done in a continuous rolled hearth-type furnace at 820 ± 20°C for approximately 2 minutes in an air atmosphere and cooled quickly to ambient temperature.

Internal residual stress inherent in HI-B strip in as-delivered condition is lower than that in CGO, thereby stress-relief annealing may be omitted in many cases.

3.9 Structural Steel

Structural steel mainly in sheet and plate form is used in the fabrication of transformer tank, radiator, conservator, clamp plate, end frame, marshalling box, cable box, roller, turret and inspection cover, etc. Normally, standard quality mild steel is conventionally used for core clamp plates. However, it is economical to use high tensile strength (HTS) steel plate, since by reducing the thickness of clamp plates, magnetic core area can be increased, resulting in improvement in the utilization factor. In such a case, lifting pins required to connect clamp plate to yoke end frame are manufactured from HTS bar (1.5% nickel-chromium-molybdenum) having a very high tensile strength in the range of 900 to 1050 MPa.

Stainless steel (austenitic chromium nickel) plates are used as cover for turrets of high current bushings to neutralize the effect of eddy current. For medium currents, magnetic circuit perpendicular to a current carrying conductor is broken by welding stainless steel inserts. These plates are titanium stabilized for weldability with

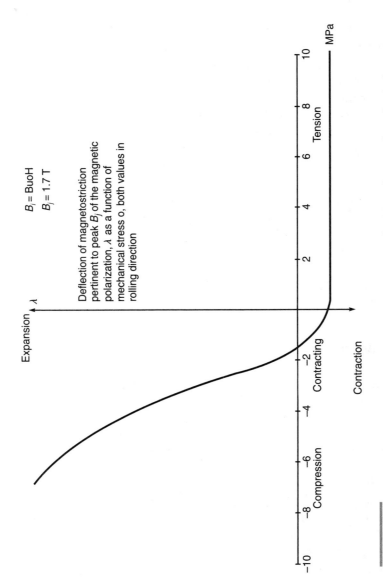

Figure 3.17 *Alternating field magnetostriction CRGO as a function of external mechanical stress.*

standard quality magnetic steels. Depending upon the requirements of application of the component, particular quality of steel is used, details of which are given in Chapter 10.

Mechanical properties of a few qualities of structural steel are given in Table 3.12.

➤ **Table 3.12** Properties of Structural Steel

Sl. No.	Material	Tensile strength MPa	Yield strength MPa (Min.)	Elongation %, (Min.)
1.	Cold rolled carbon steel sheet	275	—	—
2.	Hot rolled carbon steel sheet	330–410	205	18 (on 80 mm gauge length) 25 for thickness above 3 mm.
3.	Structural steel Std. quality Nominal thickness/diameter, mm			
	(i) 6 to 20	410–530	250	23
	(ii) 20 to 40	410–530	240	23
	(iii) Over 40	410–530	230	23
4.	Bright steel bar and section-cold drawn	430 (min.)	335	14
5.	High tensile strength structural steel plate Nominal thickness/diameter (mm)			
	(i) 6 to 16	540 (min.)	350	20
	(ii) 16 to 32	540 (min.)	340	20
	(iii) 32 to 63	510 (min.)	330	20
	(iv) Above 63	490 (min.)	280	20
6.	HTS Steel Bar (1.5% nickel-chromium-molybdenum)-hardened and tempered	900–1050	700	15
7.	Austenitic chromium-nickel-steel-titanium stabilized plate (stainless steel)	520 (min.)	—	40

3.10 Future Trends

In the recent years, intensive efforts have been undertaken to develop new transformer insulation systems which permit a more compact and thus more economical design. However, these have been, more or less, limited to small transformers up to a few MVA. Dry type transformer, silicone oil filled transformer, the vapor cooled and SF_6-insulated transformers are already available today. However these special transformers are used for a specific purpose, e.g., for weight reasons in locomotives and are normally costlier than conventional oil-cellulose transformer. The current level of knowledge and higher cost put limitations on these new insulation systems. Studies have shown that conventional oil barrier insulation system will cause no problems up to operating voltages of 2000 kV. Hence, it is very unlikely that oil impregnated cellulose insulation system will be replaced by a new insulation system for large size power transformer, at least in the near future.

To increase the short-circuit strength of windings, annealed copper conductor is already being replaced by controlled proof stress copper. Copper-silver alloy will replace high conductivity copper, specially in higher rating transformers. Under short-circuit conditions in the windings, copper-silver alloy conductor is less susceptible to annealing and is thus easily able to retain its strength. Mechanical strength of transposed conductor is further increased by bonding together all the conductors in a stack, by use of an epoxy bonding enamel over PVA enamelled strips. These changes are likely to take place faster, to cover almost all transformers. However, copper is expected to be used unless it is replaced by a new generation development of a superconducting material, since anodized aluminum strip conductors have their limitations for HV windings.

Another possible substitute for grain-oriented electrical steel for magnetic core of transformer is amorphous steel, also referred as metallic glass, which is a non-crystalline solid created by rapid quenching of metal-metalloid alloys. Amorphous steel (typical composition $Fe_{8t} B_{13.5} Si_{3.5} C_2$) of thickness 25 to 50 µm and widths up to 100 mm are now commercially available. It is expected that 175 mm wide strips would be commercially available shortly. However, there is apprehension about the technological capability of producing metallic glass in widths of 800–1000 mm to match with CRGO width.

There are numerous practical problems to be sorted out, e.g., complete transformer will have to be annealed for stress relieving, therefore pre-wound transformer coils are to be made suitable to withstand annealing temperature. Further, saturation induction of metallic glass is lower than that of CRGO and hence transformer has to be designed at lower induction. Experimental transformers of maximum rating of 100 KVA have been made so far, using amorphous steel. Nevertheless, the material holds some promise for the future, as it is expected that the problem due to its present prohibitive cost and other limitations can be overcome.

References

1. Viswanathan, P.N. et al.
   "Paper for Cable Insulation from Indigenous Wood Pulps", National Seminar on Electrical Insulation held in Nov. 1974 sponsored by the Institution of Engineers (India), Electrical Engg. Division, IEMA and ISI.
2. Philip, P.K., "Insulation Pressboard for Transformers," *Electrical India*, Vol. XIX, No. 24, pp. 5–13, Jan. 1980.
3. BICC Connolly's U.K. Publication No. 798,
   "Continuously Transposed Conductor."
4. Thyssen Grillo Funke, W.G., Brochure, *"Grain Oriented Electrical Sheet,"* 1980 edition.
5. *"Technical Data on Orient-Core,* HI-B,"
   Nippon Steel Corporation, Japan, Cat. No. EXE 367, Dec. 1976.
6. "Technical Data on Domain Refined Orient Core HI–B.," Nippon Steel Corporation, Japan. Cat. No. EXE. 706, Nov. 1987.

CHAPTER 4

Magnetic Circuit

K.N. Labh, R.C. Agarwal

In a transformer, energy is transferred from one electrical circuit to another through the magnetic field. Transformer core made of laminated sheets provides the magnetic circuit for the flow of magnetic flux mutually linking the electrical circuits. As against the air core, iron core provides a comparatively low reluctance path to the magnetic flux with consequent benefit of (a) smaller magnetizing current, (b) increase in the total flux linkage and (c) a high ratio of mutual to leakage flux resulting in reduction of stray losses. Its design, type and manufacturing methods have significant bearing on quality, transportability, operational limitations and guaranteed technical performance of the transformer.

4.1 Material

Some of the very early transformer cores were made of inferior grades of laminated steels which had inherently higher core losses and showed pronounced ageing effects, further aggravating the hysteresis component of iron-losses in the equipment. It was subsequently found that very small quantities of silicon alloyed with low carbon content steel produced a material with low hysteresis losses and high permeability. These steel sheets alloyed with silicon mitigated the problem of ageing and improved the permeability and consequently reducing the magnetizing current and core losses. In the ever-increasing pursuit of increasing the power ratings and reduction of core-losses, another innovative technique from steel manufacturers came in the form of cold rolling with orientation of the grain in the direction of rolling. This core steel known as CRGOS (cold rolled grain-oriented silicon steel) has the minimum epstein losses to the flow of magnetic flux along the directions of grain orientation and this material is universally used for the manufacture of transformer cores.

The adoption of CRGOS has brought about considerable reduction in the specific iron-losses (W/kg) over the earlier grades of core steel. However, CRGOS is susceptible to increased losses due to flux flow in directions other than that of grain orientation, effect of mechanical strain due to clamping pressure, bolt holes, jointing of limb with yokes,[1] etc. Apart from this sensitivity to the direction of rolling, CRGOS sheets are also very susceptible to impaired performance due to impact of bending, blanking the cutting. Both surfaces of the core steel sheets are provided with an insulating of oxide coating (commercially known as Carlite). The stacking factor of lamination improves by using thicker laminations, but eddy current loss goes up in proportion to square of the thickness of the lamination. For reducing the eddy current losses, thinner laminations are preferable even though the stacking factor goes down. Deburring of the laminations improves the stacking factor and minimizes the eddy losses. After machining, the material has to be annealed at 800–900°C in a neutral gas environment. The material properties of core steel are further discussed in Chapter 3.

4.2 Design of Magnetic Circuit

For a transformer design, the basic governing factors are:

(a) Rating of transformer and its performance,
(b) Operational conditions,
(c) Transport limitations (i.e., height, length, width and weight, etc.)

The design of the magnetic circuit, i.e., transformer core is also based on the above considerations and it has significant bearing on the overall economy of the transformer. For CRGOS, saturation may occur at the magnetic flux densities exceeding 1.9 tesla. Based on the input voltage and frequency variations, a suitable value of flux density can be adopted to avoid any chance of core saturation under operating conditions. By increasing the operating value of magnetic flux density (magnetic loading) the net weight of core can be reduced, but this leads to corresponding increase in the core losses. One has to compute an optimum value of the magnetic flux density, keeping in view all the above considerations.

4.2.1 Constructional Features

The type of transformer core construction depends on the technical particulars of the transformer and transport considerations. In general it is preferable to accommodate the windings of all the three phases in a single core frame. Three-phase transformers are economical over a bank of three single-phase transformers. Another important advantage of three-phase transformer cores is that component of the third and its multiple harmonics of mmf cancel each other, consequently the secondary voltage wave shape are free from distortions due to the third harmonics in mmf. However, if the three-phase ratings are large enough and difficult to transport, one has no choice but to go for single-phase transformer units.

For single-phase and three-phase transformers, the cores can be broadly classed as:
    (a) Single-phase three-limbed core
    (b) Single-phase two-limbed core
    (c) Three-phase three-limbed core
    (d) Three-phase five-limbed core.

(a) *Single-Phase Three-Limbed Core [Fig. 4.1(a)]*

The windings are placed around the central limb, also known as main limb. The main magnetic flux generated in the central limb gets divided into two parallel return paths provided by the yokes and auxiliary limbs. For the same magnetic flux density as that in the main limb, the auxiliary limbs and the yokes need to have the cross section only half of the main limb. This type of transformer core is generally preferred for single-phase transformer, as this is more economical than two limbed construction discussed below.

(b) *Single-Phase Two-Limbed Core [Fig. 4.1(b)]*

Sometimes the single-phase power ratings of transformers are so large that if the windings of full power ratings were to be placed on the central limb, its width would become too large to be transported.

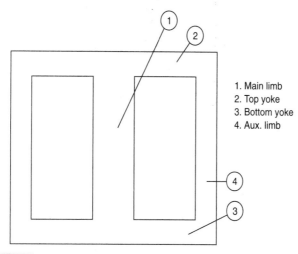

Figure 4.1 *(a) Single-phase three-limbed core: (1) main limb, (2) top yoke, (3) bottom yoke, (4) aux. limb.*

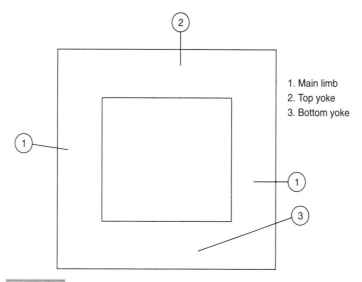

Figure 4.1 *(b) Single-phase two-limbed core: (1) main limb, (2) top yoke, (3) bottom yoke.*

To mitigate such difficulties the windings are split into two parts and placed around two separate limbs. Here the cross-sectional area of the legs (limbs) and the yokes are identical. Consequently these cores are bulkier than the single-phase three-limbed arrangements. Also the percentage leakage reactance for this type of core construction is comparatively higher due to distributed nature of the windings in the two limbs separately.

(c) *Three-phase Three-limbed Cores (Fig. 4.2)*

This type of core is generally used for three-phase power transformer of small and medium power ratings. Each phase of the winding is placed around one leg. For each phase of magnetic flux appearing in a limb, the yokes and the remaining two limbs provide the return path. If the phase fluxes are denoted as ϕ_A, ϕ_B, ϕ_C, their summation at any instant of time is identically zero, which can be mathematically stated as $\phi_A + \phi_B + \phi_C = 0$. In this type of construction, all the legs and the yokes have identical cross section.

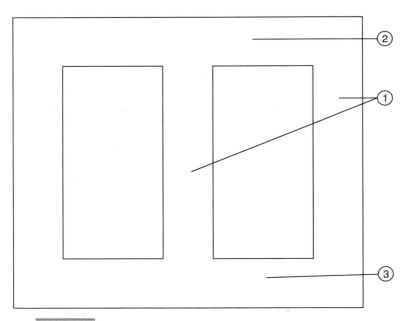

Figure 4.2 *Three-phase three-limbed core: (1) main limbs, (2) top yokes, (3) bottom yoke.*

(d) *Three-phase Five-limbed Cores (Fig. 4.3)*

For large rating power transformers, cores have to be built in large diameters. In case of three-phase three-limbed cores, the yokes

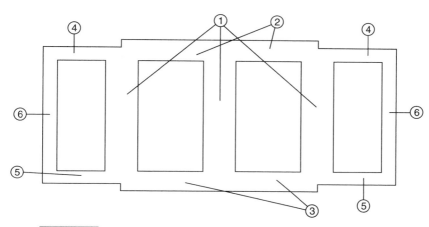

Figure 4.3 *Three-phase five-limbed core: (1) main limbs, (2) top main yokes, (3) bottom main yokes, (4) auxiliary top yokes, (5) auxiliary bottom yokes, (6) auxiliary limbs.*

have the same diameter as the limbs. In case of large diameter cores, the overall core height will go up leading to transport problem. For such cases the yoke cross-sections (and consequently yoke heights) are reduced by approximately 40% or more and auxiliary paths for the magnetic flux are provided through auxiliary yokes and limbs. The cross-section and the height of the auxiliary yokes and limbs are lower than that of the main yokes.

4.3 Optimum Design of Core

For the optimum design of magnetic core, the following aspects have to be decided.

(a) Constructional features
(b) Core cross-sectional area
(c) Number of oil ducts and location

4.3.1 Core Cross-Section

The ideal shape for the section of the cores is a circle, since this would waste no space beyond that taken up by the insulation between laminations. A perfectly circular core section, however, involves making a variation in dimensions for each successive lamination, which is possible but uneconomical. As a compromise solution, the core cross-section is made by laminations of varying widths and packet heights in such a way that the overall section approximates a circle. Such a typical core cross-section is shown in Fig. 4.4. Oil ducts are needed for cooling the core, lest the hot-spot temperature rises dangerously and their number depends on the core diameter and the specific core-losses, which is a function of operating flux density. Additionally, clamp plates made of steel are needed on either side of laminations for effectively clamping the laminations. These clamp plates should be mechanically strong enough to prevent buckling/bending of laminations and be able to withstand the lifting load of core and windings and axial short-circuit forces. The steel sheet laminations, oil ducts and the clamp plates should all lie within the core circle.

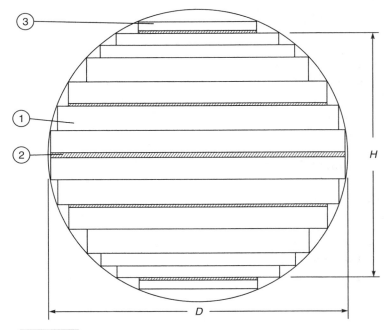

Figure 4.4 *Circular core cross-section: D – diameter of core, H – total lamination stack height (1) laminations, (2) oil duct, (3) steel clamp plates.*

The net sectional area is calculated from the dimensions of various packets and an allowance is made for the space lost between laminations (known as stacking factor) which for sheet steel of 0.28 mm thickness with carlite insulation coatings is approximately 0.96. Area is also deducted for the oil ducts. The ratio of the net cross-sectional area and the gross area of the core circle known as *utilization factor* (UF). By increasing the number of core steps UF improves. This, however, also increases the manufacturing cost. Typical cost-effective values for the number steps (i.e., one-half of cross-section from center line) lie in the range of 6 (for smaller diameters) to 15 (for large diameters). For any particular core diameter based on other design considerations, this gives out not only the optimum area and thereby reduction in the flux density and consequently iron-loss, but also helps the designer to revert to lesser value of core diameter, wherever the computational margins allow this latitude. Seen from another angle, improvement in the core utilization factor increases the core area and hence the value of volts/turn for any particular core diameter and specified flux density. This, in turn, results in the reduction in winding turns and thus reduction in copper. Therefore, core area optimization results in better economy of transformer designs.

In the following, the core optimization is discussed under (a) optimum selection of laminations (b) optimum oil ducts.

4.3.2 Optimum Selection of Laminations

For any particular core diameter, the first and foremost point to be decided is the maximum allowable height of the lamination packets. This, in turn, is determined by the design consideration of clamp plate (see Fig. 4.5) and the pertinent constraint in accommodating them inside the core circle.

For a circle of diameter D, the length of the cord at distance Y from the center is given by

$$F(Y) = 2\left(\sqrt{\left(\frac{D}{2}\right)^2 - Y^2}\right) \qquad (4.1)$$

If H is the maximum allowable packet height, the minimum allowable lamination width is given by

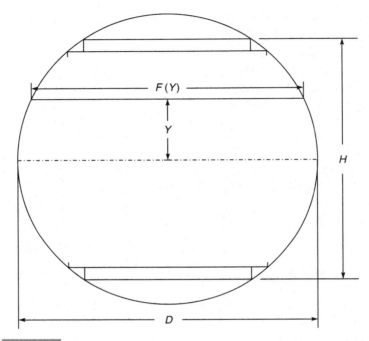

Figure 4.5 *Circular cross-section of core-dimensional relationship.*

$$L_{min.} = 2\left(\sqrt{\left(\frac{D}{2}\right)^2 - \left(\frac{H}{2}\right)^2}\right) \quad (4.2)$$

The maximum allowable width of the central packet is usually core diameter less G mm (to accommodate wooden packing)
i.e.,
$$L_{max.} = D - G \quad (4.3)$$
If the core is to be built up in N_s number of steps, where widths of laminations of individual steps are L_i, $i = 1, 2,... N_s$, these values must satisfy the following relations

$$L_{min.} \leq L_i \leq L_{max.} \quad (4.4)$$
$$i = 1, 2,...., N_s$$

For building the circular cross-section of the core in say N_s steps, we have to decide N_s different widths of laminations to be stacked one over the other as shown in Fig. 4.4. The packet height of each lamination is computed by the difference of heights of cords equal to the width of laminations and the adjacent next width below it. The gross cross-sectional area contributed by individual lamination packets is obtained by lamination widths times packet height. This is taken as the return function for selection of a particular width. The optimum selection of laminations is formulated on the method of dynamic programming. Recursive relationships are formulated for return function and state transition function for each step of decision taking. This method gives optimum selection of laminations for building in N_s number of steps.

Out of the optimum area available by the above method we must deduct areas for the oil ducts. The individual packet heights should be reduced slightly to allow for the overall manufacturing tolerance in the core built up. A computer program has been developed for the optimum selection of lamination packets based on the above algorithm. The computer program also computes (a) optimum number of oil ducts and their locations and (b) core lamination dimensions and other details required for manufacturing activities. This has automated the complete transformer core design and manufacturing informations.

4.3.3 Computation of Optimum Number of Oil Ducts

Core-losses (iron-losses) take place due to magnetic flux flow in the laminations. For CRGOS these losses are minimum for flux along the direction of grains orientation and maximum for flux flow along cross grain directions. Hot temperature spots are developed inside the core as a result of core-losses and designers must ensure that the hot-spot temperatures are well below the permissible values.

Segments of the core cross-section are rectangular as shown in Fig. 4.6. Heat dissipation takes place along the laminations (x-direction) and across the laminations (y-direction). The temperature gradients along the two directions are given by

$$T_x = W \cdot x \cdot \left(\frac{x}{2K_1} + h \right) \quad (4.5)$$

$$T_y = W \cdot y \cdot \left(\frac{y}{2K_2} + h \right) \quad (4.6)$$

where
  W = specific cross gain loss, W/kg,
  K_1 = thermal conductivity along the laminations, W/°C/mm,
  K_2 = thermal conductivity across the laminations W/°C/mm,

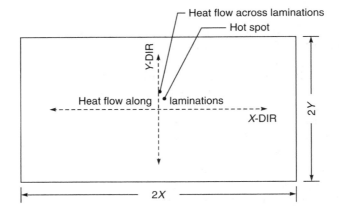

Figure 4.6 *Rectangular stack of core laminations heat flow conditions.*

h = surface heat transfer coefficient, °C/mm²/W.
T_x = temperature differential from hot spot to oil, assuming that all heat flows along the laminations.
T_y = temperature differential from hot spot to oil across lamination, assuming all heat flows across lamination.

The hot-spot temperature gradient

$$T_h = \frac{T_x \cdot T_y}{T_x + T_y} \qquad (4.7)$$

T_h (max. permissible) = maximum permissible hot-spot temperature–(oil rise + ambient temperature)

Equation (4.7) can be rewritten as

$$T_y = \frac{T_x \cdot T_h}{T_x - T_h} \qquad (4.8.1)$$

i.e.,
$$W \cdot y \left(\frac{y}{2K_2} + h \right) = \frac{T_x \cdot T_h}{T_x - T_h} \qquad (4.8.2)$$

The value of W depends on the type of core construction and remains constant for a particular type of assembly. The value of T_x increases parabolically with the value of x. An interesting condition arises when $x = x^*$ such that

$$T_x \leq T_h \qquad (4.9)$$

The equality condition of (4.9) implies that $y = \infty$, i.e., for $x = x^*$, T_h will never achieve its specified maximum value for value of y, and, along the lamination, dissipation of heat will be sufficient to maintain the value of T_h below the specified maximum permissible value. The second condition that $T_x < T_h$ leads to negative value of T_y, which is not physically possible. This also implies that along the lamination, dissipation is sufficient to contain the value of T_h below the specified limit. For such conditions as obtainable in relation (4.9), no oil duct is required for cooling.

We now examine the case when $T_x > T_h$. Equation (4.8.2) can be solved for y as

$$y^* = \frac{1}{2C_1}(-h + \sqrt{h^2 + 4C_1 \cdot C_2}) \qquad (4.10)$$

where
$$C_1 = \frac{1}{2K_2}$$

and
$$C_2 = \left(\frac{T_x \cdot T_h}{T_x - T_h}\right)\bigg/W \qquad (4.11)$$

Here, we introduce the concept of critical height $y \leq y^*$ obtainable from Eq. (4.10). For any stack height $2y$ such that $y \leq y^*$ for a specified value of x, the hot-spot temperature will never exceed the specified value of T_h. For actual stack height $H > 2y^*$, cooling surface (in the form of oil duct) is required if the hot-spot temperature is not to exceed T_h.

For any core diameter D, the minimum number of oil ducts is calculated by iteratively using the Eqs. (4.10) and (4.11) for computation of critical heights y^* starting from the central packet. Since above formulation stipulates that the maximum hot-spot temperature in the packet will be less than or equal to the specified permissible value, the above calculation gives the optimum number of oil ducts.

4.3.4 Location of Oil Ducts

For more than one oil duct, a suitable criterion is required regarding their location in the core section. For single duct, the obvious choice is its location at the center. Minimum difference of hot-spot temperatures (ideally zero) obtaining in different regions divided by the oil ducts is adopted here as the criterion of location of oil ducts. Figure 4.7 represents the case of two oil ducts in the core cross-section.

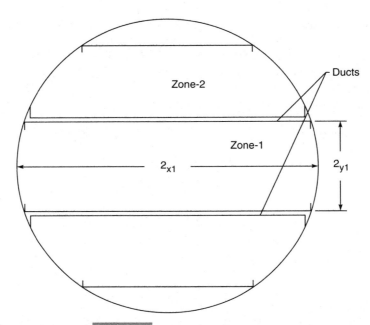

Figure 4.7 *Two oil ducts in core.*

For Zone 1

$$\frac{1}{T_{h1}} = \frac{1}{T_{x1}} + \frac{1}{T_{y1}} \qquad (4.12)$$

Where
$$\left. \begin{array}{l} T_{x1} = W \cdot x_1 \left(\dfrac{x_1}{2K_1} + h \right) \\[2mm] T_{y1} = W \cdot y_1 \left(\dfrac{y_1}{2K_2} + h \right) \end{array} \right\} \qquad (4.13)$$

and T_{h1} = Hot-spot temperature gradient of zone 1.

For Zone 2

$$\frac{1}{T_{h2}} = \frac{1}{T_{x2}} + \frac{1}{T_{y2}} \qquad (4.14)$$

where
$$\left. \begin{array}{l} T_{x2} = W \cdot x_2 \left(\dfrac{x_2}{2K_1} + h \right) \\[2mm] T_{y2} = W \cdot y_2 \left(\dfrac{y_2}{2K_2} + h \right) \end{array} \right\} \qquad \begin{array}{l}(4.15)\\ (4.15)\end{array}$$

and T_{h2} = Hot-spot temperature gradient of Zone 2.

By equating the hot-spot temperature gradients of zones 1 and 2, we obtain the following:

$$\frac{2K_1}{W \cdot x_1 (x_1 + 2K_1 h)} + \frac{2K_2}{W \cdot y_1 (y_2 + 2K_2 h)}$$
$$= \frac{2K_2}{W \cdot x_2 (x_2 + 2K_1 h)} + \frac{2K_2}{W \cdot y_2 (y_2 + 2K_2 h)} \quad (4.16)$$

By cancelling the common terms and expressing the difference of expressions of l.h.s. and r.h.s. along with a normalizing constant we compute a difference term as

$$C_0 = p \left(\frac{1}{x_1 (x_1 + 2K_1 h)} - \frac{1}{x_2 (x_2 + 2K_2 h)} \right.$$
$$\left. + \frac{K_2}{y_1 (y_1 + 2K_1 h)} - \frac{K_2}{y_2 (y_2 + 2K_2 h)} \right)$$
(4.17)

where
and
$$\left. \begin{array}{c} K_2 = K_2/K_1 \\ p = \text{normalizing constant} \\ y_1 + 2y_2 = H/2 \end{array} \right\} \quad (4.18)$$

$$x_2 = \sqrt{\left(\frac{D}{2}\right)^2 - y_1^2} \quad \Big\} \quad (4.19)$$

Since x_1 is a constant and y_2 and x_2 are related to x and y Eq. (4.19), we have

$$C_0 = f(y_1) \quad (4.20)$$

The desired value of y is the one which satisfies $C_0 \simeq 0$. This is obtained by an iterative procedure on computer. The distance of oil ducts from center of core circle are similarly computed for number of oil ducts more than two no similar principle.

In the foregoing, the broad concepts of optimum design of transformer cores of circular cross-section has been discussed. Based on the optimum core cross-sectional area and operating flux density, the windings for individual phases can be designed and from this the core window size can be worked out. This decides the overall transformer core frame dimensions for a particular type of core construction as illustrated in Figs. 4.1, 4.2 and 4.3.

4.4 Manufacturing

By adopting suitable technology for transformer cores and exercising care in handling of laminations and core building, it is possible to

achieve
- (a) higher reliability
- (b) reduction in iron-losses and magnetizing current
- (c) lowering material and labor cost
- (d) abatement of noise levels

It is necessary to apply quality checks at different stages of manufacturing to ensure quality and reliability. The core steel samples should be tested regularly for guaranteed epstein iron-loss values. The laminations should also be visually inspected and the rusted lot should be rejected. CRGOS sheet steels are susceptible to impaired losses due to cutting, punching, piercing, bending, etc. During cutting and piercing the edges develop burrs which may cut, as knife-edge, the insulation coatings on the adjacent laminations, in addition to lowering the stacking factor. It is imperative that laminations are deburred and annealed (for stress relieving), so that iron losses do not increase. For reducing the transformer noises, the laminations should be tightly clamped together and punch holes should be avoided as far as possible. The air gap at the joints can be controlled by working on tight tolerances so that value of magnetizing currents are kept to a minimum. These manufacturing aspects are discussed in more detail in the following:

4.4.1 Corner Jointing of Limbs with Yokes

Broadly speaking, the core losses can be split into (a) loss due to magnetic flux flow along the direction of grain orientation (with grain iron-losses) and (b) flux flow in cross-grain direction (cross-grain losses) occurring in the zones of jointing of limbs with yokes. The cross-grain losses depend to a large extent on the type of joints. The two must commonly used types of corner joints are (a) interleaved, (b) mitred.

(a) *Interleaved Joints (Fig. 4.8)*

Interleaved joints are the simplest, from the point of view of manufacturing. However, in the cross-grain zones the magnetic fluxes leave/enter the laminations in perpendicular direction to the grains and these losses are comparatively higher. Such type of jointings are usually preferred only for small rating transformers, where the total core-loss itself is very small.

(b) *Mitered Joints (Figs 4.9, 4.10 and 4.11)*

When the corners of the laminations are cut at 45°, the jointing is known as mitered joint. The cross-grain losses for this type of

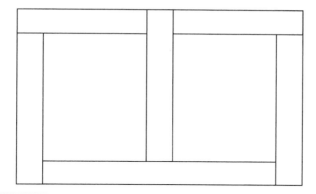

Figure 4.8 *Three-phase three-limbed core-interleaved joints*

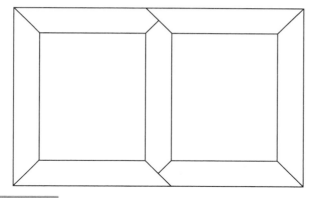

Figure 4.9 *Three-phase three-limbed core mitered joints.*

jointing is minimum, as the magnetic flux leaving/entering at the joint finds a smooth path of its flow. This, however, entails extra manufacturing cost for preparing the corner edges of the individual laminations.

Figure 4.9 shows one arrangement of laying the laminations for three-phase three-limbed cores with mitered joints. Figure 4.10 shows the laminations laying arrangement for the three-phase five-limbed cores with mitered joints.

Sometimes there may be constraints in preparation of laminations beyond certain widths. If lamination widths larger than these are required, as in case of large diameter cores, these are split into two halves, so that these can be handled easily in manufacturing. A typical core built with such split laminations is shown in Fig. 4.11.

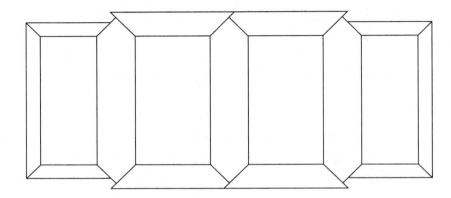

Figure 4.10 *Three-phase five-limbed core mitered joints.*

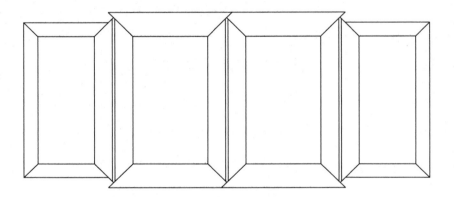

Figure 4.11 *Three-phase five-limbed core with mitered joints with split laminations for large core diameters.*

4.4.2 Preparation of Lamination Sheets

(a) *Slitting*

For building the transformer cores, lamination sheets of different widths and packet heights are needed. The manufacturing schedule may include cores of different diameters and different types of constructions necessitating slitting laminations in many widths and lengths. CRGOS rolls cannot be ordered in so many different widths and quantities. These rolls are available in standard widths of say 760, 790, 840, 915, 1000 mm, etc. For slitting operation, some widths can be combined together by suitably adjusting the cutter distances in the slitting machine.

It is evident that full width of roll cannot be utilized at any time of slitting operation and the leftover material will vary from stage to stage and depending on the widths selected in combination during the process of slitting. The meticulous care in planning is imperative to minimize wastage of core steel.

The slitting operation has been formulated based on the principle of dynamic programming to compute the optimum schedule. This provides the optimum combination of different widths to be taken together and the length for slitting, such that the scrap during slitting operation is minimum.

(b) *Cropping of Laminations*

The different shapes and sizes of laminations needed for core building are illustrated in Figs. 4.8–4.11 for different types of transformer cores.

In case of mitered core laminations, these are first cut in a trapezoidal shape. Finished shape is given by cutting off the corners (wherever needed) at the second stage of cutting by employing simple hydraulically operated guillotine machines.

(c) *Piercing Operation*

The yoke punchings (laminations) usually need holes for bolting the yoke laminations. These holes are punched after the cropping/guillotine operation by suitably adjusting the hole piercing positions in the piercing machine and selecting the right piercing tool for specified hole punch size. However, some cores are also constructed without yoke bolts.

(d) *Deburring*

During the process of slitting, cutting and piercing of laminations, the cut edges get some burrs. These burrs are removed by passing the laminations through deburring operations. Presence of burrs impairs the stacking factor. Also burrs cut into the insulation coatings and bridge adjacent laminations and thereby increase the eddy losses.

(e) Annealing/Varnishing

If the insulation coatings at the edges are scratched during deburring process or extra varnish coating is desired, the laminations are processed in a varnishing plant which provides a thin coating of varnish and quickly dried up at elevated temperature. However, varnish coating is not considered as necessary if the carlite insulation coating on the lamination is consistent. During the process of slitting, cropping, piercing, deburring, etc., mechanical strains are developed inside the laminations, which disturb the original grain orientation and thereby increase the iron-losses. This problem is mitigated by annealing the laminations in an annealing plant.

Annealing is done at an elevated temperature of 800–900°C, preferably in a neutral atmospheric zone and subsequently cooled by a blast of air.

The finished laminations are then taken to the core assembly area.

4.4.3 Core Assembly

Core building from the finished lamination sheets is done in horizontal position on specially raised platforms. The lamination sheets are susceptible to mechanical stresses of bending, twisting, impact, etc. A lot of care is exercised while handling and normally two persons are needed to hold the two ends of the laminations at the time of laying.

At first the clamp plates and end frame structure of one side of the core assembly are laid out. Guide pins are used at suitable positions for maintaining the proper alignments during core building process. Oil ducts are formed by sticking strips on lamination and put in position as required.

For each packet, the laminations are manufactured in two different lengths and these sets are laid out alternately, keeping at a time two to four laminations together. The two alternate arrangements provide overlapping at the corner joints and when the lamination packets are clamped together, these overlapping edges provide sufficient mechanical strength in holding the edges in tight grip. After laying out the complete laminations, the clamp plates, and end frame structure of the other side are laid out and the entire core-end frame structure is properly secured through bolts and steel bands at a number of positions.

The platform on which the core building takes place is of special design and the core-end frame assembly can be raised to the vertical position along with the platform which serves as a cradle. Subsequently the platform is disengaged. In this process, the core assembly is spared from the mechanical strain of lifting and raising in the vertical position. Small-size cores can however be built up without these special platforms.

Steel bands used for tightening the laminations is only a temporary arrangement and are later removed, otherwise these will form short-circuited turns. Two commonly used methods of holding the leg laminations together is their clamping by either (a) resiglass tape or (b) using skin stressed bakelite cylinders. In case of resiglass tapes, these are tightly wound around the legs at specified pitch and cured by heating. The tape shrinks after heating and provides a firm grip. The tensile strength of resiglass tapes is even higher than that of steel tapes. In the case of core legs tightened by skin stressed cylinders (base cylinder of innermost coil), these are lowered from the top and the steel bands and cut progressively. Wooden wedges are inserted along the packet corners and hammered down, so that the enveloping bakelite cylinder and the leg laminations are fitting tightly against each other.

Conventionally, the core is assembled along with all the yokes, and after assembly the top yokes are unlaced after removing the top-end frames for the purpose of lowering the windings. This takes a lot of labor and manufacturing time. The latest development is to assemble the core without top yokes and insert the top yokes after lowering all the windings in the core leg.

4.4.4 Fitting of Core in the Tank

The most commonly used method of putting the core assembly is to rest the core frame on its feet, which in turn is firmly fixed on the tank base. The extreme end feet are enclosed in a steel bracket welded on the tank base. The top portion of core assembly is also suitably locked with the tank cover, so that any possible magnification of the vibration during transit is fully arrested. An alternative to this is construction of the tank base in the shape of a channel. The laminations along with the bottom-end frames are jacket from either side against the channel walls of the tank. As such, no bolts are needed in the bottom yokes and iron-losses are comparatively lower because of the absence of these punch holes. Figure 4.12 illustrates this channel type of tank and core assembly fixing arrangement. This also has the advantage that the core is rigidly fixed to the tank bottom and no possibility of core assembly shifting exists. As a result of raising the tank base (Item 1, Fig. 4.12) the inner oil volume inside the tank is reduced. Also, the windings and the insulation rings and blocks are directly resting on the flat tank base.

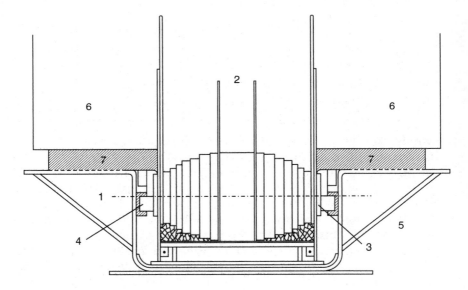

Figure 4.12 *Transformer core fitting in channel-shaped tank base*
  1. *Channel-shaped tank base*
  2. *End view of transformer core*
  3. *Bottom end frame*
  4. *Jacking of core end frame against channel walls*
  5. *Tank base stiffener*
  6. *Windings*
  7. *Insulation blocks and rims*

References

1. Brechha, H., "Some *Aspects of Modern Transformer Core Design*," Bulletin Oerlikon, No. 324, pp. 70–80.
2. Brechna, H. "*New Design Trends in Construction of Transformer Cores*," Bulletin Oerlikon, No. 326, pp. 6–14.
3. Stigant, S., and A.C. Franklin, *J and P Transformer Book*, Newnes–Butterworths, London, 10th ed., 1973.
4. Bhusan, Prabhakar, "*Optimum Design of Distribution Transformers*," M. Tech. Thesis, 1971, Dept. of Electrical Engg., IIT Kanpur.

CHAPTER 5

Windings and Insulation

M.V. Prabhakar
S.K. Gupta

Windings form the electrical circuit of a transformer. Their construction should ensure safety under normal and faulty conditions. The windings must be electrically and mechanically strong to withstand both over-voltages under transient surges, and mechanical stress during short circuit, and should not attain temperatures beyond the limit underrated and over-load conditions. For core-type transformers, the windings are cylindrical, and are arranged concentrically. Circular coils offer the greatest resistance to the radial component of electromagnetic forces, since this is the shape which any coil will tend to assume under short circuit stresses.

5.1 Types of Windings

The choice of the type of winding is largely determined by the rating of the winding. Some of the common types of windings are described below.

5.1.1 Distributed Crossover Windings

These windings are suitable for currents not exceeding about 20 A. They comprise wires of circular cross-section (Fig. 5.1) and are used for HV windings in small transformers in the distribution range. A number of such coils are joined in series, spaced with blocks which provide insulation as well as duct for cooling.

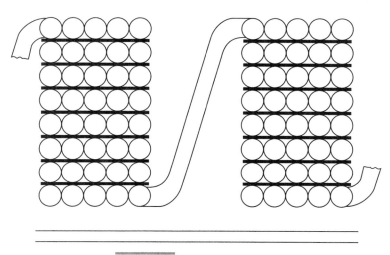

Figure 5.1 *Crossover coils.*

5.1.2 Spiral Winding

This type of winding is normally used up to 33 kV and low current ratings. Strip conductors are wound closely in the axial direction without any radial ducts between turns. Spiral coils are normally wound on a bakelite or pressboard cylinder (Fig. 5.2).

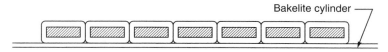

Figure 5.2 *Spiral coil (single layer, wound on flat side).*

Though normally the conductors are wound on the flat side, sometimes they are wound on the edge. However, the thickness of the conductor should be sufficient compared to its width, so that the winding remains twist-free (Fig. 5.3).

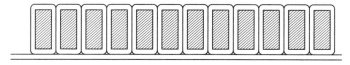

Figure 5.3 *Spiral coil (edge wound).*

Spiral windings may be made as single layer or multilayer type. Figure 5.4 shows a double-layer spiral coil where an oil duct separates the two layers. For such a coil, both the start and the finish leads lie at one end of the coil and may at times prove to be advantageous for making the terminal gear.

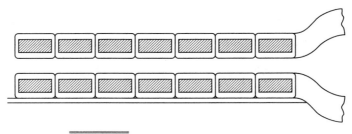

Figure 5.4 *Spiral coil (double layer).*

Normally it is not necessary to provide any transposition between the parallel conductors of a spiral winding as the lengths and the embracing of leakage flux are almost identical.

5.1.3 Helical Winding

This type of winding is used in low-voltage and high-current ratings. A number of conductors are used in parallel to form one turn. The turns are wound in a helix along the axial direction and each turn is separated from the next by a duct. Helical coils may be single-layer (Fig. 5.5) or double layer (Fig. 5.6) or multi-layer, if the number of turns are more.

Unless transposed, the conductors within a coil do not have the same length and same flux embracing and therefore have unequal impedance, resulting in eddy losses due to circulating current between the conductors in parallel. To reduce these eddy losses, the helical windings are provided with transposition of the conductors which equalize the impedances of the parallel conductors.

5.1.4 Continuous Disc Winding

This type of winding is used for voltage between 33 and 132 kV and medium current ratings. These coils consist of a number of sections

Windings and Insulation **111**

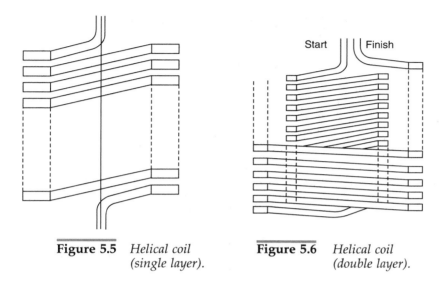

Figure 5.5 *Helical coil (single layer).*

Figure 5.6 *Helical coil (double layer).*

placed in the axial direction (Fig. 5.7), with ducts between them. Each section is a flat coil, having more than one turn, while each turn itself may comprise one or more conductors (usually not more than four or five), in parallel. The sections are connected in series, but without any joints between them. This is achieved by a special method of winding. It is not necessary to provide a cylindrical former for these coils, as these are self-supporting. Each disc is mechanically strong and exhibits good withstand of axial forces. Another particular advantage of these coils is that, each section can have either integral or fractional number of turns (for example $4\frac{15}{18}$ turns per section).

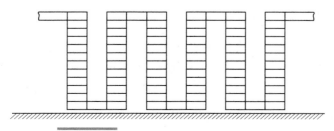

Figure 5.7 *Continuous disc winding.*

5.1.5 Transposition

(a) For helical windings, usually three transpositions are provided. The complete transposition [Fig. 5.8(a)] is provided in the middle of the windings. Two partial transpositions are provided, one at 25% of turns [Fig. 5.8(b)] and the other at 75% of turns [Fig. 5.8(c)]. In complete transposition, each conductor position is varied symmetrically, relative to the middle point, whereas in partial transpositions, two halves of parallel conductors are interchanged in the positions: the upper half becomes the lower, and vice versa. Such a transposition needs extra space in the height of the coil.

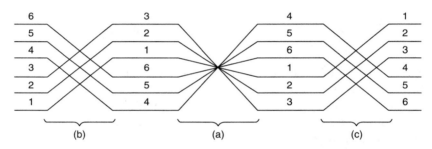

Figure 5.8 *Transpositions in helical winding (a) Complete transposition (b) and (c) Partial transposition.*

(b) With a multi-start helical winding, the transposition can be achieved by using rotary transposition. Figure 5.9 shows transposition in a double-start helical winding. By this arrangement, every conductor occupies every position by turn and thereby complete equalization of impedance is possible. Also, there is no need for extra space in the coil height.

1	12		2	1		3	2		4	3				12	11
2	11		3	12		4	1		5	2				1	10
3	10		4	11		5	12		6	1				2	9
4	9		5	10		6	11		7	12				3	8
5	8		6	9		7	10		8	11				4	7
6	7		7	8		8	9		9	10				5	6

Figure 5.9 *Rotary transposition for double-start helical winding.*

(c) For disc windings having more than one conductor in parallel, transposition is made between the conductors by changing their mutual position at each crossover from one section to another (Fig. 5.10).

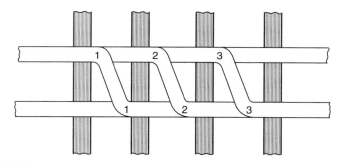

Figure 5.10 *Transposition at each crossover in continuous disc winding.*

5.1.6 Interleaved Disc Winding

A disadvantage with the continuous disc winding is that their strength against impulse voltages is not adequate for voltages above, say, 145 kV class. The impulse voltage withstand behavior of disc coils can be increased if the turns are interleaved in such a fashion that two adjacent conductors belong to two different turns. Figure 5.11 shows such a winding in which interleaving has been done in each pair of discs. It will be noticed that it is necessary to have $2n$ conductors in hand for winding when n is the number of

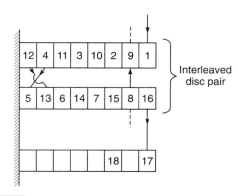

Figure 5.11 *Interleaved disc winding (2 discs per group).*

conductors in parallel. Conductors of turns 8 and 9 are joined by brazing. A crossover is given at the bottom of the disc.

Apart from interleaving between every double-disc, it is also possible to have a larger number of discs (say four) in each interleaved group (Fig. 5.12).

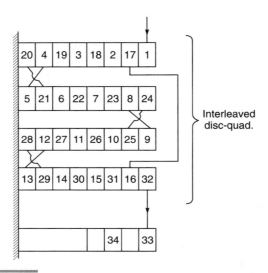

Figure 5.12 *Interleaved disc winding (4 discs per group).*

This gives further improved behavior against impulse voltage, though there are concomitant increased complexities.

Interleaved windings require more skill and labor than plain continuous disc windings. Sometimes a part of the winding is interleaved while the remaining part is plain disc, so as to combine the advantages of better impulse withstand at the high voltage end of the winding and reasonable labor cost for the winding as a whole. These are known as partially interleaved windings.

5.1.7 Rib Shielded Windings

An alternate way of increasing the series capacitance without actually interleaving is achieved in Rib Shielded Windings. Floating shields are provided inside continuous disc windings, and are comparatively easier to manufacture, when compared to interleaved disc windings. The shield wires are not conductively connected to circuit (Fig. 5.13).

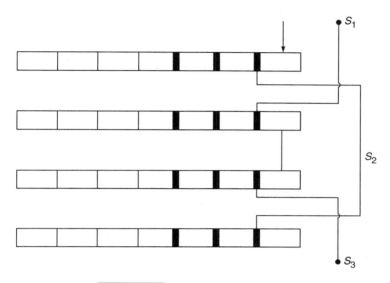

Figure 5.13 *Rib shielded winding.*

5.1.8 Shielded Layer Windings

This type of winding is generally used for star-connected transformers having graded insulation and for voltages greater than 132 kV class. The winding consists of a number of concentric spiral coils arranged in layers. The layers are graded in lengths from the longest at the neutral end (innermost layer) to the shortest at the line end (outermost layer). The layers are arranged between two concentric cylindrical shields, connected one to each end of the winding (Fig. 5.14). All these layers are connected in series, wherein two schemes are possible, viz. parallel-layer type (Fig. 5.14) and tapered-layer type (Fig. 5.15).

The layers are separated by oil ducts and unbonded paper cylinders. During winding, the latter are arranged to extend well beyond the turns of the layer and afterwards these extensions are *petalled* and bent over at right angles to form insulating flanges between succeeding layers. These flanges provide an insulation system to ground, which increases progressively from the neutral-end to a maximum for the line-end of the coil.

The winding layers and the shields form a series of capacitors and are so dimensioned that it results in substantially equal capacitances in series. This ensures a relatively uniform distribution of

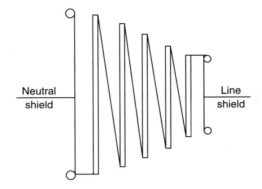

Figure 5.14 *Shielded-layer winding (parallel-layer type).*

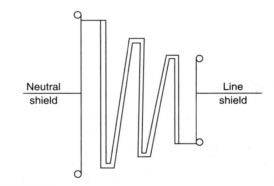

Figure 5.15 *Shielded-layer winding (tapered-layer type).*

surge voltages throughout the winding. When the winding current and the density of the leakage flux are not very high, the winding can be wound with conductors of rectangular cross-section. For large currents and higher density of leakage flux, transposed conductors are used.

5.2 Surge Voltage Behavior of Windings

In service, transformer windings are exposed to various transient over-voltages like lightning surges, switching surges, etc. These over-voltages have steep wavefronts and relatively longer tails. The transformer windings respond to surge voltages as a system of capacitance and inductance network.[1] For the first few micro-seconds after application of impulse, the winding behaves as a capacitance network. Subsequently, the self- and mutual-inductances of the winding elements come into effect and impart oscillatory nature to the voltage appearing at different parts of the winding. Thus it is worth examining separately the initial voltage distribution due to capacitive effect alone and the osciliatory behavior due to combined effect of capacitances and inductances.

5.2.1 Nature of Surge Voltages

A surge on a transmission system due to lightning discharge, either in the vicinity of the line or to the line itself may have an extremely complicated waveform. For the purpose of testing of transformers, a standard waveform has been formulated for the full-wave impulse. It has a wavefront of 1.2 μs and wavetail of 50 μs.

Impulse waves propagating along the line may occasionally create flashover at the insulators to ground, causing sudden collapse of the impulse voltage and the consequent very high rate of change of voltage in the transformer windings. This type of wave is termed chopped-wave impulse. The chopping of voltage occurs usually between 2 to 6 μs.

Another kind of over-voltage is caused due to switching on and off the loads or sources on the line and is known as switching surge. With the advent of EHV systems operated at reduced BIL, the switching surges are becoming an important factor.

The standard wave shape for the switching surges has a virtual front time of at least 20 μs, a duration above 90% of specified amplitude of at least 200 μs and a total duration to the first zero passage of at least 500 μs.

The full-wave impulse and the switching surge waveforms can be defined mathematically as the difference between two exponential functions, e.g.

$$V(t) = a_0 \, (e^{-b_1 t} - e^{-b_2 t}) \tag{5.1}$$

the chopped wave has the same waveform up to the instant of chopping and soon after that it reduces to zero. Such a waveform can be defined as

$$V(t) = a_0 \, (e^{-b_1 t} - e^{-b_2 t}) - b_0 \, (e^{-b_3 t'} - e^{-b_4 t'}) \tag{5.2}$$

where, $t' = t - t_c$
t_c = instant of chopping

5.2.2 Idealization of Transformer Winding

In a winding there exists capacitance between the adjacent turns within a disc or layer, capacitance between the adjacent discs or layers, capacitance to ground and to other windings. Similarly, there exists self- and mutual-inductances as pertaining to the individual turns, the discs/sections, one part of the winding to another or one whole winding to another. Although both the capacitance and inductance are of distributed nature, for practical computation purposes these have to be lumped in varying degrees according to the desired accuracy. Also, the effect of the winding resistance is not significant and is therefore neglected.

Figure 5.16 shows a part of a transformer winding represented as a network consisting of

  (i) series capacitances (C_s)
  (ii) ground capacitances (C_g)
  (iii) self-inductances (L)
  (iv) mutual-inductances (M)

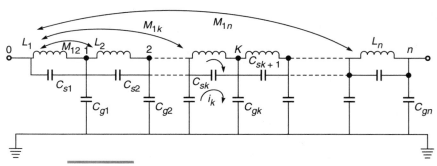

Figure 5.16 *Network representation of a winding.*

5.2.3 Initial Voltage Distribution

On the incidence of the impulse wave, the inductive elements behave like open-circuit elements and the winding can be treated as a capacitive network as shown in Fig. 5.17. The node o is impulsed, i.e., applied with voltage $v(t)$ as in Eq. (5.1) or (5.2).

Let n = number of nodes of the network
C_{sk} = series capacitance of the kth segment of the network
C_{gk} = ground capacitance of the kth segment of the network
ek = voltage appearing at the kth node

Windings and Insulation

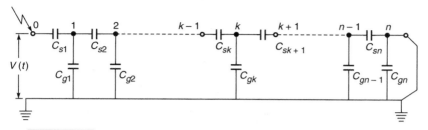

Figure 5.17 *Capacitive network representation of a winding during initial voltage distribution.*

$v(t)$ = input voltage
t = time in microseconds
$p = \dfrac{d}{dt}$ operator

The node equation for node 1 can be written as

$$(C_{g1} + C_{s1} + C_{s2})\frac{de_1}{dt} - C_{s2}\frac{de_2}{dt} = C_{s1}\frac{dv(t)}{dt}$$

and for kth node as

$$(C_{gk} + C_{sk} + C_{sk+1})\frac{de_k}{dt} - C_{sk+1}\frac{de_{k+1}}{dt} - C_{sk}\frac{de_{k-1}}{dt} = 0$$

The equations for all the nodes can be expressed conveniently by the matrix equation

$$C \cdot p \begin{bmatrix} e_1 \\ e_2 \\ \vdots \\ e_n \end{bmatrix} = \begin{bmatrix} C_{s1} \cdot p \cdot v(t) \\ 0 \\ \vdots \\ 0 \end{bmatrix} \quad (5.3)$$

or

$$p \begin{bmatrix} e_1 \\ e_2 \\ \vdots \\ e_n \end{bmatrix} = C^{-1} \begin{bmatrix} C_{s1} \cdot p \cdot v(t) \\ 0 \\ \vdots \\ 0 \end{bmatrix} \quad (5.4)$$

where

$$C = \begin{bmatrix} (C_{g1} + C_{s1} + C_{s2}) & -C_{s2} & 0\dots\dots\dots\dots.0 \\ -C_{s2} & (C_{g2} + C_{s2} + C_{s3}) - C_{s3}\dots\dots.0 \\ \vdots & & \\ \vdots & & \\ 0 & 0\dots\dots\dots\dots\dots\dots\dots.-C_{sn} \\ 0 & 0 - C_{sn-1}(C_{gn-1} + C_{sn-1} + C_{sn}) \end{bmatrix}$$

Equation (5.4) can be solved numerically (e.g., fourth order Runge-Kutta method).

For a winding having uniformly distributed series and ground capacitance, i.e., $C_{g1} = C_{g2} = \ldots\ldots = C_{gn} = C_g$ and $C_{s1} = C_{s2} = \ldots. = C_{sn} = C_s$ it has been shown that the initial voltage distribution depends on the value of $\alpha = \sqrt{C_g/C_s}$. The voltage at a distance x from the neutral end can be expressed as

$$V_x = V_m \cdot \frac{\sinh \alpha \cdot x}{\sinh \alpha} \tag{5.5}$$

The initial voltage distribution for different values of α is shown in Fig. 5.18 for windings with grounded neutral. For $\alpha = 0$, the distribution is perfectly linear. For other values of α the voltage distribution is nonlinear and the voltage gradient at the line end is α times the linear voltage gradient.

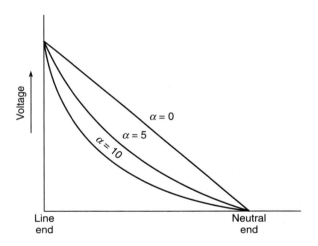

Figure 5.18 *Initial voltage distribution along a winding (grounded neutral) for different values of α.*

Due to this extra strain on the insulation between turns at the line end of a high-voltage winding, sometimes the end-turns are reinforced with extra insulating material. However, the end-turn reinforcement is a matter of careful design because inappropriately increased insulation thickness may result in undue increase of impulse voltage at these turns and thereby defeat the purpose itself.

Use of electrostatic shields at the line-end of a disc winding helps in improving the nonlinearity of the initial voltage distribution.

Sometimes these shields are used in the neutral end also. These shields are connected electrically to the end-section. Being in close proximity of the section, there exists a large capacitance between the shield and the turns of the end-section. Due to this additional capacitance, the initial voltage distribution in the end-section becomes more linear.

For plain disc windings, α generally varies from 5 to 15. An outer coil has less α compared to a similar inner coil as the latter faces two ground planes (i.e., higher C_g).

The α of the windings can be reduced either by (a) decreasing the C_g, or (b) by increasing the C_s. The latter approach is followed in case of interleaved disc windings. Referring to Fig. 5.12 we find that the voltage between two adjacent conductors is $2m \cdot (V/T)$, where m is the number of turns/section. The stored energy between the adjacent conductors is

$$Q_1 = \frac{1}{2} C \left[2m \ (V/T)\right]^2$$

$$= \frac{1}{2} \cdot 4m^2 \ C \ (V/T)^2$$

Whereas the corresponding energy between the adjacent conductors in a plain disc winding is

$$Q_2 = \frac{1}{2} C (V/T)^2$$

$$Q_1 = 4m^2 \ Q_2$$

In other words, the effective series capacitance of the interleaved disc winding is $4m^2$ times more. This will bring down the α substantially and result in nearly linear distribution.

The other approach, i.e., of decreasing the C_g is followed in shielded layer windings. The portion of the winding *seen* by earth is very small because of the concentric dispositions of the layers. Also, equalization of the inter-layer (series) capacitance, by shortening the length of layers corresponding to increased diameter, ensures nearly uniform distribution of impulse voltage between the layers (Fig. 5.19).

5.2.4 Transient Voltage Distribution

After the elapse of a few microseconds, when the rate of change of voltage comes down considerably, the inductances come into effect, in addition to the capacitances. The combined effect of inductances and capacitances gives rise to oscillatory voltages at different parts

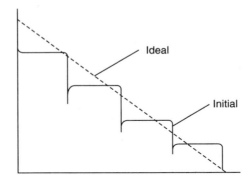

Figure 5.19 *Voltage distribution in shielded-layer windings.*

of the windings.[3,4] A multi-winding transformer can be represented by a multiple-ladder network (Fig. 5.20).

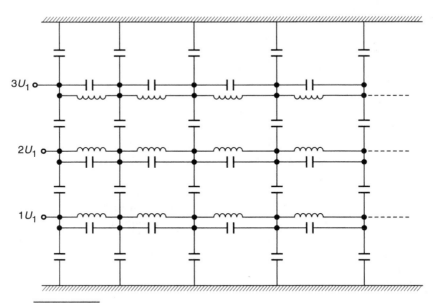

Figure 5.20 *Multiple-ladder network representation of a multi-winding transformer.*

It will be obvious that solution of such a network can be done only with the use of computers. To illustrate the principles involved for

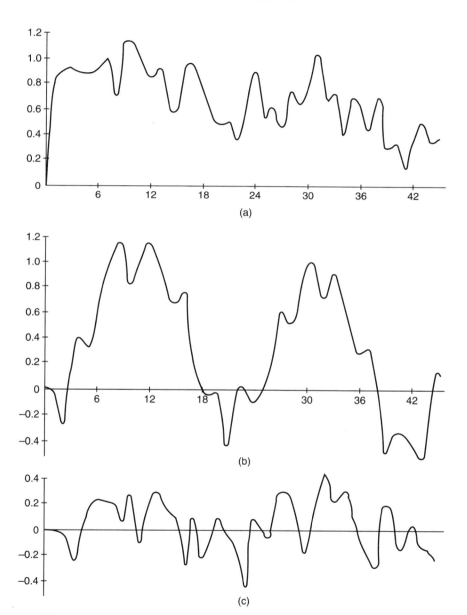

Figure 5.21 *Computer-aided analysis of impulse voltage distribution in transformer windings.*

computation, we take a simple network as shown in Fig. 5.16. The analysis presented here is in line with that of Hartill *et al*[5]. The network parameters are given as:

$$P = \begin{bmatrix} L_1 & M_{12} & M_{13} & \cdots & M_{1n} \\ M_{21} & L_2 & M_{23} & \cdots & M_{2n} \\ \vdots & & & & \\ \vdots & & & & \\ M_{n1} & M_{n2} & M_{n3} & \cdots & L_n \end{bmatrix}$$

(inductance matrix)

where L_i = Self-inductance
       M_{jk} = mutual inductance

$$S = \begin{bmatrix} \dfrac{1}{Cs_1} & 0 & 0 & \cdots & 0 \\ 0 & \dfrac{1}{Cs_2} & 0 & \cdots & \vdots \\ \vdots & & & & \vdots \\ \vdots & & & \cdots & \dfrac{1}{Cs_n} \\ 0 & 0 & & & \end{bmatrix}$$

(series capacitance matrix)

$$Q = \begin{bmatrix} \left(\dfrac{1}{C_{s1}} + \dfrac{1}{C_{g1}}\right) & -\dfrac{1}{C_{g1}} & 0 & \cdots & \cdots & 0 \\ -\dfrac{1}{C_{g1}} & \left(\dfrac{1}{C_{g1}} + \dfrac{1}{C_{s2}} + \dfrac{1}{C_{g2}}\right) & -\dfrac{1}{C_{g2}} & \cdots & 0 \\ \vdots & \vdots & \vdots & & \vdots \\ \cdots & \cdots & \cdots & & \cdots \end{bmatrix}$$

(combined series and ground capacitance matrix)

Let i_k = lower mesh current in kth mesh
     i'_k = upper mesh current in kth mesh
     $\phi_k = \int i_k \, dt$
     $\phi'_k = \int i'_k \, dt$

The solution can be written as
$$p^2\phi = (Gp^2 + H)V(t) + F \tag{5.6}$$
where $p^2 = \dfrac{d^2}{dt^2}$
$G = Q^{-1}$
$H = Q^{-1}SP^{-1}$
$F = Q^{-1}SP^{-1}(S-Q)$

Equation (5.6) can be solved numerically (say by 4th order Runge-Kutta Method)[6]. Figure 5.21 shows the responses at various points in a transformer winding under impulse voltage, as obtained by this method.

5.3 Internal Heat Transfer in Windings

The heat generated in the transformer winding is transferred to oil mainly by convection. The oil which is in contact with windings takes the heat from the latter and becomes warmer and lighter. The oil, now lighter, rises upward and finally goes to the cooling equipment, where it gets cooled. Colder oil enters the coils from the bottom of the windings and this way, continuous circulation goes on.

The heat from the inner parts of the winding is transferred to its outer parts by conduction. The thermal conductivity of copper is many times higher than that of the insulating paper on the conductor. It is thus obvious that paper tries to impede the conduction of heat and the thicker the paper, the more is this impedance. The thickness of paper covering on the conductor has to be a balance between the requirement of electrical insulation and the thermal impedance.

The cooling ducts within the windings provide a path for oil circulation. The ducts are either axial (Fig. 5.22) or radial (Fig. 5.23). The typical paths of oil flow are also shown therein. If the radial dimension of the coil is very large, it is advantageous to introduce an intermediate vertical cooling duct (Fig. 5.24). The effectiveness of these intermediate ducts can be increased by staggering their location in alternate sections.

The heat transfer can be substantially improved if the oil flow is properly directed within the coils. Figure 5.25 shows a typical arrangement wherein baffles have been provided after a few sections

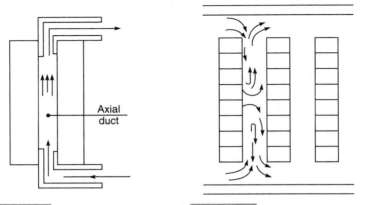

Figure 5.22 *Axial oil flow duct.*   **Figure 5.23** *Radial oil flow duct.*

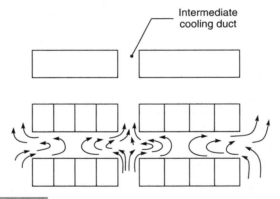

Figure 5.24 *Intermediate vertical duct improves cooling.*

to direct the flow of oil in a unidirectional manner between two baffles. Oil pumps are used externally to create increased flow of oil within the windings. Oil with increased and unidirectional flow can take away heat more effectively from the windings. This results in lower temperature gradient for the windings.

5.4 Insulation Design

For oil immersed transformers, the insulation system comprises a mixed dielectric, viz. oil and cellulosic material. The insulation

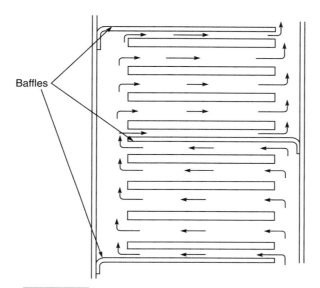

Figure 5.25 *Directed oil flow improves cooling.*

structure can be broadly categorized as per their location in transformer as follows:

(a) *Minor insulation* refers to insulations between different parts of one winding, like insulation between turns, layers, etc.

　(i) The insulation of the conductors is generally of paper, which is wrapped around the conductor. For continuously transposed conductor, the individual strands are coated with a layer of enamel before final wrapping of paper.

　(ii) Insulation between turns is provided either by conductor insulation (e.g., in spiral windings, in turns within a section of disc winding) or by conductor insulation along with radial oil ducts formed by using block (e.g., for helical coils, or for insulation between sections of a disc winding). The thickness of the insulation is determined by the voltages (power frequency as well as impulse) appearing between different conductors, whereas the thickness of the oil duct is determined from voltage as well as thermal considerations.

　(iii) Insulation between the layers of the shielded layer winding comprises vertical oil ducts and paper cylinders.

(iv) In high voltage windings, the end turns are sometimes reinforced to take care of nonlinear distribution of impulse voltage (see Sec. 5.2.3).

(b) *Major insulation* comprises insulation of windings to earth and transformer core, other windings of the same phase (e.g., HV winding to LV winding) and between one phase and another. The insulation between different windings and inner windings to core consists of pressboard cylinders separated by oil ducts. Pressboard barriers are provided between windings of different phases and between the windings and the tank. Ends of windings are insulated from yokes by adequate number of angle rings/angle washers depending upon the voltage class. For shielded layer windings, the flanges obtained by petalling the radial paper serve as the angle rings.

5.4.1 Composite Dielectric Insulation System

Some important considerations regarding design of composite dielectric insulation system are discussed below:

(a) *Oil*

(i) In a composite dielectric system, the dielectric with lower permittivity bears more than average voltage stress. The permittivity of transformer oil is nearly half that of pressboard. Therefore, the electric stress is nearly twice as great in the oil in the annular ducts as on the pressboard cylinders for the same thickness of insulation.

(ii) The electric strength of oil is substantially less than that of cellulosic material.

(iii) Oil ducts show an important characteristic of voltage withstand, that the narrower the duct, the higher is the stress withstand level (kV/mm).

Under the influence of electric field, foreign substances in oil in the form of dust, moisture, etc., have a tendency to align themselves in radial lines, giving rise to paths of low dielectric strength, with consequent danger of breakdown.

(iv) Due to the characteristic mentioned in (iii) above, the structural stability of the barriers, which form the oil ducts, is of great importance.

(b) Solid Insulation

  (i) The solid insulation is oil impregnated under vacuum. There is a considerable difference in the dielectric strength of pressboard impregnated under a relatively high and a relatively poor vacuum. Thus a high vacuum is desirable during impregnation.
  (ii) Proper placement of the insulation is the foundation of successful insulation structure. It is desirable that insulating materials are subjected only to breakdown stress, and that high creep stresses on the boundary layers of the barriers are avoided. Therefore, the boundary layers of solid insulation should correspond to the equipotential planes as far as possible. Since, however, it is not possible to entirely avoid the creep stresses, the designer has to limit the stresses within permissible values.
  (iii) Due to imperfections in their manufacture, the pressboard sheets may contain some micro-voids, which tend to lower the withstand strength. Since the probability of locations of such voids coinciding is very less, it is better to build up a required thickness of solid insulation from more than one sheet.
  (iv) Insulating materials exhibit some degree of dielectric losses when placed in an electric field. The dielectric losses are dependent on the voltage and the frequency of the field and the dielectric constant and the loss angle of the material. The heat generated due to these losses needs to be dissipated, otherwise it may lead to undue temperature rises.

(c) Partial Discharges

Partial discharges are regarded as the most hazardous phenomenon with respect to the service life of the insulation. Two types of partial discharges may occur in the solid insulations.[7]

  (i) Partial discharges of high intensities may take place on the surface of the pressboards. However, these are considered less hazardous, as the breakdown time is substantially longer.

(ii) Concentrated partial discharges may occur on the sharp edges of electrodes. These discharges gradually penetrate deeper and deeper into the pressboard and eventually lead to a breakdown.

The large oil ducts may also give rise to partial discharges when the stresses appearing in it exceed the limits of partial discharge inception voltage.

A sound design practice would be to aim at completely partial discharge free insulation structure for a given over-voltage.

5.5 Electric Field Plotting

From the above, it is clear that a sound design of an insulation system requires pre-determination of the voltage stress levels (both creep and puncture) under impulse and power frequency voltage conditions. The voltage stresses are a function of the shapes and relative dispositions of the windings, shapes of insulation structure and the dielectric media. The voltage stresses can be computed from the knowledge of electrostatic field strength at various points within the structure. Electrostatic field is Laplacian in nature[8] and the different methods of solution are:

5.5.1 Analytical Method

Where the electrode and insulation configuration is simple, a closed-form solution of the field equation can be obtained directly. For certain more complex cases, closed-form solutions can still be obtained by techniques like imaging, conjugate functions, conformal transformation, etc.

5.5.2 Graphical Method

In this method, successive graphical approximations are made on the orthogonal properties of flux and equipotential lines. However, this method becomes too tedious for any problem of practical significance.

5.5.3 Analogue Methods

Various analogues have been used by different workers, like resistance network, rubber membrane, sand models, resistance paper and electrolytic tank. The latter two are the most well known. Each of these has its own limitations, with the electrolytic tank method being the most versatile one. But the need to make expensive and time-consuming models for each problem and also the problems like polarization, surface tension, etc., seriously limit even this method.

5.5.4 Numerical Methods

In a practical problem, as in transformer, with its inherent complexities, like electrodes with odd profiles, as multitude of intervening dielectrics, the analytical methods become inadequate and this leads us to numerical methods of solution. The latter are always approximations to the true solutions, but with sufficient care yield results that are true within engineering accuracy.[9-11] The two main numerical methods are the finite-difference method and the finite-element method. A brief description of the finite-difference formulation of electrostatic field problem is given below.

5.6 Finite-Difference Method (FDM)

In this method, the continuum is replaced by a mesh system of discrete points. Consider Fig. 5.26 which shows a general point A_0 of the mesh system, surrounded by the four points A_1, A_2, A_3 and A_4. The arm lengths are p, q, r and s as shown. Each mesh has two dielectric media interfacing at the diagonal, therefore a total of eight dielectrics meet at the point $A0$. Consider a box whose cross section is the contour 1–2–3–4–5–6–7–8, passing through the midpoints of the arms, and with unit length in the direction into the paper.

The Laplace equation, $\nabla^2 \phi = 0$, in the charge-free region becomes

$$\iint \overline{D} \cdot \overline{n} \cdot dl = 0 \tag{5.7}$$

Since $D = \varepsilon E$ and that unit length into paper has been taken, Eq. (5.7) can be written as

$$\oint_{1-8} \varepsilon E \, dl = 0$$

132 Transformers

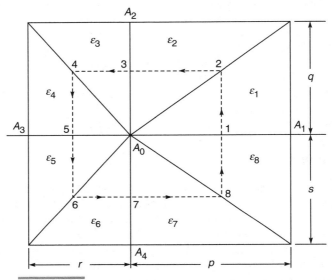

Figure 5.26 A general point A_0 in the mesh system.

This integral consists of 8 segments, e.g., 1–2, 2–3, etc., up to 8–1. Writing this integral equation in the finite difference form and noting that $E = -\,grad,\, A$, we get

$$\frac{(A_0 - A_1)}{p}\left[\frac{\varepsilon_8 s}{2} + \frac{\varepsilon_1 q}{2}\right] + \frac{(A_0 - A_2)}{q}\left[\frac{\varepsilon_2 p}{2} + \frac{\varepsilon_3 r}{2}\right]$$

$$+ \frac{(A_0 - A_3)}{r}\left[\frac{\varepsilon_4 q}{2} + \frac{\varepsilon_5 s}{2}\right] + \frac{(A_0 - A_4)}{s}\left[\frac{\varepsilon_6 r}{2} + \frac{\varepsilon_7 p}{2}\right] = 0 \quad (5.8)$$

Putting
$$C_1 = [\varepsilon_8 s + \varepsilon_1 q]/2p$$
$$C_2 = [\varepsilon_2 p + \varepsilon_3 r]/2q$$
$$C_3 = [\varepsilon_4 q + \varepsilon_5 s]/2r$$
$$C_4 = [\varepsilon_6 r + \varepsilon_7 p]/2s$$

The Eq. (5.8) can be written as
$$A_0 (C_1 + C_2 + C_3 + C_4) = A_1 C_1 + A_2 C_2 + A_3 C_3 + A_4 C_4$$

or
$$A_0 = \frac{\sum_{i=1}^{4} A_i C_i}{\sum_{i=1}^{4} C_i}$$

where A = scalar electrostatic potential function
D = electric flux density
E = electric field intensity
ε = permittivity of dielectric medium

This equation for the potential A_0 is successively applied to every node in an iterative manner, till sufficient accuracy is obtained. It is to be noted that this equation is true in charge-free regions only. Thus, whenever a node which is part of a conductor with definite potential, is encountered in the iterations, the computation is skipped for that node.

The electrostatic field problem is essentially a boundary value problem and the nature of the boundaries are to be known a priori. The two types of boundaries met are Dirichlet and Neumann boundaries. For the former, the potential on the boundary is a constant, whereas the Neumann boundary is characterized by constant gradient condition. Any axis of lateral symmetry in the problem region can be thought of as a Neumann boundary.

5.6.1 Use of Digital Computers

A generalized computer program based on the above algorithm can cater for almost all types of insulation structures. The grid lines are so chosen that all interfaces between dielectrics and conductors, etc., coincide with grid-lines. The curved lines can be approximated by segmenting them into a number of straight lines. The grid system can be closely spaced where higher accuracy is desired.

The problem is defined through a set of input data as follows:

(a) Number of horizontal and vertical grid-lines and their distances
(b) Position of the insulation and the conductors
(c) Permittivities of the media.

5.6.2 Special Features

Since the formulation is in terms of potentials at discrete points within the region, it is possible to consider both point conductors and conducting foils exactly as they occur in the problem. This is of particular importance in the case of transformers where electrostatic shields, which are essentially foil-conductors, are used to a large extent. A special feature incorporated in the program is the computation of potential acquired by a floating conductor in the region of the field.[12]

5.6.3 Application of Field Plotting Technique to Transformer Insulation Design

To illustrate the nature of the fields encountered in transformers, a typical field plot of the coil-end region of a high-voltage transformer is shown in Fig. 5.27. The different coils have been shown along with the equipotential lines in steps of 10%. According to the nature of the field, three zones have been marked as A, B and C.

The zone A, or the vertical zone has a nearly uniform field. The number and thickness of the barriers is decided by the stress-withstand of the oil gaps. The zone B has nonuniform field and is generally of horizontal configuration. In this zone the oil gaps are larger than in zone A. The placement of barriers and static rings tend to make the field more uniform.

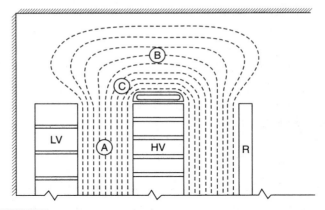

Figure 5.27 *Typical field plot of the coil-end region in a high-voltage transformer.*

The zone C is the transition zone between zones A and B and is the most important from insulation design consideration. It is a zone of high gradients and is subject to both puncture and surface creep stresses. It is relatively easy to guard against puncture, but minimization of surface creep stresses requires the knowledge of the exact positions of the equipotential lines. If the insulation profiles are designed to conform to the shape of equipotential lines, creep stresses are minimized.

Figure 5.28 shows another example of electrostatic field plotting for an EHV transformer.

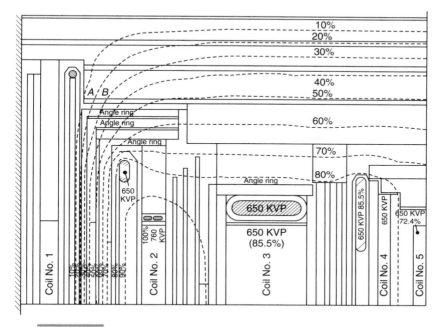

Figure 5.28 *Electrostatic field plot inside transformer top end of coils under impulse test.*

Once the knowledge of the field is obtained, it is necessary to ensure that the stresses appearing at various parts of the insulation structure are contained below the limits of partial discharge inception.[13] For achieving this, the designer resorts to any one or all of the following methods to bring the stresses within limits.

(a) Electrodes having very sharp corners produce undesirably high stresses around the corners. It is therefore imperative to avoid such sharpness in electrodes. The minimum allowable radius of electrodes is governed by their relative disposition.

(b) Though it may suffice from thermal considerations to use a small size conductor for low current, high-voltage leads, it may sometimes be necessary to go in for a conductor of a larger diameter, as the latter gives rise to lower electric stresses.

(c) In oil immersed electrode systems, where the stress is considerably high on the surface of the electrode and falls off exponentially, the electrode is covered with solid insulation to take advantage of its higher withstand capability and to push the oil insulant to a safe point where it is subjected to allowable stress limits.

(d) Should the above technique be not viable, withstand capacity of oil is considerably increased without lowering the stress to which it is subjected to, by dividing the highly stressed oil zone into a number of thin laminae by using barriers of solid insulation.

(e) In some important areas, use is made of moulded insulation items whose contours are made in accordance with the shape of the equipotential lines. Thus surface creep stress can almost be avoided and a very compact insulation system can be evolved.

(f) Another source of partial discharge inside the transformer is the cables and the terminal gear. Here also, it is necessary to ensure that the electrical stresses at the copper-paper interface or paper-oil interface are within limits. Provision of too much insulation on the copper, however, impedes the heat transfer.

REFERENCES

1. Heller and Veverka, *Surge Phenomenon in Electrical Machines*, Ileffe Books Ltd, London.
2. Blume, L.F. et al, "*Transformer Engineering*," John Wiley & Sons, New York, 1965.
3. Lovas-Nagy et al., "A Matrix Method of Calculating the Distribution of Transient Voltages in Transformer Winding," *Proc IEE*, Vol. 110, p 1663, 1963.
4. Miki, Hosoya, and Okuyama; "A Calculation Method for Impulse Voltage Distribution and Transferred Voltage in Transformer Windings," *Trans IEEE*, PAS-97, No. 3, May/June 1978.
5. Dent, B.M., Hartil et al., "Method of Analysis of Transformer Impulse Voltage Distribution Using a Digital Computer," *Proc IEE*, Vol. 105, Part A, 1958.
6. Smith, G.D., *Numerical Solution of Partial Differential Equations: Finite Difference Methods*, Oxford Press.
7. Moser, H.P., "Transformer board," *Scientia Electrica*.

8. Binns, K.J., and Lawrenson, P.I, *"Analysis and Computation of Electric and Magnetic Field Problems,"* Pergamon Press, 1973,
9. Galloway, R.H., Ryan, H.M. and Scott, M.F., "Calculation of Electric Fields by Digital Computers," *Proc IEE*, Vol. 114, pp 824–829, 1967.
10. Ryan, H.M., Mattingley, J.M. and Scott, M.F., "Computation of Electric Field Distribution in High Voltage Equipment," *Trans IEEE*, PAS, pp 148–154, 1971.
11. Cermak, I.A., and Silvester, P. "Boundary-Relaxation Analysis of Rotationally Symmetric Electric Field Problems," *Trans IEEE*, PAS-89 (5/6), pp 925–932, 1970.
12. Mittal, M.L., Sarkar, S.C. and Kumar, P.R., *Computerised Electric Stress Analysis of Insulation Structures in Transformers,"* International workshop on EHV and UHV insulation (Instac-82), Institution of Engrs (India) Ltd., 1982.
13. Mittal, M.L., and Sarkar, S.C., *"Recent Design Practices in Power Transformers,"* International conference on transformers, *IEMA*, New Delhi, 1982.

CHAPTER 6

Voltage Regulation and Tapchanger

B.L. Rawat
A.K. Ekka

Voltage variation in electrical systems is a normal phenomenon, because of rapid growth of industries and distribution network. It is very essential to maintain the system voltage within prescribed limits for the better health of electrical equipments. Voltage of the system can be varied by changing the turn ratio of transformer. The device tapchanger is used for adding or cutting out turns of primary or secondary winding of the transformer. Basically tapchanging equipment can be divided in two categories:

(a) Off-circuit tapchanger,
(b) On-load tapchanger.

6.1 Off-Circuit Tapchanger

The cheapest method of changing the turn ratio of a transformer is the use of off-circuit tapchanger. As the name implies, it is essential to de-energize the transformer before changing the tap.

An off-circuit tapchanger, as shown in Fig. 6.1, consists of principally the following three parts:

(a) Operating handle projecting outside the transformer
(b) Fixed contact with connecting terminal
(c) Insulating shaft with moving contact system

The basic transformer winding circuit arrangements using

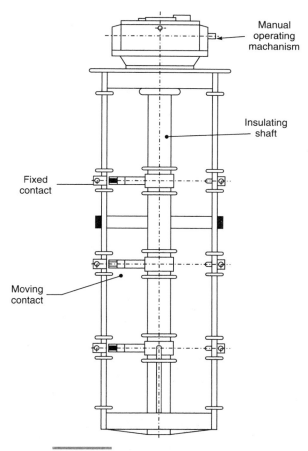

Figure 6.1 *Off-circuit tapchanger.*

off-circuit tapchanger are as shown in Fig. 6.2. They are:
(a) Linear
(b) Single-bridging
(c) Double-bridging
(d) Series-parallel
(e) Star-delta

Depending upon the requirement, any of the above arrangements of the winding can be made use of to get desired voltage regulation. To prevent unauthorized operation of an off-circuit tapchanger, a mechanical lock is provided. Also to prevent inadvertent operation,

140 *Transformers*

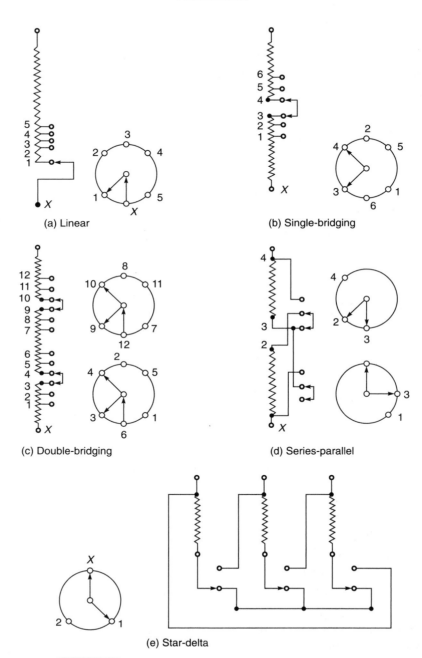

Figure 6.2 *Off-circuit tapchanger connection circuits.*

an electromagnetic latching device or microswitch is provided to open the circuit breakers to de-energize the transformer while operating the handle of tapchanger before movement of contacts on tap switch.

6.2 On-load Tapchanger (OLTC)

On-load tapchangers are employed to change turn ratio of transformer to regulate system voltage while the transformer is delivering normal load. With the introduction of on-load tapchanger, the operating efficiency of electrical system has considerably improved. Nowadays, almost all the large power transformers are fitted with on-load tapchanger.

All forms of on-load tapchanging circuits possess an impedance, which is introduced to prevent short circuiting of tapping section during tapchanger operation. The impedance can be either a resistor or center-tapped reactor. The on-load tapchanger can in general, be classified as resistor or reactor type.

6.2.1 Reactor Transition Type OLTC

In early designs, the use of center-tapped reactor as tapchanging impedance, in general was more popular, in spite of the inevitable shorter contact life. One of the principal advantages of mid-point reactor transition is that, twice as many active working positions as that of transformer tappings could be obtained. This can be of considerable advantage where large numbers of tapping positions are required. Reactor transition type on-load tapchangers are manufactured and used only in the United States of America. In other parts of the world, resistor transition type on-load tapchangers are being manufactured and used.

6.2.2 Resistor Transition Type OLTC

Resistor transition has a considerable advantage of longer contact life, due to relatively short arcing time associated with unity power factor switching. With the introduction of high speed resistor transition tapchanging, it is possible to break the arc at first to current zero. Furthermore, the introduction of contacts using copper/tungsten alloy arcing tips has brought about a substantial improvement in contact life.

Transition resistor type tapchangers can be divided into two types, those which carry out selection and switching on the same contacts, and those which have tap selectors and a separate diverter switch. The first category is known as single-compartment type, while the second one as double-compartment type.

6.2.3 Single-Compartment Type OLTC

The single-compartment type employs a rotary form of selector switch with single-transition resistor or double-transition resistor. Figure 6.3 illustrates the switching sequence in moving from one tap to the next tap employing single-transition resistor. This switching cycle is known as "asymmetrical pennant cycle." Tapchangers constructed with one-transition resistor are suitable for power flow in one direction only. This particular contact arrangement is not suitable for power flow in reverse direction.

Figure 6.4 illustrates the switching sequence in moving from one tap to the next tap employing two transition resistors. This switching cycle is known as *Flag cycle* and the arrangement is suitable for bi-directional power flow. Single compartment tapchangers available presently are suitable for currents up to 600 A and 66 kV voltage class (with certain limitations up to 132 kV class) winding of transformers.

6.2.4 Double-Compartment Type OLTC

On larger transformers, the on-load tapchanging equipment is more usually arranged with separate tap selectors and divertor switches. The tap selectors are generally arranged in a circular form. The divertor switches have contacts operating in rapid sequence with usually four separate make and break units.

Figure 6.5 shows typical selector and switching arrangements with on-load tapchanger on neutral end of the star-connected winding of a transformer.

Figure 6.5(a) shows linear selector arrangement for 16 steps and 17 positions. In the diagram, the on-load tapchanger is shown connected to tapping 4. While changing from tapping 4 to tapping 5, M_1 is opened first and this transfers the load current via A_1 with

Voltage Regulation and Tapchanger 143

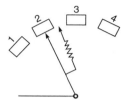
(a) Tap position N. 2. the main contact is carrying the load current. The transition contact is open and rests between the fixed contacts 2 and 3.

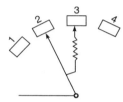
(b) Transition contact makes on the fixed contact 3, the transition resistor bridges 2 and 3 and carries circulating current.

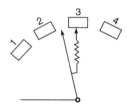
(c) The main switching contact breaks and the transition resistor carries the load current.

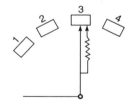
(d) The main switching contact makes on contact 3 and carries the load current.

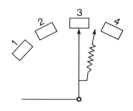
(e) The transition contact opens. the tap change operation complete.

Figure 6.3 *Switching sequence.*

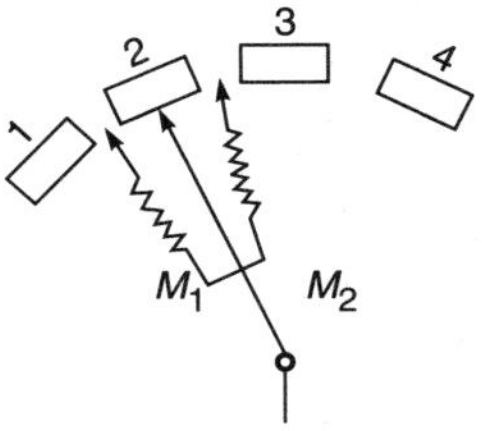

Transformers

(a) Tap position No. 2. the main contact is carrying the load current. The resistor contats M_1 & M_2 are open, resting between fixed contacts.

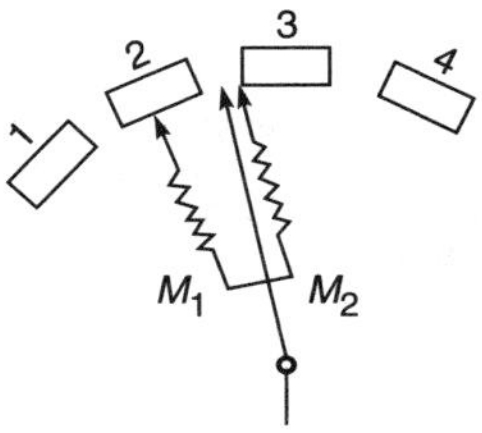

(b) The resistors contact M_1 has made on the fixed contact 2 and the main contact has broken the transition resistor contact M_1 carries load current.

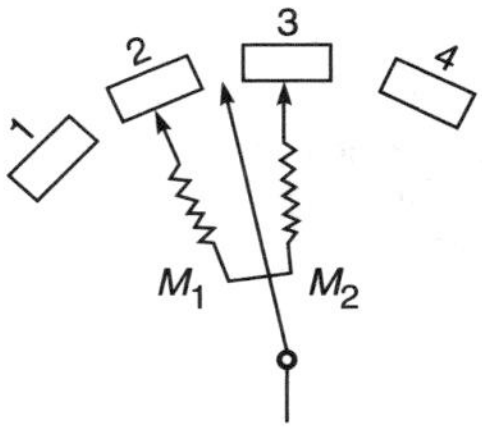

(c) The resistor contact M_2 Has made on the fixed contact 3. The load current is divided between resistor contacts M_1 & M_2. The circulating current is limited by resistors.

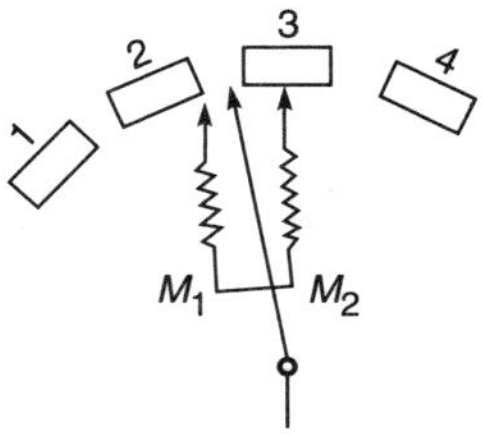

(d) The resistor contact M_1 has broken from the fixed contact 2. The transition resistor contact M_2 carries the load current.

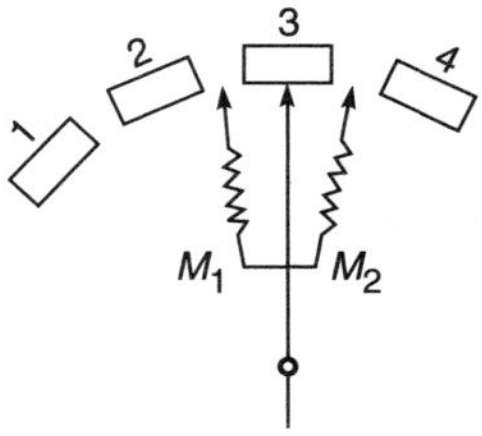

(e) Position No. 3 the main contact has made on fixed contact 3. The resistor contact M_2 has broken from the fixed contact 3. The main contact is carrying load current

Figure 6.4 *Switching sequence.*

Voltage Regulation and Tapchanger 145

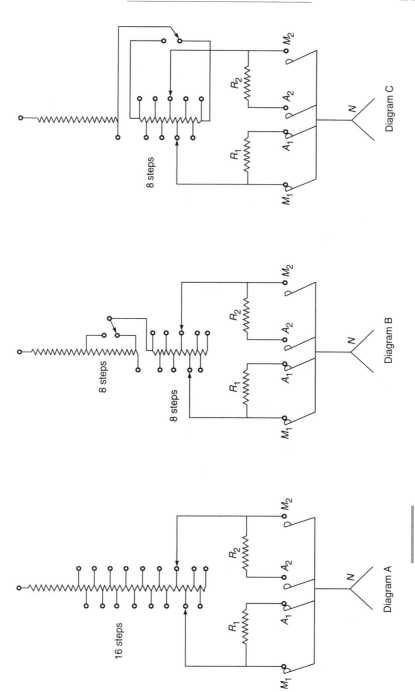

Figure 6.5 Linear coarse-fine and reversing switch-type connection diagram.

resistor R_1 in series. Then A_2 closes and two resistors R_1 and R_2 are in series across tappings 4 and 5. A circulating current will flow through these resistors because of step voltage between tappings 4 and 5. The load current is divided passing through each of the resistors to each of the tappings. A_1 then opens and interrupts the circulating current and the load current is transferred to tapping 5, passing through resistor R_2.

Finally M_2 closes and takes load current, completing the tapchange. This sequence of tapchange in most of the designs takes 40 to 80 ms. For a tapchange in the opposite direction the sequence is reversed. For successive tapchange in the same direction, first the tap selector contacts movement takes place. However, for the first reversal, the tap selector does not move. This feature is obtained by using a lost motion coupling in the mechanical drive of tap selector.

6.2.5 OLTC with Changeover Selector

When the tapping range is large it is advantageous to halve the length of tapping winding and to introduce a reversing or changeover selector. In Fig. 6.5(b) the tapped portion of the winding is shown divided into eight sections and a further untapped portion has a length equal to eight sections. In Fig. 6.5(c), the tapped section of the winding itself is reversed to get double steps. Decision regarding use of on-load tapchanger to adopt am of the two circuit designs will depend on the design of the transformer.

With tappings on the neutral end of HV winding of three-phase transformer, only one three-pole tapchanger is required per transformer. However, for large regulating transformers, single pole tapchanger for individual phase shall be required. Three such tapchangers are mechanically coupled together and driven by a common driving gear.

6.2.6 Methods of Potential Connections

Figure 6.6 shows a typical selector and switching arrangement generally followed by transformer design engineers depending upon transformer specification requirements. During changeover operation of reversing switch or coarse tap selector, the tapping winding gets temporarily disconnected from the main winding. During this period tap winding acquires a potential determined by the winding capacitances. The difference in voltage will, therefore, appear during contact separation of the changeover selector as a recovery voltage. If the calculation results show recovery voltage more than the specified limit declared by the tapchanger manufacturer for the particular design, the tapping winding should be connected continuously or during changeover operation to a fixed potential.

This is possible by the following methods shown in Fig. 6.6.

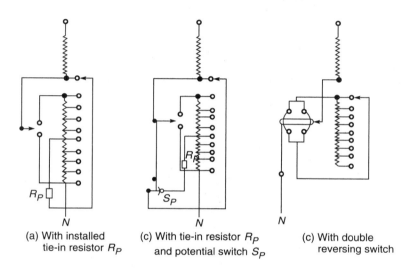

Figure 6.6 *Method of potential connection.*

(a) Potential connection by a continuously connected ohmic resistor (tie-in resistor).
(b) Potential connection by an ohmic resistor which will be inserted during changeover operation by means of a potential switch.
(c) Double reversing switch (available with a particular special design only).

6.2.7 OLTC with Coarse-Fine and Reversing Arrangement

Comparing reversing with coarse-fine arrangement, the latter gives lower copper-losses at the minimum number of turns position. However, coarse-fine arrangement requires a more sophisticated winding layout from dielectric point of view. Further, the wide variation of impedance over the tapping range can be avoided by the use of coarse-fine arrangement.

6.2.8 OLTC for Delta-Connected Winding

If the tappings are to be provided on the delta-connected winding, usually three isolated single-pole design tapchangers are employed. The connection diagram is shown in Fig. 6.7. Special arrangement of tappings shown in diagram (c) can employ one two-pole tapchanger and one single-pole tapchanger, mechanically coupled and operated by single driving gear. From economic consideration, it is preferable to employ tapchanger on the neutral end of star-connected winding. Presently, tapchanger employed on delta-connected winding are of 132 kV class or less.

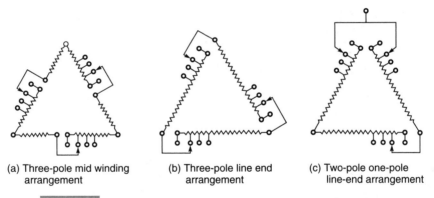

(a) Three-pole mid winding arrangement

(b) Three-pole line end arrangement

(c) Two-pole one-pole line-end arrangement

Figure 6.7 *Tapping arrangement on delta-connected winding.*

6.3 Constructional and Operational Features of OLTC

As described earlier, the on-load tapchangers mainly fall in two categories:

(a) Single-compartment design
(b) Double-compartment design

6.3.1 Single-Compartment Design

In single-compartment design, same contacts are employed for selection and transfer of current. All the contacts are housed in the same compartment. The insulating oil of this compartment is kept isolated from the main transformer oil. Two designs are available in single-compartment type on-load tapchangers.

In one design the electrical contacts, transition resistors and mechanical gear arrangement are housed in a pressure tight insulation cylinder which is housed alongwith the transformer in the same tank. Motor drive for manual/electrical operation is mounted on the tank outside.

In the other design, electrical contacts, transition resistors and geneva gears are housed in a steel tank, which in turn is mounted on the side of transformer tank. The connection between transformer tapping leads and tapchanger is through terminal board of epoxy resin moulding. This terminal board is mounted either on transformer tank or tapchanger housing, depending on the design and it also acts as a barrier to separate the insulating oil of transformer and tapchanger.

6.3.2 Double-Compartment Design

Large transformers of high voltage class employ double-compartment type on-load tapchangers. The compartment which houses make-break contacts and transition resistors is called diverter switch compartment. The tap selector which is the other part of the tapchanger is housed in a second compartment. The modern design based on Jenson principle presently available uses insulating pressure-tight cylinder to house make-break switches, transition resistors and energy accumulators. The tap selector with geneva wheel gear is without housing. The diverter switch and tap selector are mechanically coupled and housed alongwith transformer in the same tank. The tap selector contacts for each phase are arranged in two circles, one having even numbered contacts and the other odd numbered contacts. To achieve alternating movement of tap selector contacts, geneva wheel gear train (Fig. 6.8) is employed.

As the tap selector is housed in the same tank along with transformer, the Buchholz relay of transformer takes care of tap selector also. However, for diverter switch a separate oil surge relay is provided because the oil in diverter switch compartment is not in contact with transformer tank oil. This oil surge relay trips the

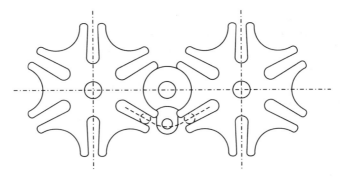

Figure 6.8 *Geneva wheel gear train.*

transformer for any electrical fault that takes place in the diverter switch compartment.

6.4 Manual and Electrical Operation of Tapchanger

A motor drive unit for operation of on-load tapchanger is supplied and fitted on the side of the transformer tank. Motor drive housing is designed for outdoor duty. The motor drive housing contains all mechanical and electrical parts necessary for operating the tapchanger.

To facilitate manual operation of tapchanger, a hand crank kept in the motor drive unit is to be pushed on the shaft so that it engages the gearing. While pushing the crank on the shaft, safety switch operates and breaks and supply to motor and the control circuit. After the tapchanger reaches end position, further manual operation results in the interruption, disengaging the drive shaft after about two to three revolutions of the crank handle. The gears get re-engaged after operation of crank handle in the opposite direction.

To electrical control of motor drive follows the step-by-step principle, i.e., after energization of electrical control and starting of driving motor, the tap change operation is accomplished independent of whether the push button or switch have been operated during the running time of motor drive. Another switching operation is possible only when the control system is again in the rest position. Shortly before the motor drive arrives in the end position, the limit switch opens so that the motor and its control can no longer be energized.

6.5 Automatic Control of Tapchanger

Automation requires the use of automatic voltage control at substation, so that a predetermined constant or compensated busbar voltage can be maintained. In general, a tapchanger is provided on a transformer for maintaining a predetermined outgoing voltage where the incoming voltage is subjected to variation due to voltage drops and other system variations. The output of the voltage transformer connected to controlled voltage side of the transformer, is used to energize automatic voltage regulating relay.

The voltage regulating relay does not issue a control signal, as long as the voltage to be controlled remains within preset limits. Once, however, the voltage deviation exceeds the set limit, a control signal RAISE or LOWER is given to initiate tapchanger operation. It is usual to introduce a time delay element within the relay itself, to prevent unnecessary operation or hunting of the tapchanger during transient voltage change. As an additional feature, load compensation is frequently added. The load drop compensator serves for compensating the resistive and the reactive voltage drops of an outgoing feeder. This device keeps the voltage on-line end constant, regardless of the load (amplitude and phase angle of current). One tapped resistor and a tapped reactor, provided in the load compensator of voltage regulating relay, are fed from the secondary of a current transformer. The primary winding of this current transformer carries the load current and is arranged to subtract a voltage drop proportional to the load on main transformer from voltage applied to the voltage regulating relay. By suitable adjustment of the resistance and reactance components, which will depend upon the outgoing line characteristics, it is possible to obtain constant voltage at some distant point on a system, irrespective of the load or power factor. Figure 6.9 shows the principle of the regulating relay with compensator.

6.6 Tapchanger Selection

While selecting a tapchanger for a particular transformer, the following points are to be considered:
— Voltage class of transformer winding and its rating.

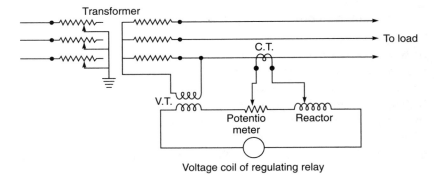

Figure 6.9 *Automatic voltage regulating relay with line drop compensation-connection diagram.*

— Percentage voltage variation required.
— Maximum through current.
— Step voltage between adjacent contacts.
— The switching capacity (maximum through current, step voltage).
— Insulation level to ground and between various contacts.
— Number of steps and basic connection (linear, reversing or coarse-fine).
— Temporary overloads.
— Short-circuit strength required.
— Number of operations required (any special duty).

6.7 Latest Trends in Tapchanger Design

At present tapchangers are available for the highest insulation level of 1475 kVp impulse and 630 kV power frequency voltage. Efforts are being made to develop tapchangers suitable for still higher insulation level class. Further efforts are being made for developing tapchangers smaller in size and having high reliability and performance. The use of vacuum switches in the diverter switch of tapchanger is being tried to increase its performance. Also, thyristorized tapchanger will be available in the future for very special applications where excessively high numbers of operations are required.

Reference

1. Stigant, S., and A.C. Franklin, *The J & P Transformer Book*, Newnes–Butterworths, London, 10th ed., 1973.

CHAPTER 7

Electromagnetic Forces in Power Transformers

M.V. Prabhakar

T.K. Ganguli

The most severe mechanical stressing occurs in a transformer, when it is subjected to a sudden short circuit. Since the currents flowing through the windings at that time are enormous, the forces generated are also enormous. Much work has been done to analyze and calculate these forces. The problem does not yield easy solution due to the transient and dynamic nature of the phenomenon.

7.1 Leakage Flux in a Typical Two Winding Transformer

The typical leakage flux pattern in a two winding transformer of core-type is given in Fig. 7.1. The flux lines are almost axial at the middle of the winding and start bending in the radial direction as we move towards the winding ends. In other words, the axial component is predominant along the height of the winding, except at winding ends (top and bottom), where the radial component becomes significant. The direction of current in the winding conductors is perpendicular to the flux lines, thus resulting in the axial and radial component of forces.

7.2 Nature of Forces

Forces are produced due to the interaction of the current and the magnetic flux density vectors. Thus, in general, the force vector can have any direction. Nevertheless, it is easier to speak of the *radial*

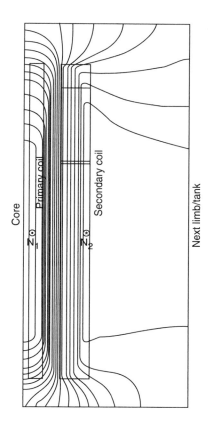

Figure 7.1 *Typical leakage flux pattern in a two winding transformer.*

forces and the *axial* forces in a transformer, since these two components of the force can be calculated and analyzed independently. Also, the two components have influence on different parts of the total transformer and it is necessary to obtain the two components for design purposes. The nomenclature *axial* and *radial* is applicable to concentric wound core type transformer and is assumed in the following discussion.

7.2.1 Radial Forces

Radial forces are those that act in the radial direction and are generated by the interaction of the current and the *axial* component of the leakage flux density. They tend to squeeze the inner winding and burst the outer winding. The calculation of radial forces does not present much difficulty since the axial component of the leakage flux density is calculable fairly accurately. Calculation of the stresses due to this force acting on the conductors is more complex, especially for inner windings. The compressive strength of winding is influenced by the radial thickness of conductors, its work hardening strength, number of blocks per circle, and winding dimensions, etc.

Similarly, the outer winding experiences outward radial force which results in tensile stress. The tensile strength of winding depends on the work hardening strength of the conductors.

7.2.2 Axial Forces

Axial forces are those that act in the axial direction and are generated by the interaction of the current and the *radial* component of the leakage flux density. These forces tend to bend the conductors in the axial direction, and their sum total act on the coil-clamping ring and other clamping structures. The calculation of axial forces presents some problems since the radial component of the leakage

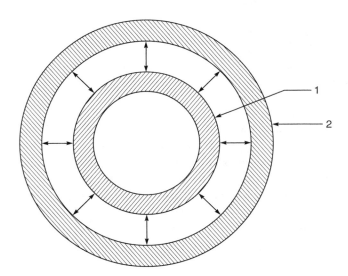

Figure 7.2 *(a) Radial forces.*
1. Inner winding 2. Outer winding.

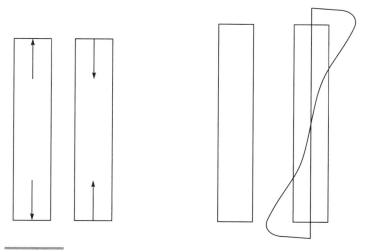

Figure 7.2 (b) *Axial forces.*
(c) *Most common pattern of radial component of leakage flux.*

flux density is difficult to calculate accurately. When ampere turns are perfectly balanced so that the leakage flux pattern is symmetrical, then the leakage field is axial over the major part of the coil height. But due to the flux lines dispersing in the radial direction in the vicinity of the winding ends, the axial flux density tends to decrease, and the resultant flux density at the ends can be resolved into the radial component causing axial forces. These axial forces are unequally distributed between the outer and inner winding, due to the presence of core.

The axial forces at the top and bottom are in opposite directions as the currents are in the same direction. In case the ampere turns are perfectly balanced and the leakage flux pattern is symmetrical, the resultant force on the winding would be zero. Any axial displacement between the magnetic centers of HV and LV windings will result in a net axial force, tending to increase the displacement even further. However, in practice, a complete balance of all elements of the winding cannot be achieved entirely for a number of reasons like provision to tappings, HV line lead exit from the center of winding, dimensional accuracy and stability of windings, etc.

7.3 Basic Formula and Methods for Force Evaluation

7.3.1 Basic Formula for Force Calculation

The basic equation for the calculation of any electromagnetic force is

$$\vec{F} = \vec{i} \times \vec{B} \times l \tag{7.1}$$

where

$\vec{B}$ is the flux density due to leakage flux at mean radial depth of winding,
i the short-circuit current, and
l the length of the conductor.

An examination of Eq. (7.1) shows that, since F is the cross-product of i and B, a radial flux will give rise to axial force while an axial flux will give rise to radial force.

The right-hand side of Eq. (7.1) involves only three quantities, viz. i, B and l. Of these, the current i and the length l are known exactly. The flux density B is due to the leakage flux, since the conductors lie in the field of the leakage flux. The calculation of the radial and axial components of the leakage flux density is the most difficult task and forms the crux of the problem.

7.3.2 Methods of Force Calculations

Attempts in solving the force problem were made by various designers. The main thrust was towards obtaining a closed-form solution, so that the designer would directly apply a set of formulae and graphs to his particular design for calculating the forces. With the advent of the digital computer, field plotting techniques were developed and today numerical methods of force calculation using the digital computer are common.

Calculation of radial forces requires evaluation of the axial leakage-flux density. Formulae for the axial leakage-flux density are readily available and are the same as used for leakage reactance calculations. (Refer para. 9.14 of Chapter 9). Hence, the radial forces are easily calculated to good accuracy. On the contrary, evaluation of the radial leakage-flux density, required for axial force calculation, is quite complex. Various investigators have given different empirical formulae and constants based on their experiments, for the calculation of the axial forces. Such formulae are to be used with due care, as to their applicability to the case in hand.

7.3.3 Analytical Methods

(a) Billing[1,2] has given formulae and curves for the calculation of radial and axial short-circuit forces in transformers. His formulae are derived considering the windings as current sheets. These are fairly accurate for close-wound coils (spiral coils).

(b) Waters[3,4] has given formulae and constants for the calculation of radial and axial forces. He uses the residual ampere-turn method to obtain the unbalanced ampere-turns from which the axial forces are calculated using certain empirical constants. The method is suitable for transformers with only two distinct windings or coils.

7.3.4 Numerical Methods

The most well-known methods for force calculation in this category are due to Roth.[5,6] Roth's method is basically an analytical method. He has used the method of images to obtain the magnetic flux density as a double-Fourier series. In rectangular coordinates, the solution is in the form of algebraic equations; in cylindrical coordinates, the solution involves Bessel and Struve functions. The method is most suitable for programming on a digital computer. A detailed exposition of the method can be found in Ref. 7.

Rabins[10] introduced a simpler form of the field solution using a single Fourier series as opposed to the double-Fourier series of Roth. Rabins developed this for reactance calculations, but the expression for the vector potential which he obtained, when differentiated, leads to the flux density and hence the electromagnetic forces. This method also is most suitable for digital computers, though the basic formulation leads to an analytic expression. Other numerical methods involve obtaining a point-to-point solution for the magnetic field and then applying Eq. (7.1) systematically. Such field solutions can be obtained in a number of ways; finite difference methods and finite element methods are the most popular of these.

7.4 Radial Forces

A simple formula for the average radial force can be derived by evaluating the axial component of the leakage flux density and applying Eq. (7.1). An assumption is made that all the leakage flux is axial only and passes between top and bottom yokes in straight lines. This assumption is also made in calculation of leakage reactance and will lead to a slightly higher value of radial force.

The flux density at the mean radial depth of the winding is given by

$$B = \frac{1}{2}\mu_o \times NI/L \qquad (7.2)$$

The average radial force is

$$F_r = B \times i \times l$$

$$= \frac{1}{2}\mu_o \frac{NI}{L} \times I \times \pi D_m \text{ N}$$

$$= \frac{1}{2}\mu_o \times \frac{(NI)^2}{L} \times \pi D_m \text{ N}$$

$$= 19.739 \times 10^{-7} \frac{(NI)^2}{L} \times D_m \text{ N} \qquad (7.3)$$

Due to the assumption that all the flux is axial in nature, Eq. (7.3) will result in a force larger than that due to any other formula. It is used in all stress calculations and in the design of windings. The maximum force per turn will occur at the inner turn of innermost layer of the outer winding and the outer turn of the inner winding and will be given by

$$F_m = \frac{2F_r}{N} \qquad (7.4)$$

If the turn experiencing this force has small radial dimension and is unsupported by any adjacent turns, the stress calculations are done using Eq. (7.4). Otherwise the *average* force of Eq. (7.3) is used for stress calculations.

7.5 Axial Forces

The electromagnetic forces arising due to a short circuit are oscillatory in nature, and act on an electric system immersed in oil and consisting of winding conductors, insulation components, and clamping structure. Such forces, dynamically transmitted to various parts such as conductors, end supports, press plate, and clamps may be quite different, both in magnitude and in waveshape from the internally generated electromagnetic forces depending on the relationship between the excitation frequency and the resonant frequency of the system.

When evaluating axial forces, winding misalignments, caused by workshop tolerances need to be considered.

The design force calculations are performed both for the symmetrical configuration and the displaced configuration of windings.

In case of a symmetrical winding arrangement in axial direction having uniform current distribution, there is no resulting force against the yoke, and the winding tends to be compressed in the axial direction only.

Figure 7.3(a) is a typical case of a symmetrical winding arrangement with no forces towards yoke. Different yoke distances and tapping in the main winding and uneven current distribution in the axial direction can cause the force integral to reach a final value greater than zero. For example in Fig. 7.3(b), the residual force, Fr is towards the bottom yoke, while in Fig. 7.3(c) it is seen that the integral has passed the zero line once and obtained a negative value at the bottom yoke resulting in residual force towards top yoke.

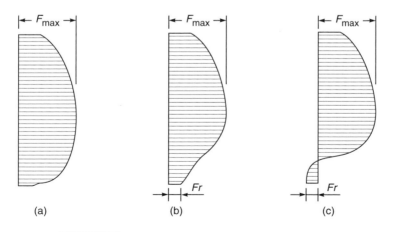

Figure 7.3 (a) No forces on yoke.
(b) Force towards bottom yoke.
(c) Force towards top yoke.

While a simple formula could be obtained for radial forces, such is not the case for axial forces. The two major questions encountered are:

(a) What is the value of ampere-turns causing the radial leakage flux?
(b) What is the effective length of path for the radial flux?

An approximation to the ampere-turns causing radial leakage-flux is the residual ampere-turns caused due to tappings, gaps in windings, unequal winding heights, etc. (see Fig. 7.4). This is only an approximation, since even perfectly balanced windings will still have fringing of flux at the coil ends, giving rise to radial component of flux. The approximation is on the conservative side, since the peak value of the residual ampere-turns are used for calculations. Such a peak will not occur in practice, or will occur only at one point.

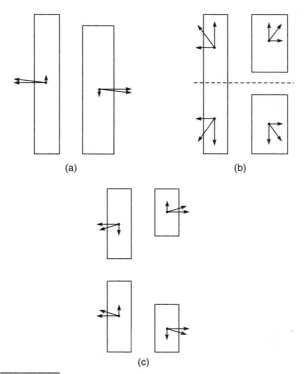

Figure 7.4 *Unbalanced windings causing axial forces.*

Considering the second question about the effective length of path for the radial flux, even an approximation is not easily forthcoming. The flux lines are influenced by the core legs, yokes, the clamping structure, the tank walls and all other magnetic structures within the tank, and follow quite complex paths for closure. An exact evaluation of the effective length of path is extremely difficult, if not impossible.

Attempts have been made to obtain empirical equations for the effective length-of-path of the radial leakage-flux. Billig[2] proposed a value based on Rogowski's work on sandwich windings. It is argued that this gives wrong results in certain cases. Waters[3] has given values for the factor $\pi D_m/l$ for various cases. Forgestad[8] has given some empirical formulae for the effective length, based on his experiments. Knaack[9] has given certain other formulae from his experiments.

The simplest of all these approaches is that of Waters[3]. Considering that there is an unbalance of NI ampere-turns, the mean radial flux B_r is given by

$$B_r = \frac{\mu_o}{2} \times a \times \frac{NI}{l_{eff}}$$

The axial force on the winding is

$$F_a = B_r \times i \times l$$

$$= \frac{\mu_o}{2} \times a \times \frac{NI}{l_{eff}} \times I \times \pi D_m \, \text{N}$$

$$= \frac{\mu_o}{2} \, a(NI)^2 \, \frac{\pi D_m}{l_{eff}} \, \text{N} \tag{7.5}$$

Reference (3) gives the value of the factor $(\pi D_m/l_{eff})$ for various residual ampere-turn configurations and two values of the ratio of window height to core circle. The paper opines that these two values of the ratio suffice for all normal power transformers.

7.6 Roth's Method of Force Calculation

A brief exposition of Roth's method of force calculation is given here, since it is an important and accurate method of evaluation. Roth's solution in rectangular coordinates is treated.

The transformer window region is considered and all the four boundaries are assumed infinitely permeable. This rectangular region contains an arrangement of a number of rectangular conductors (p)

carrying current, as shown in Fig. 7.5. The field solution is obtained by using the method of images. Equation (7.1) is then applied to obtain the forces.

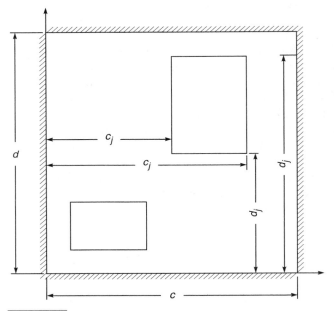

Figure 7.5 *Transformer window region with conductors.*

From a consideration of the method of images, the p conductors will form an infinite number of images both in the x and y directions. Hence the field function must be periodic with both coordinates, and will be given by

$$A = \sum_m \sum_n B_1 \cos mx \cdot \cos ny + \sum_m \sum_n B_2 \cos mx \cdot \sin ny$$
$$+ \sum_m \sum_n B_3 \sin mx \cdot \cos ny + \sum_m \sum_n B_4 \sin mx \cdot \sin ny \quad (7.6)$$

Applying the boundary conditions and simplifying the vector potential function will have the form

$$A = \sum_{h=1}^{\infty} \sum_{k=1}^{\infty} B_{h,k} \cos (h-1) \frac{\pi x}{c} \cdot \cos (k-1) \frac{\pi y}{d} \quad (7.7)$$

Electromagnetic Forces in Power Transformers

Equation (7.7) must satisfy Poisson's equation if it is to be the true field solution. Applying this condition, the following relations for the constant $B_{h,k}$ are obtained:

(a) When $h \neq 1$ and $k \neq 1$

$$B_{h,k} = \frac{4\mu_o}{cd} \frac{1}{m_h^2 + n_k^2} \sum_{j=1}^{p} J_j \frac{\sin m_h c_j' - \sin m_h c_j}{m_h}$$

$$\frac{\sin n_k d_j' - \sin n_k d_j}{n_k} \quad (7.8)$$

(b) When $h = 1$ and $k > 1$

$$B_{1,k} = \frac{2\mu_o}{cd} \sum_{j=1}^{p} J_j (c_j' - c_j) \frac{\sin n_k d_j' - \sin n_k d_j}{n_k} \quad (7.9)$$

(c) When $h > 1$ and $k = 1$

$$B_{h,1} = \frac{2\mu_o}{cd} \sum_{j=1}^{p} J_j (d_j' - d_j) \frac{\sin m_h c_j' - \sin m_h c_j}{m_h} \quad (7.10)$$

(d) When $h = k = 1$, $B_{1,1}$ is a constant, which can be ignored because A has an arbitrary origin.

In the above

$$m_h = 2(h-1)\frac{\pi}{2c}$$

$$n_k = 2(k-1)\frac{\pi}{2d}$$

To calculate the forces, it is necessary to evaluate the flux density acting within the conductor. The flux density is simply the gradient of the vector potential function of Eq. (7.7). This is evaluated and the force calculated for a filament of area $dx \times dy$. This is then integrated over the area of the conductor. The final equations for the force is given by

$$F_{xj} = -J_j \sum_{h=1}^{\infty} \sum_{k=1}^{\infty} B_{h,k} \frac{(\cos m_h c_j' - \cos m_h c_j)(\sin n_k d_j' - \sin n_k d_j)}{n_k}$$

$$(7.11)$$

$$F_{yj} = J_j \sum_{h=1}^{\infty} \sum_{k=1}^{\infty} B_{h,k} \frac{(\sin m_h c_j' - \sin m_h c_j)(\cos n_k d_j' - \cos n_k d_j)}{m_h}$$

$$(7.12)$$

Here $B_{h,k}$ is given by Eqs (7.8), (7.9) or (7.10) as the case may be. A fairly accurate solution is obtained when h and k are varied from 1 to 20.

7.7 Modes of Failure of Windings and Design of Windings

7.7.1 Failure Due to Radial Forces

(a) *Outer Windings*

The outer of two concentric windings will experience a simple hoop-stress under short-circuit. This is so, since the outer winding has no extraneous radial support and all the force is to be withstood by the conductors. The average force per turn from Eq. (7.3) will be

$$F_{rt} = \frac{F_r}{N} \tag{7.13}$$

The pressure due to this force is

$$P_{rt} = \frac{F_r}{N} \frac{1}{\pi D_m t_a} \tag{7.14}$$

The hoop-stress due to this pressure is

$$P_{ht} = \frac{F_r}{N} \frac{1}{\pi D_m t_a} \frac{D_m}{2 t_r}$$

$$= \frac{F_r}{2 N a_c} N/m^2 \tag{7.15}$$

The above is true when there is more than one turn in the radial direction and each such turn supports the other, so that all conductors may be assumed to have the same mean stress. This is not so if the outer conductors do not support those with a smaller radius. In such a case, the innermost conductor nearest to the duct will experience twice the average force as given by Eq. (7.4). It is unlikely that an inner turn is subjected to bursting force and remains unsupported by the turn over it. First, the coil is wound under tension to form an integral and compact winding.

Secondly, a consideration of plastic behavior will lead to the conclusion that very slight plastic deformation will bring the second and subsequent turns into the analysis. Thus, the mean stress is a good indicator of the hoop-stress on the outer winding.

A recent paper by Patel[12] shows that the maximum hoop-stress may be about 20% higher than that calculated from Eq. (7.3). He has applied the theory of elasticity to the problem, and by using a *force generation rate*, has set up a second-order differential equation. The solution yields the hoop-stress.

The designer has to ensure that the conductor section chosen is adequate to withstand this hoop stress. If inadequate, the cross-section has to be increased or a more work-hardened copper has to be used.

(b) *Inner Windings*

A much more difficult problem is presented by inner windings, whether disc or helical. Normally, inner windings are supported from the core by spacers at a number of points along the circumference. The simplest approach is to treat the conductor as a simply-supported beam between two such spacers. This will give an erroneously large value for the stresses. The winding resembles a tube supported internally and subjected to external pressure. Failure can be of two types: bending of the conductors between supports as in Fig. 7.6(a) or by buckling as in Fig. 7.6(b). The designer must ensure that neither of these occur.

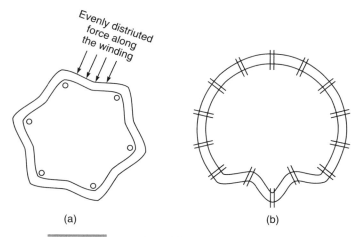

Figure 7.6 *Modes of failure of inner windings.*

In general, the problem of the collapse of the inner winding is difficult to solve. Solutions have been obtained considering the winding as a tube supported internally and subjected to external pressure, but others have experimentally proved these formulae to be conservative. Treatment of the winding as a curved beam or in general as a ring subjected to external pressure and supported internally at a number of points has also been done. Kojima et al[11] have presented a finite-element analysis of the problem and claim good agreement with experimental results.

7.7.2 Failure due to Axial Forces

Assuming that the axial spacers are rigid and do not fail, axial forces will tend to bend the conductors in an axial direction between the axial spacers. Again, an approximate method is to treat each segment as a simply-supported beam. Analysis as a continuous beam can also be done.

A typical failure in disc coils is when the conductors tilt in a zig-zag pattern as in Fig. 7.7. Waters[4] has given an analysis of this type of failure, and his expression for what he calls the critical load W_{cr} is

$$W_{cr} = \frac{\pi E_c N a_c}{6} \frac{t_a}{R} \qquad (7.16)$$

The failure of windings due to axial forces depends more on the strength of the clamping structure and the insulating spacers. Once the clamping structure gives way even slightly, the resultant displacement of the winding enormously increases the forces and leads to a complete collapse of the winding. All manufacturers of transformers use some kind of prestressing of the windings in order to obviate movement of coils due to short-circuits and shrinkage of insulation spacers in service. The design of clamping structures to withstand short-circuit force is a specialized area by itself and is discussed in the following section.

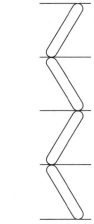

Figure 7.7 *Failure of disc winding.*

7.8 Strengthening of Coils to Withstand Short-Circuit Forces

In spite of adequate design strength of the conductor, a coil will still fail under short-circuit if it is permitted to move under the effect of short-circuit forces. The very nature of these forces is such that they tend to increase the unbalance causing them. Thus, any movement of the coil greatly increases the force causing the movement and ultimately leads to failure.

Some of the means adopted for strengthening the transformer against short-circuit forces are given below:

Adequate radial supports are provided for inner coils.

Precompressed insulation material is used in manufacture in order to minimize shrinkage in service.

Proper processing and prestressing of coils is done to obtain dimensional stability of windings.

Depending upon prevalent design practices, devices like springs or hydraulic dampers are used to absorb the impact load due to short-circuit forces.

High tensile steel is also used in places where ordinary steel is not adequate.

7.9 Design of Clamping Structures

The purpose of a well-designed clamping structure (Fig. 7.8) is to prevent any movement within the windings. This can be achieved

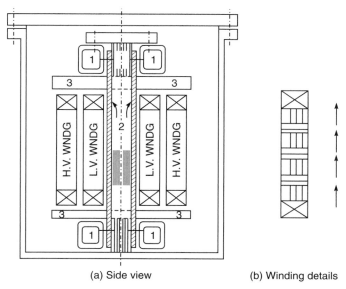

(a) Side view (b) Winding details

1. End frame
2. Clamp plate
3. Clamping ring

Figure 7.8 *Typical construction.*

by any clamping device which puts the coils permanently under a pressure, higher than those produced by the short-circuit forces. Thus, the first requirement for clamping structure is that they are designed for exerting short-circuit forces without permanent deformation.

7.9.1 Stresses in Top Clamping Ring

For maintaining a constant pressure on windings, the force is applied by a set of bolts or spring loaded jacks. These bolts are tightened against top-end frame for providing the required pressure to windings through top-end ring. Refer to Fig. 7.9 for top clamping ring. A, B, C and D are the locations of bolts/jacks. Under short-circuit force, the winding will try to bend the free span of the clamping ring between the two consecutive bolts. The stresses in top ring are

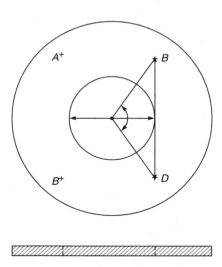

Figure 7.9 *Top clamp ring.*

Bending moment

$$M = \frac{Wl^2}{10} \quad (7.17)$$

Section modulus of ring

$$Z = \frac{1}{6}bt^2 \quad (7.18)$$

Bending stresses

$$f_b = \frac{M}{Z} \quad (7.19)$$

The results obtained by Eq. (7.19) are on the conservative side, because of the assumption made for boundary condition of fixing bolts as average of simply supported and fixed beam conditions. The allowable bending stresses further depend on allowable factor of safety, which is generally very low for short-circuit forces.

7.9.2 Stresses in Clamp Plates

During the period when short-circuit forces are developed, axial forces are also developed in windings, which in turn transfer to top and bottom end frame and these try to separate out from the core. Hence, to keep them under position, clamp plates are used. These also take the static load of transformer during lifting. The tensile stresses are developed at section AA and shearing stresses at BB as shown in Fig. 7.10.

  (a) Tensile stresses

$$f_t = \frac{P}{(X - D_1) \times t} \quad (7.20)$$

  (b) Shearing stresses

$$f_s = \frac{P}{2 \times H_1 \times t} \quad (7.21)$$

  (c) Bearing stresses

$$f'_b = \frac{P}{D_1 \times t} \quad (7.22)$$

The stress values computed from Eqs. (7.20) to (7.22) should be less than the yield point of the clamp plate material.

7.9.3 Stresses in Top End Frame

The stresses in top end frame are developed due to the following:

  (a) The force due to the clamping pressure applied for tightening the coils.
  (b) The end thrust created by electromagnetic forces in the windings due to an unbalance in ampere turns.
  (c) The lifting load of transformer.

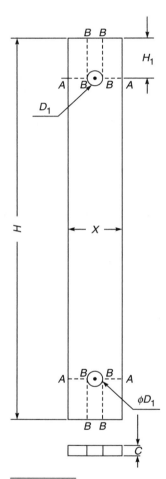

Figure 7.10 *Clamp plate.*

In general, the short-circuit force condition is most critical. Figure 7.11 shows the top end frame. The end frame shown is for single-phase transformer, where A is the core bolt position while C is the pin used for connecting top end frame with clamp plate. Since the end frame is symmetrical along center line, only half portion shall be analyzed for stresses. Figure 7.12 shows the short-circuit forces, acting locations and support position. The core bolt points are assumed as support locations.

The bending moment at support is given by

$$M_s = F_1 \times L_1 + F_2 \times L_2 + F_3 \times L_3 + F_4 \times L_4 \quad (7.23)$$

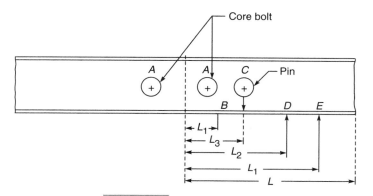

Figure 7.11 Top end frame.

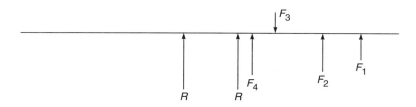

Figure 7.12 Free body diagram.

The section modulus of end frame is calculated by

$$Z_s = \frac{b_1 \times h_1^2}{6} - \frac{(b_1 - t_1)(h_1 - 2 \times t_1)}{6h_1} \qquad (7.24)$$

(See Fig. 7.13, section of end frame.) The bending stresses are given as

$$f_b = \frac{M_s}{Z_s} \qquad (7.25)$$

The maximum bending moment will be at C, where the force sign is changing. Bending moment at C is given by

$$M_c = F_1 \times L_1 + F_2 \times L_2 \qquad (7.26)$$

Section modulus is calculated from Eq. (7.24) and maximum bending stresses should be less than yield point of end-frame material with sufficient factor of safety.

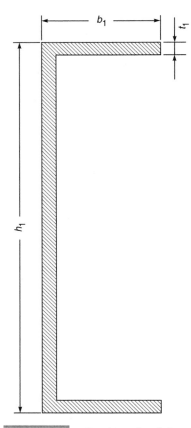

Figure 7.13 *Section of end frame*

7.9.4 Bottom End Frame

The same philosophy is applicable to the bottom end frame. However, the flanges of bottom over which windings rest are critical areas. Sufficient stiffening of flanges shall be done by providing vertical stiffeners below flanges.

7.10 List of Symbols

All units are SI

μ_0 Permeability of free space = $4\pi \cdot 10^{-7}$.
a Unbalance between windings expressed in per-unit.

a_c	Cross-sectional area of turn.
A	Vector potential.
B	Flux-density at mean radial depth of winding.
B_r	Mean value of radial leakage flux-density.
B_1, B_2, B_3, B_4	Constants
D_m	Mean diameter of main intercoil duct.
E_c	Modulus of elasticity of conductor material.
$\vec{F}$	Force vector.
F_m	Radial force on conductor adjoining the main inter-coil duct.
F_r	Radial force at mean radial depth of winding.
F_{rt}	Average radial force per turn.
$\vec{i}$	Current vector.
I	Asymmetric peak of the short-circuit current.
l	Length of conductor.
L	Length of path for axial leakage flux.
l_{eff}	Effective length of path for radial leakage flux.
N	Number of turns in winding.
P_{ht}	Hoop-stress on turn.
P_{rt}	Average radial pressure per turn.
R	Radius of turn.
t_a	Axial dimension of turn.
t_r	Radial dimension of turn.
W_{cr}	Critical load.
F_a	Axial force on the winding.
W	Force per unit length due to short-circuit forces.
1	Pitch between two consecutive bolts in cm.
b	Width of top ring in cm.
t	Thickness of top ring in cm.
P	Total short-circuit force on one clamp plate in kg.
X	Width of clamp plate in cm.
D_1	Hole diameter for pin in cm.
H_1	Height of clamp plate from center of hole to top/bottom surfaces in cm.
F_1, F_2, F_3 and F_4	Short-circuit forces intensities acting on end-frame in kg.
L_1, L_2, L_3 and L_4	Locations of forces from center line in cm.

References

1. Billig, E., *Mechanical Stresses in Transformer Windings*, Critical Resume, E.R.A. Report Q/T 101, 1943.
2. Billig, E. Mechanical Stresses in Transformer Windings, *Journal I.E.E.*, Vol. 93, Part II, June 1946.
3. Waters, M., The Measurement and Calculation of Axial Electromagnetic Forces in Concentric Transformer Windings, *Journal IEE*, Vol. Part II.
4. Waters, M., *The Short-Circuit Strength of Power Transformers*, Book, Macdonald and Co, London, 1966.
5. Roth, E., Analytical Study of the Leakage Field in Transformers and of the Mechanical Forces Acting on the Windings, *General Electric Review*, Vol. 23, p. 773, May 1928.
6. Roth, E., *Magnetic Leakage Inductance in Transformers with Cylindrical Windings and Forces Acting on the Winding*, RGE, Vol. 40, August 1936.
7. Binns, K.J., and P.J., Lawrenson, *Analysis and Computation of Electrical and Magnetic Field Problems*, 2d edition, Pergammon Press Ltd., Oxford, 1973.
8. Fergestad, R., Electromagnetic Forces in Core-type Transformers with Concentric Windings, Paper 114, *CIGRE*, Vol. II, 1956.
9. Knaack, W., The Mechanical Stressing of Transformer Windings under Short-Circuit, Paper 135, *CIGRE*, Vol. II, 1956.
10. Rabins, L., Transformer Reactance Calculations with Digital Computers, *Communications and Electronics*, Vol. 75, p. 261, 1956.
11. Kojima, H., H. Miyata, S. Shida and K. Okuyama, Buckling Strength Analysis of Large Power Transformer Windings Subjected to Electromagnetic Force under Short-Circuit, *IEEE Trans. on PAS*, Vol. 99, No. 3, p. 1288, May–June 1980.
12. Patel, M.R., *Hoop-Stress in Transformer Coils under Radial Short-Circuit Forces*, IE(I), Electricial Engineering Division. Vol. 61, June 1981.

Chapter 8

Cooling Arrangements

C.M. Sharma

In power transformer, the oil serves a dual purpose as an insulating medium as well as a cooling medium. The heat generated in the transformer is removed by the transformer oil surrounding the source and is transmitted either to atmospheric air or water. This transfer of heat is essential to control the temperature within permissible limits for the class of insulation, thereby ensuring longer life due to less thermal degradation.

8.1 Various Types of Cooling

8.1.1 ONAN Type Cooling

The generated heat can be dissipated in many ways. In case of smaller ratings of transformers, its tank may be able to dissipate the heat directly to the atmospheric air, while bigger ratings may require additional dissipating surface in the form of tubes/radiators connected to tank or in the form of radiator bank. In these cases, the heat dissipation is from transformer oil to atmospheric air by natural means. This form of cooling is known as ONAN (oil natural, air natural) type of cooling.

8.1.2 ONAF Type Cooling

For further augmenting the rate of dissipation of heat, other means such as fans blowing air on to the cooling surfaces are employed. The forced air takes away the heat at a faster rate, thereby giving better cooling rate than natural air. This type of cooling is called ONAF (oil natural, air forced) type of cooling. In this cooling arrangement, additional rating under ONAN condition viz. after shutting off fans, is available, which is of the order of 70–75%.

8.1.3 OFAF Type Cooling

Still better rate of heat dissipation could be obtained if in addition to forced air, means to force circulate the oil are also employed. The oil can be forced within the closed loop of transformer tank and the cooling equipment by means of oil pumps. This type of cooling is called OFAF (oil forced, air forced) type of cooling. Mixed cooling transformers of radiator type can have two or three ratings available, one for each type of cooling, viz. OFAF, ONAF and ONAN.

OFAF cooling can also be obtained by using OFAF compact coolers. These coolers offer the advantage that they occupy less space. The disadvantage of OFAF compact coolers is that, there is no ONAN rating available in case of failure of fans and pumps, thereby necessitating provision of a stand-by cooling equipment for switching in, immediately upon receipt of failure signal. Continuity of auxiliary supply to fans and pumps is to be ensured for uninterrupted power flow.

8.1.4 OFWF Cooling

Since the ambient temperature of water is always less than the atmospheric air, it is possible to use water as a better heat-transfer media. Such an arrangement employs oil to water heat exchangers. A prerequisite for such an arrangement is the availability of a source of sufficient quantity of water. In most of the transformers for hydropower stations, this type of cooling is used. Such a cooling is called OFWF (oil forced, water forced) type of cooling.

8.1.5 Forced Directed Oil Cooling

Additional means of improving the heat dissipation rate are also employed on higher ratings for transformer. These comprise arrangements which direct the transformer oil in the windings through predetermined paths. This directed oil flow type of cooling is utilized with advantage in case of forced oil system. The cool oil entering the transformer tank from the cooler/radiators is passed through the windings in a pre-decided manner ensuring faster rate of heat transfer. This type of cooling is called ODAF (oil directed, air forced) or ODWF (oil directed, water forced) type. Figure 8.1 shows a typical arrangement used for directing the oil in the winding.

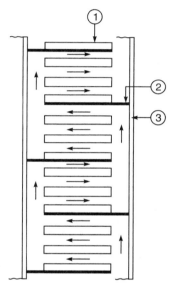

1. Winding conductors
2. Directed oil flow washers
3. Insulation cylinders

Figure 8.1 *Directed oil flow.*

8.2 Cooling Arrangements

Depending upon the type of cooling and rating of the transformer, the cooling equipments can be arranged in various ways. Some of the commonly used ones are described hereunder.

8.2.1 Arrangement with Radiators

Radiators are commonly used for ONAN, ONAF/ONAN and OFAF/ONAF/ONAN types of coolings. Figure 8.2 shows a typical radiator. Radiators consist of elements joined to top and bottom headers.

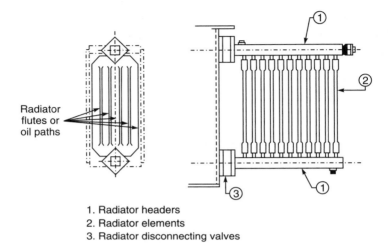

1. Radiator headers
2. Radiator elements
3. Radiator disconnecting valves

Figure 8.2 *Typical five flute radiator.*

Elements shown in Fig. 8.2 are made by welding two previously rolled and pressed thin steel sheets to form a number of channels or flutes through which oil flows. These radiators can be either mounted directly on the transformer tank or in the form of a bank and connected to the tank through the pipes. The surface area available for dissipation of heat is multiplied manifolds by using various elements in parallel. As oil passes downwards, either due to natural circulation or force of a pump in the cooling circuit, heat is carried away by the surrounding atmospheric air.

(a) *Tank Mounted Radiators*

Tank mounted radiators are used for smaller ratings of transformer for ONAN or ONAF/ONAN type of coolings. The radiators are connected to the tank with an interposing valve at the top and the bottom. The hot oil enters the radiators from the top and after dissipating the heat to surrounding air it goes back in the tank from the bottom. The radiators are arranged on the tank in such a manner that the center line of radiator is always at a higher level than the transformer winding center line. This difference in the heights is called *thermal head* and is responsible for producing thermosyphonic action, enabling the oil to circulate. Figure 8.3 depicts the available *thermal head* in case of tank mounted radiators. It is

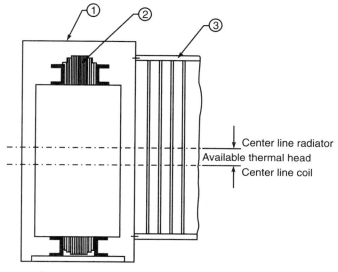

Figure 8.3 *Available thermal head.*

obvious that it is governed by the height of the tank and the length of the radiator.

(b) *Banked Radiators*

Radiators can also be arranged to form a bank. This is necessitated when transformer rating is higher and the total number of radiators, required to dissipate the transformer losses, are more. Under such circumstances it is not possible to physically arrange the radiator on to the transformer tank. The radiators are mounted on to the headers, which are supported from the ground. Interconnecting pipework is required to be provided with isolating valves for maintenance and erection purposes. Separately mounted radiator bank is generally called for in case of OFAF type of cooling, since the interconnecting pipework is required to mount a pump for force circulating the oil.

In case of tank mounted radiators, the available *thermal head* cannot be adjusted much, whereas in case of banked radiators, supports for the headers can be adjusted in height to give adequate *thermal head* enabling better cooling under natural circulation also. The cooling fans required for ONAF and OFAF types of coolings can be arranged either underneath the radiators or on the sides.

The total number of radiators required for cooling of transformer can be arranged in many ways. Usually one bank of radiators for 100% capacity is sufficient. However, if desired 2–50% banks can be formed by dividing the number of radiators required into two equal parts, each part being capable of dissipating 50% losses. Each 50% part can be arranged in a completely independent bank with separate auxiliary cooling equipments like fans and a pump. An alternative arrangement having 2–50% capacity radiators with one 100% capacity pump is also specified. It is obvious, that because of the additional pump and bank supports, etc., the fully independent alternative shall be costlier.

To cater to the exigencies of failure of auxiliary pumps and fans, provision of a standby pump and a fan is also sometimes specified. In case of failure of the running pump and fan, the standby pump and fan can be brought into operation. It is desirable to specify bare minimum standby capacity, that too only for cases where continuous operation at full load is envisaged. In a few cases, 3–50% radiator banks have also been specified, but in our opinion one complete 50% bank as standby in case of mixed cooling banks is not necessary. This is due to the fact that all mixed cooling transformers have a self-cooled (ONAN) rating of at least 50% inherently available.

8.2.2 Arrangements Using Coolers

Compact OFAF coolers for dissipation of heat to atmospheric air and OFWF coolers for dissipation to water are also in common use. These coolers are connected to the transformer tank through interconnecting pipework and pumps for oil circulation.

(a) *Oil to Water Heat Exchangers (OFWF Coolers)*

This type of cooling equipment is commonly employed where water is available in abundance. Figure 8.4 shows a cutaway section of such a heat exchanger. Cylindrical shell (1) is generally made of cast iron or mild steel which houses a removable tube stack (2). The tube stack is generally made of copper, admiralty brass or stainless

Cooling Arrangements 183

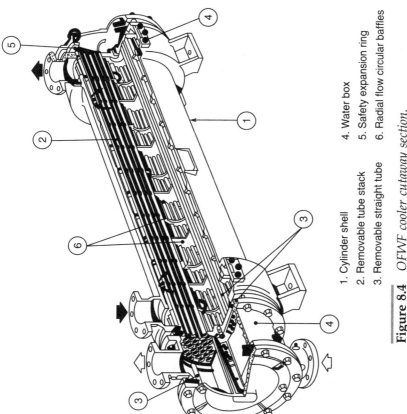

1. Cylinder shell
2. Removable tube stack
3. Removable straight tube
4. Water box
5. Safety expansion ring
6. Radial flow circular baffles

Figure 8.4 *OFWF cooler cutaway section.*

steel depending upon the impurities present in cooling water. The tubes are assembled by roller-expanding the tube-ends in the drilled holes of the tube plates. One tube plate is fixed between the shell and water box (4). The other plate is free to move, to allows expansion of the tube stack as a whole. A safety expansion ring (5) prevents intermixing of oil and water. Water enters through water box and flows through the tube stack while the oil flows over the tubes guided by circular baffles (6) to give maximum cooling efficiency.

Water cooled oil coolers are available suitable for horizontal as well as vertical mounting. In certain cases it is possible to mount these coolers on the tank supported on brackets. Generally 2–100% OFWF coolers are provided which includes one complete 100% cooler as standby. It is very important to ensure that under working conditions, the oil pressure in the coolers is always more than the water pressure, so that possibility of water leaking into oil is eliminated. OFWF coolers can be arranged independent of each other. But for better flexibility, usually both the coolers are specified to be interconnected in such a manner, that by regulating the various interposing valves it is possible to use either cooler with either of the pumps. Such an arrangement is schematically shown in Fig. 8.5. This arrangement requires two coolers and two pumps only.

(b) *Oil-to-air Heat Exchanger (OFAF Coolers)*

In this type of coolers, tubes with fins are employed to increase the surface area. Figure 8.6 shows such a heat exchanger. The principal components are ferrous or non-ferrous tubes threaded through spaced aluminum or copper fins, to which they are metal bonded for strength and heat transfer. These tubes are fitted into brass plates, which in turn are attached to brass headers.

Axial-flow motor driven fans are attached to the cooler body rigidly to reduce vibrations. These coolers can be mounted either horizontally or vertically. These coolers are mounted directly on transformer tanks in vertical position on small rating transformers.

These coolers are compact and as such occupy less space, and therefore these are most suited where available space is constraint.

These coolers can be arranged in many combinations, viz. $1 \times 100\%$, $2 \times 50\%$, $4 \times 25\%$, $5 \times 20\%$, etc. Usually, in addition to coolers required for 100% rating, one complete unit cooler is kept as a standby. A number of small unit coolers offer an additional advan-

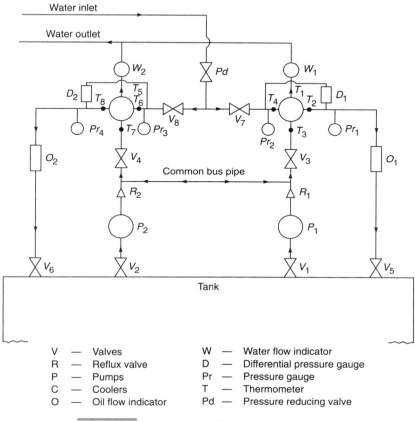

Figure 8.5 2–100% OFWF cooler schematic.

tage, that under light load conditions some of the unit coolers can be cut off, thereby saving the energy consumed by auxiliary fans and pumps.

8.3 Propeller Type Fans

Fans constitute an important part of the radiator type cooling equipment. The fans are suitable for outdoor duty and are required to give large air deliveries at slow or moderate speeds with low power consumption and noise level.

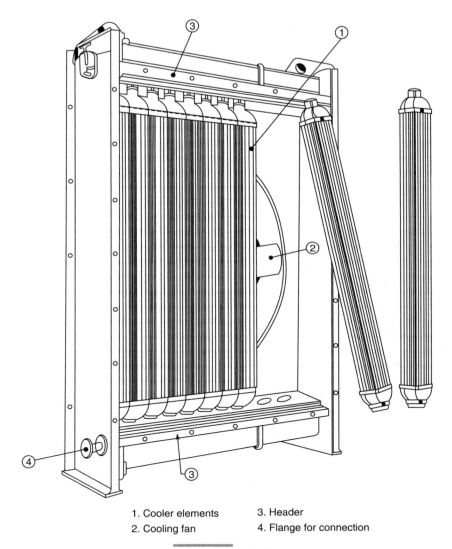

1. Cooler elements
2. Cooling fan
3. Header
4. Flange for connection

Figure 8.6 *Cooler.*

Figure 8.7 shows the main components of a fan with motor. The fan blades are bolted on to the hub and the assembly called impeller is properly balanced to minimize the wear of bearings, vibration and noise level. The direction of air flow, also referred to as *form of running*, is generally away from the motor. The blades are covered

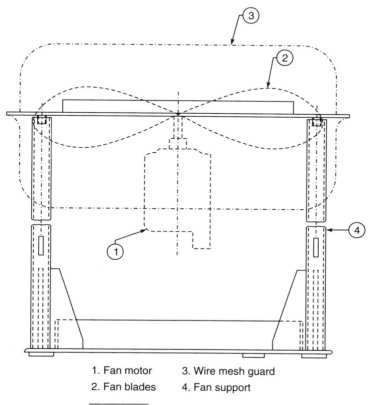

1. Fan motor 3. Wire mesh guard
2. Fan blades 4. Fan support

Figure 8.7 *Fan with its support.*

from both top and bottom sides with wire guards, and the whole assembly is supported on a steel structure when fans are mounted below the radiators.

8.3.1 Fan Mounting Arrangements

Fans can be mounted underneath the radiators, blowing air from the bottom to the top. This arrangement permits use of bigger sweep fans. The air flow cone encompasses sufficient surface of the radiators to give the required amount of heat dissipation. The radiator surface not within the cone is natural cooled. These fans can be supported either directly onto the radiators or mounted separately from the ground. The latter case requires provision of foundations for fan supports. A typical fan and support for ground mounted fan is shown in Fig. 8.7 and some fans are mounted directly on to radiators. It should be ensured that fans mounted on the radiators do not produce undue vibrations.

Another method of mounting the fans is on the side of the radiators. This is with a view to ensure that the air blown cools elements of radiators where oil temperature is higher. This arrangement is used when a number of small capacity fans are used. To accommodate a number of fans, rows of fans are formed in the height of the radiators.

8.4 Transformer Oil Pump

OFAF and OFWF type of cooling methods require oil circulating pumps in the cooling circuit of a transformer. These pumps circulate comparatively cold oil through the windings and carry away the heat generated quickly, to keep hot-spot temperature in the windings within permissible limits. This is a closed-circuit operation and the pumps are required to develop enough pressure, to overcome the frictional head loss during the flow of oil in pipe work, cooling equipment and winding, etc.

8.4.1 Constructional Features

Transformer oil pumps employ special features of construction, in which the pump and motor is an integral unit. The pump impeller is screwed and locked on the driving motor shaft. The motor unit is embodied inside the pump casing, completely immersed in transformer oil, and is cooled by the surrounding oil. As the motor is attached to the pump without the intervention of a stuffing box, the equipment forms a leakproof glandless close-coupled unit suitable for outdoor installation. No shaft seal is required, thus eliminating the dangers of oil leakage or entry of air. The heavy-duty ball and roller bearings carrying the shaft are lubricated by internal circulation of transformer oil. The casing of pump is made of cast iron or aluminum, with flanged suction and delivery for bolting to the transformer oil-pipe circuit. The impeller is made of aluminum or bronze.

Normally two different designs are employed depending upon the duty requirements.

(a) Axial Flow-in-line Type

Single-stage axial-flow pumps are used to circulate large volumes of oil against comparatively low frictional head losses. These pumps are compatible with ONAN cooling mode of a mixed cooling (OFAF/ONAF/ONAN) transformer employing radiator bank, as they offer minimum thermosyphonic resistance to oil flow. Figure 8.8 shows a cross-sectional view of such a pump.

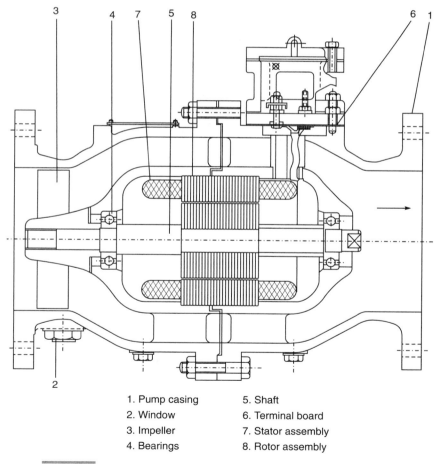

1. Pump casing
2. Window
3. Impeller
4. Bearings
5. Shaft
6. Terminal board
7. Stator assembly
8. Rotor assembly

Figure 8.8 *Cross-sectional view of an axial flow-in-line type.*

(b) Radial Flow-in-line or Elbow Type

These pumps are used on transformers where oil is required to circulate against large frictional head as encountered in oil-to-water and oil-to-air heat exchangers. Figure 8.9 shows a cross-sectional view of a radial flow elbow type pump.

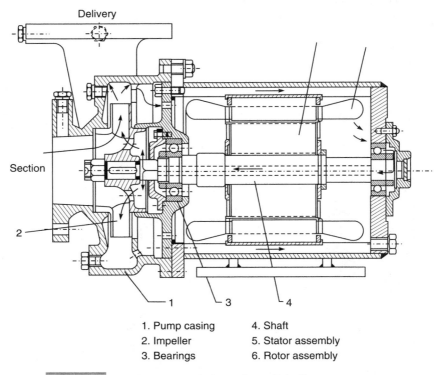

1. Pump casing
2. Impeller
3. Bearings
4. Shaft
5. Stator assembly
6. Rotor assembly

Figure 8.9 *Cross-sectional view of a radial elbow type pump.*

8.5 Flow Indicators

It is essential to monitor continuous flow of oil and water in the cooling circuit of transformers for its safe working. Flow indicators employed for this purpose perform the following two functions:

— indicate the full flow in proper direction.
— provide signal by operation of a switch when flow drops below a preset limit.

8.5.1 Construction and Principle of Operation

A suspended vane is used as a sensor. Its surface is kept at right angles to the direction of flow in the pipe. Figure 8.10 shows a schematic diagram of a flow indicator. When liquid starts flowing through the pipe, the vane gets deflected along with vane shaft. This deflection is transmitted through a magnetic coupling to an indicating needle on the dial. The needle indicates the flow of liquid corresponding to vane deflection. When the flow in the circuit drops to 70–80% of full flow, a switch provided in the instrument closes with the help of a cam and initiates the alarm.

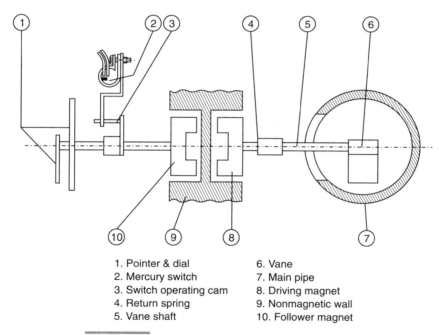

1. Pointer & dial
2. Mercury switch
3. Switch operating cam
4. Return spring
5. Vane shaft
6. Vane
7. Main pipe
8. Driving magnet
9. Nonmagnetic wall
10. Follower magnet

Figure 8.10 *Schematic-diagram of flow indicator.*

8.6 Stress and Hydraulic Analysis of Pipework

The analysis of piping may be divided into three categories:
  (a) Stress analysis of pipe under internal pressure.
  (b) Frictional head-loss analysis of complete piping system including pipe bends, etc., under ONAN cooling.

(c) Frictional head-loss analysis of complete piping system, including pipe bends, etc., under OFAF cooling.

8.6.1 Stress Analysis of Pipe

To check the leakage, etc., pipes are tested under very high oil pressure. However, under normal conditions the internal pressure is very low. Hence, to use the economical size and thickness of pipe, it is imperative to do stress analysis of the pipe. The stresses in the pipe under internal pressure may be computed from the following formula.

$$f = \frac{PD}{2t} \tag{8.1}$$

It has been observed that the medium thin pipes are sufficient for withstanding internal pressure.

8.6.2 Frictional Head Analysis of Pipes Under OFAF Cooling

A typical schematic diagram showing the complete piping system with the fittings and pump is shown in Fig. 8.11. Under OFAF cooling condition, the pump is required to work against a particular pressure head and flow of oil for dissipation of heat through radiators. In general, this pressure head may be divided into the following:

(a) *Velocity head* This is the height through which a body must fall in a vacuum to acquire the same velocity as that of oil flowing into the pipe and equals $V^2/2G$.
(b) *Entry head* Head required to overcome the frictional resistance at the entrance to the pipe.
(c) *Frictional head* This is due to the frictional resistance to flow within the pipe. This includes pipe bends, etc.
(d) *Frictional head in fittings* This is due to the frictional resistance to flow through the fittings, such as valves, expansion joints, radiators, headers, etc.
(e) *Frictional head in windings* In the case of directed flow, the oil is forced to circulate through the ducts in windings, etc., which in turn provides frictional resistance to flow.

All the above five elements of pressure heads added together give the total head against which pump is required to generate the pressure head. It is evident from pump characteristic, if the pressure

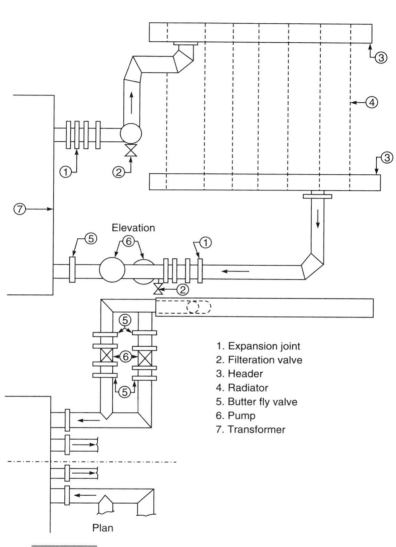

Figure 8.11 *Schematic diagram of complete piping system.*

head developed by pump is lower than the frictional head, the flow of oil will reduce, which in turn will reduce the heat dissipation.

8.6.3 Frictional Head Analysis of Pipes Under ONAN Coolings

In the ONAN cooling system, the pump to force oil is not used. The available thermal head due to top and bottom temperature and dimensional difference is responsible for flow of oil under natural cooling condition. This *thermal head* is also required to overcome the frictional losses in the cooling system. The frictional head losses are dependent on diameter of pipe, flow velocity and hence to decide the correct size of pipe diameter, the overall study of head losses is required to be done as illustrated for OFAF cooling.

8.6.4 Design Calculations (OFAF Cooling)

For finding out the total head-loss in the complete pipework, one should know the type of flow which may be found out by calculating the Reynold's number. In 200 mm and 150 mm diameter pipes, the Reynold number is very high and is within the turbulence zone. Hence, the flow in cooling pipes is always turbulent. After finding out the type of flow, the hydraulic losses in the following parts are computed.

  (a) Straight length of pipe
  (b) 90° bends
  (c) Valves
  (d) Expansion joints
  (e) Main header
  (f) Sub-header
  (g) Nipple
  (h) Elements of radiators
  (i) Due to contraction and enlargement of sub-header, nipple, elements, etc.
  (j) Pipe required for directing the oil into windings for directed flow
  (k) Exit from transformer tank
  (l) Reducers, etc.

The frictional head loss in straight length pipe would be

$$H_p = \frac{4\,FLV^2}{2\,GD} \tag{8.2}$$

The frictional head loss in other components would be

$$H_c = K\frac{V^2}{2G} \tag{8.3}$$

The factor K depends on many factors, such as diameter, Reynold number and shape of components.

Some typical results are compiled and given in Tables 8.1 and 8.2 for OFAF and ONAN cooling respectively. Depending upon the length of the pipework, number of bends and type of joints and other fittings, the size of the pipework is selected for each type of cooling. Depending on the velocity of oil in the pipe, the head losses in various components may be computed from Eqs. (8.2) and (8.3), which are summarized in Table 8.2. However, the driving force required to make the oil flow inside pipe against frictional head is computed from the following formula

$$\text{Driving force} = H \times \frac{\delta_{LT} - \delta_{HT}}{\delta_{MT}} \tag{8.4}$$

The above given procedure is applicable to all types of cooling.

As most of the jobs have mixed cooling arrangements, in order to decide the diameter of pipe and the optimum number of elements, the ONAN cooling conditions become the governing criteria.

NOMENCLATURE

f	Stress in kg/cm^2
D	Internal diameter of pipe in cm
P	Working pressure in kg/cm^2
t	Thickness of pipe in cm
H_p	Frictional head loss in pipe length
H_c	Frictional head loss in components
F	Frictional constant
L	Length of pipe in meters
V	Velocity of oil in pipe in m/s
G	Gravity constant in m/s^2
H	Thermal head in cm
δ_{LT}	Density of oil at lower temperature
δ_{HT}	Density of oil at higher temperature
δ_{MT}	Density of oil at mean temperature

Table 8.1 Summary of Pressure Head Losses in Pipe Assembly in OFAF Cooling

Sl. No	Description	20 Cms Dia.	Q GPM	V Cm/Sec.	Turbulent or Laminar	Remarks
1.	Pressure head loss in 1500 cms length pipe	20.0	600	145	Turb.	20 cm 15 cm
2.	Head loss in 90° bends (total 8 bends)	21.7	600	145	Turb.	Reynold No.
3.	Head loss in butterfly valve (2 Nos.)	1.6	600	145	Turb.	36125 48172
4.	Head loss in expansion joints	9.9	600	145	Turb.	
5.	Main header 20 radiators with 28/36 elements	0.66	600	50	Turb.	
6.	Sub-header 28/36 elements	2.07	36	34	Turb.	
7.	Head loss in nipples	0.86	1	10	Laminar	
8.	Head loss in elements	0.26	1	2.4	Laminar	
9.	Head loss due to contraction enlargement of sub-head or nipple in radiators	0.11	1	100	Laminar	
10.	Head loss at joint of sub-header to mark header	0.75	1	34	Turb.	
11.	Enlargement from pipe to tank	—	—	—		
12.	Contraction from 20 cm to 15 cm	—	—	—		
	Total	67.91 cm		810.4		
	For 20 cm dia.	Without directed flow 67.91 cm With directed flow 67.91 + 1.53 + 28.1 = 97.54 cm				

Cooling Arrangements

Table 8.2 Summary of Pressure Head Losses in Pipe Assembly in ONAN Coolings

Sl. No.	Description	20 cm dia.	Q GPM	Remark
1.	Pressure head loss in 1500 cm length pipe	0.821	121.9	With directed flow
2.	Head loss in 90° bend (total 8 bends)	1.3767	121.9	
3.	Head loss in butterfly value (2 Nos.)		0.06	121.9
4.	Head loss in expansion joints	0.40	121.9	For 20 cm Head loss = 41.23 cm
5.	Main header	0.0244	121.9	
6.	Sub-header	0.058	6.07	
7.	Head loss in nipple	0.0397	0.217	
8.	Head loss in elements	0.009	0.217	
9.	Head loss due to contraction/enlargement of sub-heads, nipple in radiator	0.006	0.217	Extra loss due to direct flow = 3.36 cm
10.	Head loss at joint of sub-header of mainheader	0.0336	0.217	
11.	Head loss in entering the tank	0.4	121.9	
12.	Reducer from 20 cm to 15 cm		121.9	
		3.1507 cms		
	For 20 cm dia.	Without directed flow 3.1507 cms With direct flow 3.36 cms		

CHAPTER 9

Design Procedure

R.C. Agarwal

9.1 Specifications of a Transformer

Transformers have large number of variables and their requirements differ a great deal. Complete specifications as stipulated in Chapter 15 must be furnished for designing the transformer to suit the specific requirements. In the following paragraphs design procedure, selection of various parameters and their effect on the various performance criteria have been discussed.

9.2 Selection of Core Diameter

Core diameter of a transformer depends upon a number of factors like rating, percentage impedance between windings, basic insulation level, transport height, overfluxing requirements, type of core and quality of core steel. Hence, it is fairly complicated to derive a universal and exact formula for determining diameter. In practice, the core diameter is selected by the designer keeping into view the limiting conditions as mentioned above. Based on this, guaranteed parameters, viz percentage impedance and losses are worked out. Core diameter is adjusted to meet the guaranteed parameters.

9.2.1 Influence of Varying Core Diameter

Increasing core diameter increases area of cross-section, thereby increasing voltage per turn, which reduce the number of turns in various windings. The percentage reactance between windings is directly proportional to number of turns and diameters of various coils and is also inversely proportional to volts per turn and coil depth. In order to have specified reactance, increased core diameter necessitates reduction in coil depth and increase in coil dimensions in lateral direction, which leads to reduction in core height and increase in core leg centers. Inspite of reduction in core height due to increase in core diameter, overall weight of core steel increases, which also increases no-load loss of transformer. Also, reduced number of turns in windings even with larger length of mean turn results in reduced copper weight, which in turn also reduces the load loss of transformer. Similarly, reduced core diameter results in reduced core steel weight and no-load loss and increased copper weight and load loss.

9.2.2 Core Area

Stepped core construction is used to obtain an optimum core area within circumscribing circle of a core. Core area depends upon number of steps, grade of core steel, insulation on the laminations, i.e., varnished or unvarnished, and method of core clamping. As the number of steps increases, core area also increases, however it needs extra labor to cut the various sizes of laminations. Optimum number of steps are used to give overall economy. Large rating transformers are usually provided with high tensile steel clamp plates for clamping the core laminations, which provide increased core area for a particular core diameter. Depending upon the flux density in the core, adequate number of ducts are provided to keep its hot-spot temperature within permissible limits. (For more details on hot-spot temperature and cooling ducts in core, refer to Chapter 4.)

9.3 Selection of Flux Density

Value of flux density is chosen to suit the guaranteed performance. Normally flux density is chosen near knee point of magnetization curve, however, adequate margin should be kept to take care for system conditions like overfluxing, frequency and voltage variations. In certain cases value of flux density is reduced to limit the noise level of transformers.

9.3.1 Influence of Varying Flux Density

Keeping the other parameters the same, increase in the value of flux density in the core results in higher volts per turn. Hence number of turns in various windings are reduced. Effect of increase of flux density on reactance is similar to that of increase in core diameter. In order to meet the requirement of specified reactance, coil depth is reduced and lateral dimensions of coils are increased. In spite of small increase in core leg centers, reduced core height results in lower core-steel weight. Increased flux density, with corresponding reduced core-steel weight results in higher no-load loss of transformer. Also reduced number of turns in windings results in lower copper weight and load losses. Similarly, reduction in the value of flux density causes increased core-steel weight, lower no-load loss and increased copper weight and load loss.

9.4 Selection of Type of Core

The cores are made of CRGO steel, which is available in various grades having different properties. Suitable grade of steel is chosen to suit the performance requirement and overall economy. Different types of cores are used as described in Chapter 4. Selection of type of core depends upon the rating of transformer and transport limitations. For large three-phase transformers, five-limb construction is adopted to overcome the problem associated with the transport height. For single-phase transformers, one center-wound limb with two return limbs is a common configuration, though sometimes both limb wound cores are also adopted. For very large ratings, core construction having two wound limbs with two return limbs may also be adopted.

9.5 Selection of Leg Length

Maximum leg length available for designing the windings, is dependent upon maximum transport height, type of wagon, to be used for transportation and type of core selected. By proper shaping of the transformer tank for well wagon, the leg length can be increased. Generally for larger rating transformers maximum value of leg length gives the overall economy. Whereas, lower rating transformer may have reduced leg length to give economical design. By reducing leg length, reactance of transformer increases, and vice versa.

Influence of core diameter, flux density and leg length over various parameters while maintaining the % reactance as constant is tabulated below:

Parameter	Increased core diameter	Increased flux density	Increased leg length
Copper weight	Decrease	Decrease	Increase
Load loss	Decrease	Decrease	Increase
Core steel weight	Increase	Decrease	Decrease
No-load loss	Increase	Increase	Decrease

9.6 Selection of Type of Windings

The windings along with its insulations form the electric circuit of the transformer. Due care must be taken while designing the windings to ensure its healthiness during normal as well as fault conditions. The windings must be electrically and mechanically strong to withstand both over-voltage under incidence of surges and mechanical stresses during short-circuit conditions. The temperature of windings at rated, over-load and short-circuit conditions should be within the limits, ensuring the proper life of the transformer.

The power transformers are manufactured for a very wide range of outputs and voltages and realization of these requirements is possible only by using different kinds of windings. The following types of windings are used in power transformers:

(a) Spiral winding
(b) Helical winding
(c) Reversed section winding
(d) Parallel layer winding
(e) Tapered layer winding
(f) Interleaved disc winding

Spiral winding is a medium-current and low-voltage winding. Tertiary winding, of star/star/delta connected power transformer used for stabilizing purposes and sometimes feeding small loads, could be generally spiral winding.

Helical winding is a high-current and low-voltage winding. Normally, it is used for LV coils of large generator transformers.

When number of turns preclude the use of helical winding, the reversed section winding is used. It is generally used for high-voltage and low-to medium current rating. Reversed section (disc) winding is usually used up to 132 kV class windings.

Higher voltage windings above 132 kV class are mostly multi-layer or interleaved disc winding. Number of layers of layer-type windings are generally five to nine. Ordinary reversed section disc winding is not suitable for voltage above 132 kV because of impulse distribution characteristics of winding. However, by doing the inter-leaving of conductors/turns, impulse distribution characteristics improve and therefore, interleaved disc windings are used for HV coil above 132 kV class.

Constructional details of various types of windings have been dealt within Chapter 5.

9.7 Selection of Tapchanger

To cater for the voltage regulation in the transformer and system voltage variation, off-circuit/on-load tapchanger is provided. Off-circuit tap switch is provided when tapchanging is required only occasionally. Generally generator transformers are provided with off-circuit tap switch. When tapchanging is required under loaded conditions, on-load tapchangers are provided. Generally, unit auxiliary transformers, station transformers, and system transformers are provided with on-load tapchangers. The choice of tapchangers is governed by the following factors:

(a) Tapping range
(b) Number of steps
(c) Step voltage
(d) Current rating
(e) Location of tapping
(f) Design of tapping winding, i.e., linear, reversing, coarse and fine
(g) Insulation level

(h) Type of voltage variation, i.e., constant flux, variable flux and mixed type

(i) Power flow requirement, i.e., unidirectional or bi-directional

Constructional details and various types of tapchangers have been described in Chapter 6.

9.8 Calculation of Number of Turns

emf equation of transformer can be written as under:
Volts per turn,

$$E_t = 4.44\, B_m\, Af \tag{9.1}$$

where B_m = Maximum flux density in tesla
A = net cross-sectional area of core in m^2
f = frequency in Hz

Therefore, after fixing the core area and flux density, volts per turn is calculated by Eq. (9.1). Then LV (lowest voltage winding) turns are calculated. LV turns are rounded off to nearest integer and volts per turn are adjusted and new value of flux density is worked out by back calculation. The number of turns in other windings are determined by the new value of volts per turn.

9.9 Selection of Conductor and Current Density

Phase values of currents for different windings are calculated. Selection of current density in any winding depends upon type of winding, loss level and temperature rises permitted. Current density in any winding should be such that winding gradient and hot-spot temperature are within limits. Sometimes, current density in a winding is restricted due to short-circuit-force withstand point of view. Generally current density in various windings is chosen so as to meet the load-loss requirement and also such that their gradients are close to each other, thus resulting in an economical cooling equipment.

After deciding the values of current densities, copper area required for each winding is determined. Now suitable number of parallel conductors or pretransposed cables in both radial and axial direction and their width and thickness are selected, such as to meet the desired performance values. For large current ratings, pretransposed cables offer the advantage. Conductors of chosen dimensions must be strong enough to withstand short-circuit forces; also they should not have excessive eddy currents.

9.9.1 Influence of Varying Current Density

Increased current density reduces copper weight but leads to higher load loss, gradient and forces in the winding. Reduced copper weight results in lower core dimensions causing reduction in core-steel weight and no-load losses.

9.10 Insulation Design

Transformer windings have insulation within the winding, between windings and windings to earth. Insulation within the winding is generally paper insulation, however helical and disc type of windings have duct between turns or discs. Paper thickness of conductor should be such that it should be able to withstand various voltage stresses appearing during normal and transient conditions. Sometimes paper thickness are increased on pre-transposed cables of a large cross-sectional area, to increase the mechanical strength of paper insulation. Electrical clearances between windings of various voltage class and windings to earth depend upon their BIL and insulation arrangement adopted. Various clearances and disposition of solid insulation should be such that adequate cooling ducts are available to have effective cooling of windings. Also voltage stresses are controlled within limits.

Details of insulation design, method of improvement of voltage stress, etc., have been discussed at length in Chapter 5.

9.11 Calculation of Lateral and Axial Dimensions of Coils

Based upon the number of parallel conductors in radial direction and covered thickness of conductor, lateral dimensions of various coils can be calculated. If there are a number of layers in a coil, the lateral dimension of layer can be worked out and by taking into account the gap between layers, total lateral dimensions of coil can be worked out. Axial dimension of a coil depends upon covered axial dimensions of conductors and total number of conductors per turn in axial dimension. If cooling ducts are provided between conductors, these are also to be accounted for. Extra space is provided for dummies and transpositions if required.

9.12 Ampere-Turn Balancing

HV and LV winding are disposed axially in such a way that perfect ampere-turn balancing throughout the length of coil is achieved. Location and type of tapping coil disturb the ampere-turn balancing at different tap positions. Inter-wound type of tapping coil, in which each step occupies the whole axial length of winding, is ideal from ampere-turn balance point of view. Ampere-turn balancing between windings is essential from short-circuit force point of view. Unbalanced ampere-turn may give excessively high forces. Also, if ampere-turn balancing is not proper, it may give considerable increase in stray losses due to radial component of leakage flux.

9.13 Reactance Calculation

The estimation of reactance is primarily the estimation of the distribution of the leakage flux and the resulting flux linkages with primary or secondary coils. The distribution of the leakage flux depends on the geometrical configuration of the coils and of the neighboring iron masses and also on the permeability of the latter. The diagrams in Fig. 9.1 show typical leakage flux distribution. In case of cylindrical-core type coils of equal length [Fig. 9.1 (a)], the leakage field is packed in the space between HV and LV windings and it runs parallel to the core for nearly the full length of coils. Where there is an inequality in the coil lengths, the field is considerably altered, as shown in Fig. 9.1(b).

9.13.1 Cylindrical Concentric Coils (Equal Length)

For this case the actual leakage field, e.g., Fig. 9.1(a) is assumed to consist of a longitudinal flux of uniform and constant value in the

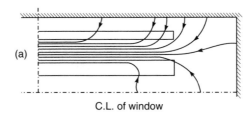

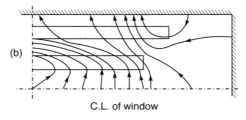

Figure 9.1 *Leakage flux.*

inter-space between primary and secondary, and a field crossing the conductors, reducing linearly to zero at the outer and inner surfaces. Further, the permeance of the leakage path external to the coil length is assumed to be so large as to require the expenditure of a negligible mmf, i.e., all the mmf is expended on the coil length. The effect of the magnetizing current in unbalancing the primary and secondary ampere-turn equality is neglected. Considering the Fig. 9.2(a), percentage reactance (X) between HV and LV winding can be calculated by the following equation:

$$X = \frac{2\pi f \mu_0 L_{mt}(AT)}{L_c E_t}\left(a + \frac{b_1 + b_2}{3}\right) \times 100 \qquad (9.2)$$

where
- f = Frequency (Hz)
- μ_0 = Magnetic space constant = $4\pi \times 10^{-7}$
- L_{mt} = Mean length of turn of primary and secondary together (meter)
- AT = Ampere-turns per limb of either coil
- L_c = Coil length (meter)
- E_t = Volts per turn
- a = Gap between HV and LV windings (meter)
- b_1 = Lateral dimension of HV winding (meter)
- b_2 = Lateral dimension of LV winding (meter)

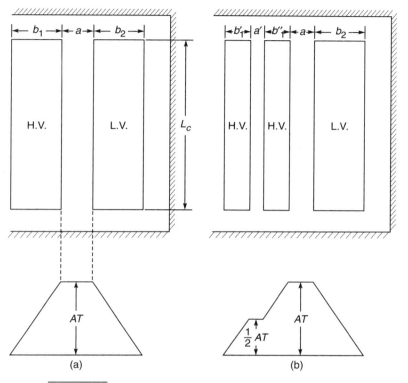

Figure 9.2 *Ampere-turn diagram for cylindrical coils.*

When HV winding is split as in Fig. 9.2(b), percentage reactance can be calculated by the equation

$$X = \frac{2\pi f \mu_0 L_{mt}(AT)}{L_c E_t}\left(a + \frac{b_1' + b_2 + b_1''}{3} + \frac{a'}{4}\right) \times 100 \quad (9.3)$$

where b_1' and b_1'' stand for lateral dimension of layers of HV windings and a' is gap between two layers.

9.13.2 Cylindrical Concentric Coils (Unequal Length)

The leakage field depends on the proportional difference in length, and on where the difference occurs (e.g., at one or both ends or in the middle, etc.). The case is very common, and it may be produced by end-turn reinforcement, tappings or by the normal small difference of coil length. The effect of divergences on the reactance requires investigation for each individual case. Figure 9.3 shows one method

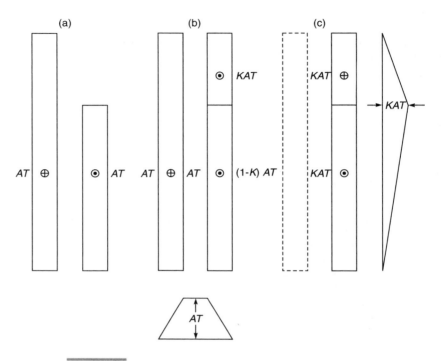

Figure 9.3 *Analysis of asymmetrical cylindrical coils.*

of treating the problem. The coils, each of ampere-turns (*AT*) are shown in the section. The actual arrangement (a) can be considered as equal to the sum of (b), a symmetrical system of longitudinal ampere-turns amenable to the treatment leading to Eq. (9.2), and (c) an asymmetrical transverse system of ampere-turns producing a cross flux. Having determined the flux distribution, due to each of the systems (b) and (c), the reactance is determined from the emf's induced in actual windings due to each flux distribution.

9.13.3 Sandwich Coils

Figure 9.4 shows the simplified case where the end coils have one-half the turns of the remainder. There will be n HV coils, $(n-1)$ LV coils of equal ampere-turns and two LV half-coils. Percentage reactance between HV and LV can be calculated by the equation

$$X = \frac{\pi f \mu_0 L_{mt}(AT)}{W^n E_t}\left(a + \frac{b_1 + b_2}{6}\right) \times 100 \qquad (9.4)$$

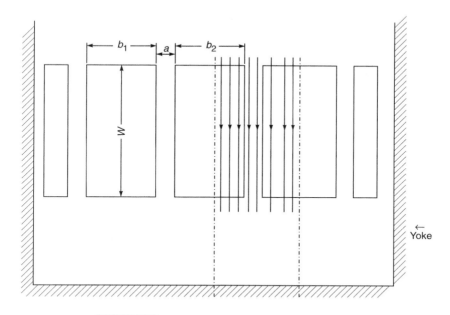

Figure 9.4 *Leakage of flux in sandwich coils.*

The percentage reactance can be calculated, for any coil arrangement, with the help of an ampere-turn diagram, as discussed above. Desired value of reactance can be achieved by suitably altering the various parameters.

The percentage of impedance and reactance are practically the same for large transformers, as they have very low winding resistance. The regulation in the system fixes the highest acceptable impedance and the requirements of transformer protection, which fixes the lowest acceptable impedance value. It is of interest to note that variation in impedance affects the cost of transformer. It is difficult to give a specific picture of how transformer economics are affected by variation in impedance, because this depends upon the relative price of copper and core material as well as the rates at which core-loss and copper-loss are capitalized. In general, for a given material content, a high impedance value would result in higher copper-loss and lower iron-loss. In any case, very high impedances are undesirable, as they may give substantially high stray losses, resulting in hot spots. Low impedance values are desirable where regulation is regarded as most important and on the other hand, they pose considerable problems for designing the transformer suitable for large mechanical forces developed under short-circuit conditions. Also, low impedance transformers would require large core diameter, resulting in higher iron-loss, lower copper-loss and higher transport weight of the largest packet.

9.14 Iron Weight and Losses

Transformer no-load loss occurs because of flow of main flux in the core. It depends upon grade of steel, frequency, flux density, type and weight of core and manufacturing techniques (i.e. mitered or interleaved joints, bolted or boltless, core construction, etc.). Modern core-steel used has different magnetic properties in the direction of grain orientation and cross-grain direction. Cores are designed such that flux travels mostly in the direction of grain orientation to give maximum advantage. However, at joints, flux travels in cross-grain direction, giving increase in losses. For accurate estimation of no-load loss, core weight for cross-grain and along-grain portion are calculated separately. Steel characteristics furnished by their suppliers give the iron-loss in watt/kg for both direction of flux travel (i.e., along the grain orientation and across the grain orientation) No-load loss is computed with these parameters.

9.15 Copper Weight and Losses

Winding conductor weight depends upon the specific gravity of conductor material, number of turns in the winding and cross-sectional area of conductor and this can be calculated by the following equation:

$$G = DALN \qquad (9.5)$$

where G = weight of conductor material (kg)
D = density (kg/m^3) = 8900 kg/m^3 for copper
A = area of cross-section (m^2)
L = mean length of winding turn (M)
N = number of turns.

By putting the corresponding values for different windings in Eq. (9.5), their conductor material weights can be worked out.

The alternating current flow in the windings is associated with load-loss comprising I^2R-loss and eddy current loss in the windings and stray losses in tank, clamping gear, etc. The resistive loss in the windings, i.e., I^2R-loss depends on resistivity of conductor material, number of turns, and cross-sectional area and length of mean turn of conductor.

The resistance (R) of winding is calculated by the following equation:

$$R = \frac{\rho \, LN}{A} \quad (9.6)$$

where ρ is specific resistivity of conductor material in Ω-meter.

By knowing the resistance of each winding, I^2R-loss in various windings can be worked out. Eddy current loss in conductors depends upon their configuration and stray loss in tank and clamping structure, etc., depends upon leakage flux, and can be suitably estimated with reasonable accuracy. The load-loss of transformer is a function of temperature and generally expressed at a reference temperature of 75°C.

9.16 Stray Losses in Transformer

The load losses in the transformer consist of losses due to the ohmic resistance of winding and stray losses. These stray losses take place in the loaded and unloaded windings, clamping framework and transformer tank and depend upon leakage flux and magnetic field surrounding the leads. The total stray loss in the transformer is generally made up of the following components.

(a) Stray loss in the transformer tank due to leakage flux from the windings.
(b) Stray loss in the tank due to leads.
(c) Stray loss in the clamping framework due to leakage flux from the windings.
(d) Stray loss in the clamping framework due to leads.
(e) Stray loss in the loaded windings.
(f) Stray loss in the unloaded windings.

9.16.1 Leakage Flux

Some amount of leakage flux will always exist in all types of transformers. The total leakage flux increases approximately as the square root of MVA rating of transformer. Therefore, increased capacity of transformer results in larger values of leakage flux and this gets further amplified with increased value of reactance of large transformers, causing increased stray losses. This leakage flux can sometimes lead to local overheating. By controlling the leakage flux path, stray losses can be reduced considerably, and thus resulting in reduced local heating.

9.16.2 Calculation of Leakage Flux

Leakage flux is generated by the winding while carrying current and is the resultant of the field surrounding the windings. This leakage flux is carried in the gap between HV and LV windings. The leakage field in the gap is fairly constant along the length of windings but at ends of winding, fringing of flux takes place. Various paths followed by leakage flux are shown in Fig. 9.5

The component of flux parallel to the axis of coil is termed as axial flux and that perpendicular to the axis of coil is termed as radial flux. The axial and radial flux densities at various regions can be calculated by various methods such as image method, Roth method, Rabins method, etc.

9.16.3 Losses in Metallic Parts due to Leakage Field

As shown in Fig. 9.5, leakage flux cuts the various metallic parts namely tank walls, core clamping plates, etc., resulting in eddy current losses in them. These eddy current losses and the losses in the region of high intensity field can lead to local overheating. The regions of these losses and overheating are: the tank walls near the winding ends, core clamping plates at winding ends, clamp plates on core limbs, etc.

Leakage flux going back to core, does not give significant losses and temperature rise. However, the flux impinging upon flat surface of core, where the flux cuts perpendicular to the lamination, additional losses will take place and can lead to high local temperatures.

The losses due to leakage field emerging from winding depend upon the amount and intensity of flux, permeability and resistivity

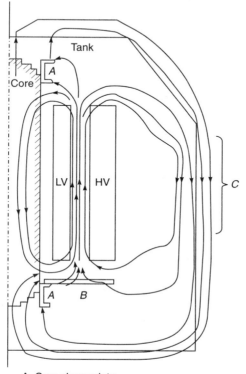

A. Core clamp plate
B. Base plate
C. Flux density greatest in this region of tank surface

Figure 9.5 *Principal leakage flux path.*

of metallic materials. Low resistivity materials will give rise to higher eddy currents and higher losses. The losses also depend upon the distance of metallic parts from the source of leakage flux, as with the increased distance the amount and the intensity of flux is reduced.

9.16.4 Stray Losses in the Winding

Major portion of stray losses takes place in the winding. This consists of eddy current loss and circulating current loss. Eddy current loss is proportional to the square of the flux field in which winding conductor is lying and to the power of four of the thickness of the conductor perpendicular to the direction of flux, and also to the square of total number of radial conductors in the winding.

In large transformers, a turn in winding consists of large number of conductors. In such an arrangement, unequal voltages are induced in different conductors due to varying leakage field in radial direction and variation in length of each conductor. The unequal voltages in parallel strands (conductors) give rise to circulating currents. The losses due to circulating current within a few stands of turn may not reflect much in overall losses, but still can cause over-heating of strands.

Eddy current and circulating current losses can also take place in idle windings, not connected in load circuit but lying in leakage field.

9.16.5 Stray Losses due to Current Carrying Leads

Leads carrying current induce magnetic field around the leads. If field is strong and in close vicinity to metallic parts, it produces eddy currents and losses. Intense field can lead to local hot-spots. The resultant losses are not very important, but the hot spots can lead to gasification of oil or decomposition of insulation. The effect of current in the leads depends upon the magnitude of the current, distance of leads from metallic parts and resistivity of material.

9.17 Stray Loss Control

A minimum stray loss design can be achieved by analyzing systematically, the source of leakage flux and its path. Main leakage field in the transformer will always exist and losses arising out of this can be reduced by various methods given below:

9.17.1 Magnetic Yoke Shields

Magnetic shields, made up of core laminations are used under yokes as shown in Fig. 9.6. A large proportion of the axial leakage flux is fed back into the yokes. The yoke clamp assembly is shielded and reduction of radial flux to tank side is also achieved. Magnetic shunts can be conveniently used for three-phase, five-limbed and single-phase three-limbed transformers. For three-phase, three-limbed transformers, the flux collection from outer limbs and feeding to yokes is difficult and needs special transfer technique.

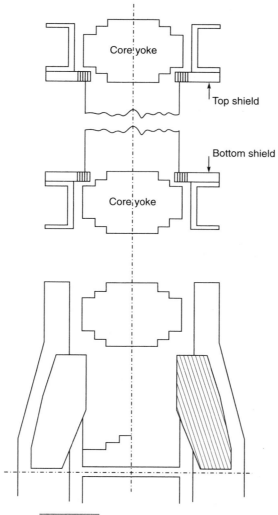

Figure 9.6 *Magnetic yoke shields.*

The reduction in stray loss is considerable in the case of large transformers.

9.17.2 Magnetic Shunts

The magnetic shunts consisting of packets of core laminations are fixed inside the tank to absorb stray flux. The thickness of lamination packets is decided by the flux density used. The typical arrangement of magnetic shunts is shown in Fig. 9.7. The reduction in stray loss in a tank is considerable in large transformers.

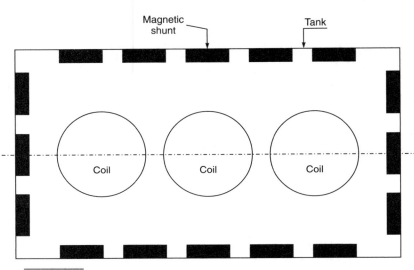

Figure 9.7 *Fixing of magnetic shunts on three-phase transformer.*

9.17.3 Electromagnetic Screens

High conductivity materials like aluminum or copper are used as screens on tank walls. The stray flux induces eddy currents in these screens, which suppresses the original field and relieves the tank walls from stray flux. The screens are effective beyond a critical distance and placement of screens within critical distance may result in increased losses.

9.17.4 Losses in Clamp Plates and Core Packets

The core limb clamp plates are very near the leakage field area and are subjected to intense radial field at the ends of windings. This results in eddy losses on clamp plate ends and can lead to local hot-spots. Stray loss and heating is further increased due to high current inner winding leads. The outside packet of lamination on core limbs are also subjected to similar conditions. The control in temperature rise and losses can be achieved by using clamp plates of non-magnetic materials like stainless steel. Substantial reduction in eddy losses and temperature rise can be achieved by providing slots on the clamp plate.

The clamp plates made of thin lamination stacks and placed edgewise are ideal for large transformers. Outer packets of core are also sometimes subdivided to reduce temperature rise on core plates for large rating transformers.

9.17.5 Losses in Winding

Subdivision of conductors radially reduces the eddy current loss due to axial leakage field. Similarly, subdivision of conductors axially reduces the eddy current loss due to radial component of leakage field. To reduce radial component of leakage flux, it is essential that ampere-turn of HV and LV windings shall be perfectly balanced, which in turn gives the reduced eddy current loss due to radial component of leakage flux. It is worth mentioning here that large unbalance in ampere-turns may lead to very high stray loss. To eliminate the circulating current between parallel strands (conductors) of a turn, transposition is essential. This leads to positioning of strands of a turn, such that flux linkage is same, thus equalizing the induced emf in each strand. Methods of various transposition have been discussed in Chapter 5.

The use of transposed cables for high-current windings result in considerable reduction in stray losses. In high-voltage winding with moderate current, requiring two or three conductors in parallel, bunched conductors can be used to improve winding space factor, and to provide subdivision of conductors.

9.17.6 Losses due to Leads

Losses due to high-current leads can be reduced by spacing them suitably from metallic structures. The field effect of leads can be eliminated by positioning together the current carrying leads of opposite direction (go-return arrangements). In three-phase connection, the leads of all the three phases can be grouped together so that the net vectorial effect of field is minimum. Losses due to leads can be reduced by shielding the nearby surfaces by non-magnetic material like aluminum. High-current bushing mounting plate has high eddy current losses. This can be reduced by putting the non-magnetic inserts to break eddy current path or by using the plate of non-magnetic material like aluminum or stainless steel for mounting the bushings.

9.18 Impulse Calculation

The voltage distribution in power frequency-voltage tests is substantially uniform between turns and coils. Under impulse test, however, the distribution can be far from uniform and initial voltage distribution is governed solely by capacitance network. It depends upon value of α, which can be expressed by the following equation:

$$\alpha = \sqrt{\frac{C_g}{C_s}} \qquad (9.7)$$

where
C_g = winding to ground capacitance
C_s = series capacitance of winding

The determination of initial voltage distribution is merely the calculation of the various capacitances of windings. The greater the value of α the greater is the divergence from uniform voltage distribution from line to ground. It is evident that value of α is high for a small transformer. For transformers of higher rating and especially for higher line voltage, C_g decreases because it is determined largely by clearances, while C_s increases because of greater radial depth of the winding. Thus, α is lower and initial voltage distribution is closer to uniform. Due to stress concentration at the ends of winding, a few end turns are provided with extra insulation. To reduce stress concentration at the ends, disc coil windings are provided with stress rings that act as radial shields, although they do not materially improve the axial distribution. Axial improvement can be gained by the addition of shields or by several other means of controlling the electric field distribution. Also refer to Chapter 5, Clause 5.2 for surge behavior of windings.

9.19 Mechanical Forces in Windings

When transformer is loaded, the flow of currents in its primary and secondary windings gives rise to leakage flux and mutual forces between windings. The mechanical repulsive force with normal load currents is low compared to the strength of coils. Under fault conditions, the forces produced may be increased many times. The design of transformer windings, insulating cylinders, coil clamping rings, clamping plates, etc., should be such that they are able to withstand these forces. The forces produced are radial as well as axial. Radial component of force tends to burst outer windings and crush inner windings and axial component gives bending and compressive stresses. Various forces and their effect have been discussed at length in Chapter 7.

9.20 Temperature Gradient and Cooling Calculation

Difference between mean winding temperature and mean oil temperature is termed as winding gradient. It depends upon the losses in the winding and surface area available for cooling. Winding design, i.e., selection of current density, size of cooling ducts, etc., should be proper, resulting in low gradients.

Gradient of the windings can be reduced by providing the directed oil flow, in the case of forced oil cooled transformer. The gradient in various windings of a transformer are adjusted by suitably altering the winding design to result in economical cooling equipment and thus overall economy. The design of cooling equipment is to be such that it shall be capable of delivering the rated MVA, under specified conditions, without exceeding the guaranteed values of top oil temperature rise and mean winding temperature rise. Top oil temperature rise depends upon mean oil temperature rise and half of the difference in oil temperatures at the inlet and the outlet of the cooling equipment. Mean winding temperature rise depends upon mean oil temperature rise and winding gradient. Various types of cooling have been discussed in Chapter 8.

9.21 Typical Design Calculations for Two Winding and Auto-Transformers

Example 9.1 Typical design calculations for a 20 MVA, 132/33 kV, star/star, three-phase, 50-Hz, power transformer with on-load tapping on HV winding for voltage variation of LV from − 5% to + 15% in 16 steps. Percentage impedance 10% on 20 MVA base.

Select, core diameter = 515 mm, say
B_m = 1.524 tesla, say
Type of core: Three-limbed construction
$A = 0.18167$ m^2 (Core utilization factor 87.2%)
$f = 50$ Hz.

Substituting the values in Eq. (9.1),
$$E_t = 61.46$$

LV turns per phase $(N_1) = \dfrac{33000}{\sqrt{3} \times 61.46} = 310$

HV turns per phase $(N_2) = \dfrac{1,32,000}{\sqrt{3} \times 61.46} = 1240$

HV turns (max.) $= \dfrac{1240}{0.95} = 1305.3$
$= 1305$ (say)

HV turns (min.) $= \dfrac{1240}{1.15} = 1078.3$
$= 1078$ (say)

Tap winding turns $= 1305 - 1078 = 227$
(Linear tapping winding foreseen)

H.V. winding design

Type of winding: Reversed section winding
Number of coils in parallel
$= 2$ (mid-entry of line lead)
Number of discs in each coil
$= 50$
Conductor size $= 2.1 \times 5.4$ (mm)
Radial paper insulation over conductor
$= 0.60$ mm
Cooling duct between two discs
$= 6.50$ mm (compressed)
Maximum number of turns in a disc
$= 22$

Phase current $= \dfrac{20,000}{\sqrt{3} \times 132}$
$= 87.48$ A
Current density $= 3.98$ A/mm^2 (say)
Conductor area $= 21.96$ mm^2

Axial dimension of coil

$= \dfrac{0.60 \times 2}{1.20}$

10% paper compression $-\ 0.12$

Design Procedure

$$\begin{array}{r} 1.08 \\ 5.40 \\ \hline 6.48 \times 50 \\ 324.0 \end{array}$$

Cooling ducts $6.50 \times 49 = 318.5$

$$\begin{array}{r} \hline 642.5 \times 2 \\ 1285 \end{array}$$

Top and bottom insulation

$$\begin{array}{r} 235 \\ \hline \end{array}$$

Leg length $\quad 1520$ mm

Radial dimension of coil

$$= \begin{array}{r} 0.60 \times 2 \\ 1.20 \\ 2.10 \\ \hline 3.30 \times 22 \\ 72.6 \end{array}$$

Say 74 mm

Tapping winding design

    Type of winding: Reversed section winding
    Number of coils in parallel = 4
    Number of discs in each coil = 32
        Conductor size = 1.9×3.2 (mm)
    Radial paper insulation over conductor
        = 0.60 mm
    Cooling duct between two discs
        = 3.70 mm (compressed)
Maximum number of turns in a disc = 9
        Copper area in coils = 22.88 mm^2 (for current density of 3.82 A/mm^2, (say)

Axial dimension of coil

    Calculating in the same manner as for HV coil, axial dimension/coil for 32 sections = 252.00
    The coils shall be placed concentrically one over the other.

Radial dimension of coil

    For 9 turns per section, calculating in the same manner as for HV, radial dimension of coil = 29 mm (inclusive of building tolerance).

LV winding design

Type of winding: Reversed section winding (single coil conceived)

Number of discs = 110
Conductor size = 5112.8 × 7.6 (mm)
Radial paper insulation over conductor
   = 0.225 mm
Cooling duct between two discs = 3.70 mm (compressed)
Maximum number of turns in a disc = 3

$$\text{Phase current} = \frac{20{,}000}{\sqrt{3} \times 33} = 349.92 \text{ A}$$

Copper area in coil = 103.65 mm²
Current density in coil = 3.38 A/mm²
Calculating as for HV coil the axial dimension of coil
   = 1520 mm (inclusive of 235 mm top and bottom insulation)
Radial dimension of coil = 50 mm (inclusive of building tolerance)

Reactance calculation at normal tap (Fig. 9.8)

$$\% X = \frac{K(AT)\pi}{EtL} \times \Sigma\, DM \times 10^{-5} \tag{9.8}$$

K = constant
   = 3.86
L = reactance length
   = 1368 (including fringing effect)
$\Sigma\, DM\, LV$ = 10,117
Gap = 30,844
HV = 24,055
   ─────
   65,016

$$\% X = \frac{3.86 \times \pi \times 349.92 \times 310 \times 65016 \times 10^{-5}}{61.46 \times 1368}$$

= 10.2 (Guaranteed impedance 10%)

Core weight and loss calculations

Weight of legs = 7.65 × 3 × 1816.7 × 152 × 10⁻³	= 6337
Weight of yokes = 7.65 × 4 × 1816.7 × 107.5 × 10⁻³	= 5976
Weight of corners = 7.65 × 2 × 1816.7 × 50 × 10⁻³	= 1390
	13,703 kg

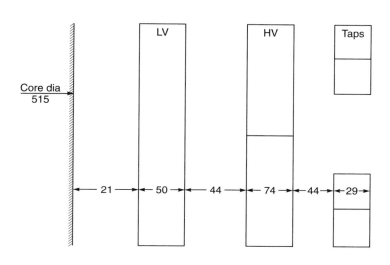

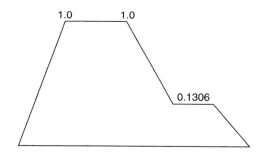

Figure 9.8 *Ampere turns diagram at normal tap position.*

At 1.524 tesla, specific watt/kg of M4 steel
$$= 0.9176$$
$$\text{No-load loss} = 1.15^* \times 0.9176 \times 13{,}703 \times 10^{-3}$$
$$= 14.5 \text{ kW}$$

Copper weight calculations

$$\begin{aligned}
\text{LV} &= 8.89 \times \pi \times 607 \times 310 \times 103.65 \times 10^{-6} &= 544.8 \\
\text{HV} &= 8.89 \times \pi \times 819 \times 1078 \times 21.96 \times 10^{-6} &= 541.7 \\
\text{Taps} &= 8.89 \times \pi \times 1010 \times 227 \times 22.88 \times 10^{-6} &= 146.5 \\
& & \overline{1233.0} \text{ kg}
\end{aligned}$$

---

* Building factor from experience of similar transformers.

Total weight for three phases = 1233.0 × 3
= 3699.0 kg

Resistance calculations
LV = 21.05 × π × 607 × 310 × 10^{-6}/103.65 = 0.1201 Ω
HV = 21.05 × π × 819 × 1078 × 10^{-6}/21.96 = 2.6590 Ω
Taps = 21.05 × π × 1010 × 227 × 10^{-6}/22.88 = 0.6627 Ω

Total HV resistance at normal tap

$$= 2.6590 + 0.6627 \times \frac{162}{227}$$
$$= 3.1319 \ \Omega$$

Copper loss calculations at normal tap
LV = $(349.92)^2$ × 0.1201 × 10^{-3} = 14.7
HV = $(87.48)^2$ × 3.1319 × 10^{-3} = 24.0
                                      ───────
                                      38.7 kW

Total I^2R loss for 3 phases = 116.1 kW

Eddy current loss and stray losses based on similar past design as 5 kW and 10 kW respectively

Total copper loss = 131.1 kW

Impulse distribution across different sections of coils, temperature gradients of windings and mechanical forces are further done to ascertain the design adequacy vis-à-vis test requirements.

Example 9.2 Typical design calculations for 20 MVA, 132/66/11 kV, Star/Star/Delta, 3 phase, 50 Hz auto transformer with onload tappings on HV winding for voltage variation of HV from – 10% to + 10% in 16 steps. Percentage impedance 12.5% on 20 MVA, base.

Core diameter – 405 mm, say

B_m = 1.684 Tesla, say

Type of core: Three-limbed construction

A = 0.11145 sq. meter (core utilization factor 86.5%)
f = 50 Hz

Substituting the values in Eq. (9.1),

Et = 41.66

LV turns per phase (N1) = $\frac{11000}{41.66}$ = 264

Design Procedure

IV turns per phase (N2) = $\dfrac{66000}{\sqrt{13} \times 41.66}$ = 914.7

Say 914.5

HV turns per phase (N3) = $\dfrac{132000}{\sqrt{3} \times 41.66}$ = 1829.4

Say 1829

Common Winding turns = 914.5

Series Winding turns = 1829 − 914.5
= 914.5

Trapping turns = 1829 × 0.1
= 182.9   Say 184 (reversing taps)

Current calculations

LV phase current = $\dfrac{20{,}000}{3 \times 3 \times 11}$

= 202.02 A

IV phase current = $\dfrac{20{,}000}{\sqrt{3} \times 66}$

= 174.96 A

HV phase current = $\dfrac{20{,}000}{\sqrt{3} \times 132}$

= 87.48 A

Series winding phase current
= 87.48 A

Common winding current = 174.96 − 87.48
= 87.48 A

Series and common windings can be reversed section coils. Tapping winding can be inter-wound spiral coil placed between common and series windings or elsewhere depending upon the impedance variation required. LV winding can be spiral coil.

Design of these coils can be done in a similar manner. While calculating impedance and other performance, auto factor is suitably used as applicable.

REFERENCES

1. Say, M.G., "Electrical Engineers Reference Book."
2. "The J & P Transformer Book".
3. Blume, L.F., A. Boyajian, G. Camilli, T.C. Lennox, S. Minneci, and V.M. Montsinger, "Transformer Engineering" a Treatise on the Theory, Operation and Application of Transformers," Book, Chapman and Hald Ltd., London, 1951.
4. Say, M.G., "Performance and Design of AC Machines," Book, Sir Isaac Pitman & Sons, 1962.
5. Narke, D.V., and R.K. Talwar, "Paper on Stray Losses in Power Transformer," BHEL, Bhopal.
6. Khosla, A.K., Paper on "Recent Trends in the Design of Large Size Transformers," BHEL, Bhopal.

Chapter 10

Structural Design of Transformer Tank

M. K. Shakya
S. G. Bokade

Transformers are housed in metallic tanks, which are structurally robust enough to withstand the loadings such as full vacuum during processing of transformers, oil pressure and concentrated point loads of lifting, hauling, jacking, etc. The tank sizes reach the transportable limits and call for a lot of ingenuity in the design to meet such stringent conditions as minimum electrical clearances from high voltage points of windings and leads, and proper shaping to reduce oil quantity, transportable profile suitable for loading on rail wagons, transportable weight, etc. From these design considerations, in general, and for large power ratings, transformer tanks are structurally quite complicated. Though, for medium size, plain tanks are also used quite often for the sake of ease, and economy of cost of fabrication.

The structural design of transformer tanks comprises the computation of the combined behavior of plate and shells with stiffeners, which involves a realistic estimate of boundary conditions. For computing the stresses and displacements at a few selected points the classical method is handy, however for computing the stresses and displacements in global sense, the classical method is not sufficient and one has to make use of rigorous methods such as finite element method.

10.1 Types of Tank Constructions

Rectangular tanks are simpler in fabrication. However for large rating transformers, shaping of tanks becomes necessary to conform to transportable profile. Shaping is provided by rounded corners at the ends, truncation of lower portion of walls from consideration of loading in well wagon girder and on the covers to reduce the height. To minimize the tank oil, the tank profile may closely follow the electrical clearances along the coils. As is evident, shaping gives saving in tank material and oil but increases complexity and fabrication cost. From the standpoint of structural strength, shaping is advantageous and this aspect is discussed later in the text.

Transformer tanks may be classified as–
— plain tanks
— shaped tanks
— bell-shaped tanks
— corrugated tanks
— stub-end type tanks

10.1.1 Plain Tanks

Plain tanks are rectangular box type in shape. These tanks may sometimes also have rounded corners. These are commonly used for small and medium rating transformers. This type of construction may however be uneconomical for larger sized tanks, particularly if the welding is done manually.

10.1.2 Shaped Tanks

In order to make more economical transformers, it is possible to shape the profile of the tank body suitably, so that the inside volume of the tank is less. It is suitably truncated in the lower portion for ease of loading inside the well of the wagon and it also ensures the required minimum electrical clearance at all points. Usually many shaped tanks have the tank walls shaped towards the HV-leads side, where electrical clearances are comparatively much larger, and a flat wall towards the LV side. Structurally the curved portions in the tank walls act as stiffener, which is an added advantage. The shaping in the tank construction is decided by the electrical layout considerations of transformer windings and terminal gear/ tapchanger mounting arrangements.

10.1.3 Bell Shaped Tanks

Sometimes it is desirable to construct the tank in two parts, so that if the top portion is removed, the height of the lower part is such that the core and the winding of the tank are easily accessible for inspection and maintenance. This arrangement is preferable at site, where it is not possible to lift the core and windings together from the tank for inspection and maintenance work. Such tanks which are made into two separable parts are known as bell type. The profile of the tank walls can be either rectangular or bell shaped.

10.1.4 Corrugated Tanks

An alternative for providing vertical ribs welded to the plates is to form corrugation on the plates by suitably folding the plates. These corrugations play the role of stiffeners and this process reduces welding by replacing the welding vertical ribs on the tank walls. The usual type of tank construction with corrugated plates for tank walls employs box type of horizontal stiffeners to provide necessary structural strength against bending stresses. Another advantage of corrugated tank walls is that it provides additional cooling area on the tank walls and tank weight reduction.

10.1.5 Stub-End Wagon Type Tanks (Fig. 10.1)

It may not be possible to transport very large rating power transformers on any conventional wagon. The height of the girder on which the consignment rests adds to the overall transport height. As a remedy, such large-size transformers are not supported on such girders. The tanks are suitably designed such that they can be

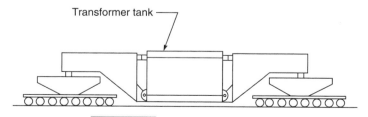

Figure 10.1 *Stub-end type tank.*

supported from either end by stub-end wagons and the transformer hangs in the vertical position, leaving the minimum necessary clearance between the bottom of the tank and railway track. These tanks are of special construction and designed to withstand the dynamic loads during transit in addition to static loads.

10.2 Structural Design of Transformer Tanks

10.2.1 Classical Method

Structural design is necessary to guarantee against material failure under design loads. For the sake of clarity and ease of formulation, the analysis has been provided for a plain rectangular tank. The whole tank structure is suitably subdivided into a number of plate panels. The shaped transformer tanks can be similarly subdivided in such plate panels and the present analysis with some approximation would also be valid for shaped tanks.

For the plates with stiffeners in either direction (i.e., transverse and longitudinal), it has been seen that the behavior of the transformer tank wall is very near to the average behavior between simply supported and encastered beams.

10.2.2 Theoretical Formulation

Simplified rectangular tank with horizontal box stiffeners and vertical ribs (stiffeners) is illustrated in Fig. 10.2. For the purpose of analysis the entire face is subdivided into a number of panels, whose boundaries are edges of the face and center lines of stiffeners. This is illustrated in Fig. 10.3. These plate panels are subjected to design loads like vacuum, oil pressure, concentrated point loads, etc.

10.2.3 Role of Stiffener

Figure 10.4 shows the rectangular plate supported all around. For such a plate loaded under uniformly distributed load (udl) p considering a simply supported beam along longitudinal direction (i.e., in the direction of l), the bending moment at the center is given by

Structural Design of Transformer Tank

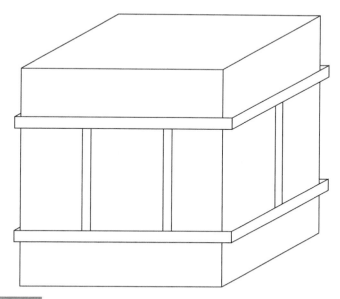

Figure 10.2 *Rectangular tank with horizontal and vertical stiffeners.*

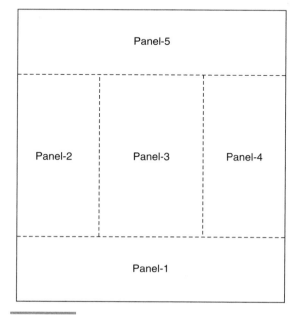

Figure 10.3 *Subdivision of tank wall into panels.*

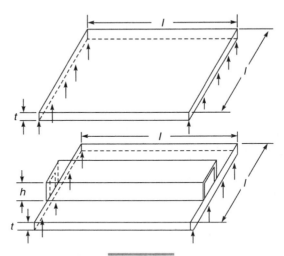

Figure 10.4

$$M = \frac{pbl^2}{8} \tag{10.1}$$

The stresses and deflections may be computed by the following formulae

$$Z_p = \frac{1}{6} bt^2 \tag{10.2}$$

$$f_{ss} = \frac{M}{Z_p} \tag{10.3}$$

$$= \frac{pbl^2}{8 \times Z_p} \tag{10.4}$$

$$\delta = \frac{5\,pbl^4}{384\,EI} \tag{10.5}$$

Referring to Fig. 10.5, the combined section modulus of plate and stiffener will now be taken in the above formula as

$$Z = Z_p + Z_s \tag{10.6}$$
$$I = I_p + I_s \tag{10.7}$$

Consequently, the maximum stresses and the deflection will be proportionately reduced and these can be computed from the above formulae by substituting the combined section modulus and moment of inertia.

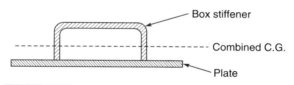

Figure 10.5 *Combined section plate with stiffener.*

Similarly, the bending moment, stresses and deflections for a encastered beam may be given as

$$M = \frac{pbl^2}{24} \tag{10.8}$$

$$f_{EB} = \frac{pbl^2}{24.Z} \tag{10.9}$$

$$\delta = \frac{pbl^4}{384.EI} \tag{10.10}$$

By considering stiffeners as shown in Fig. 10.5, the combined section modulus of plate and stiffeners will be taken as in Eq. (10.6).

Consequently, by substituting these values of section modulus and moment of inertia, the values of stresses and deflection will come down proportionately. In the case of a transformer tank, the walls have some average behavior of the two cases, namely simply supported and encastered.

$$f = (f_{ss} + Kf_{EB})/2 \tag{10.11}$$
$$\delta = (\delta_{ss} + K\delta_{EB})/2 \tag{10.12}$$

where K is an empirical value and will depend on the weld size, the rigidity of the corners and on stiffener sizes.

10.2.4 Behavior of Continuous Plates with Intermediate Flexible Supports

The bending of each span of the plate can be computed by combining the known solution for laterally loaded, simply supported rectangular plates, with those of rectangular plates bent by moments distributed along the edges.

The expressions for stresses and displacements of plates and plates with stiffeners, considering the bending effect in both the direction, are given as

$$f_1 = \frac{pb^2l}{12\,I_{xx}} \left(\frac{1}{1+K_1 \dfrac{I_{yy}}{I_{xx}}\left(\dfrac{b}{l}\right)^3} \right) \qquad (10.13)$$

$$f_2 = \frac{plb^2}{12\,I_{yy}} \left(\frac{1}{1+K_1 \dfrac{I_{xx}}{I_{yy}}\left(\dfrac{l}{b}\right)^3} \right) \qquad (10.14)$$

$$f'_1 = \frac{pb^2}{2t^2} \left(\frac{1}{1+K_1 \left(\dfrac{b}{l}\right)^4} \right) \qquad (10.15)$$

$$f'_2 = \frac{pl^2}{2t^2} \left(\frac{1}{1+K_1 \left(\dfrac{l}{b}\right)^4} \right) \qquad (10.16)$$

Subscript 1 for shorter side and 2 for longer side have been used in the above expressions.

$$\delta' = \frac{pl^4 b}{K_2 E I_{yy}} \left(\frac{1}{1+\dfrac{I_{xx}}{I_{yy}}\left(\dfrac{l}{b}\right)^4} \right) \qquad (10.17)$$

$$\delta'' = \frac{pl^4}{K_3 E t^3} \left(\frac{1}{1+\left(\dfrac{b}{l}\right)^4} \right) \qquad (10.18)$$

where Eqs. (10.17) and (10.18) are for the deflection of stiffened and unstiffened plate respectively.

In case of actual transformer tanks, the dimensions are fairly large and unless one uses very thick plates, one has to stiffen the different faces by welding stiffeners of suitable sizes, either along longitudinal or transverse or both the directions, such that the resulting panel widths are safe values. The different arrangements are illustrated in Figs. 10.6 – 10.9 and these are self-explanatory. It may be pertinent to mention here that for pressure-loadings on transformer tanks, the horizontal stiffeners are more effective as these bear most of the bending moments compared to vertical stiffeners. But horizontal stiffeners alone may not be sufficient except for smaller size tank, and a combination of these stiffeners along with vertical ribs are most effective and most economic from design standpoint. Here the variable parameters are the plate thickness, sizes of horizontal and vertical stiffeners and the cost of fabrication.

In the case of curved tank structures, the vertically curved surfaces act like vertical stiffeners. However, the bending moment distribution along the profile of the tank walls are very complicated.

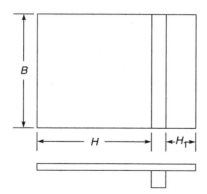

Figure 10.6 *Structural considerations in tank design.*

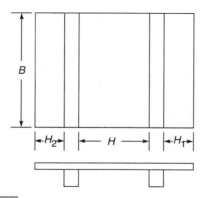

Figure 10.7 *Structural consideration in tank design.*

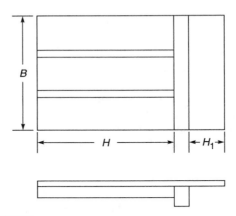

Figure 10.8 *Structural considerations in tank design.*

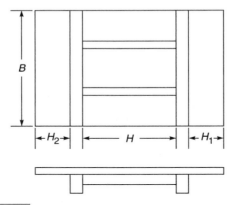

Figure 10.9 *Structural considerations in tank design.*

As such, shaped tanks are more complicated for structural analysis by classical methods. However, some simplified assumptions can be made based on experience, such as the resulting subdivision of plate panels, the boundary conditions along the edges of these panels and the formulae discussed earlier can be applied without significant loss of computational accuracy.

10.2.5 Finite Element Method

Any analytical method is based on certain simplifying assumptions, like a particular type of boundary condition such as simply supported or fixed. The panel behavior is further analyzed as a beam problem or a plate problem. It is difficult to account for structural irregularities, such as cutout or openings or bracket type of projections. Some of these zones may be areas of high stresses and need to be investigated. It is very difficult to account for such complexities in the tank structure by any classical analytical method. Nevertheless, it is necessary to investigate the stress contours in the global sense, in order to improve reliability in the design and parametric optimization based on the results of stress mapping under design loads. A rigorous and more accurate method of stress and displacement analysis is based on the formulation of finite element method, which is discussed in the following section.

In finite element formulation, the region of interest is subdivided into subregions known as finite element. Within each subregion the variables are assumed in the form of a function. The functions assumed for unknown variables are defined by interpolation functions associated with the nodal values of the unknown parameters. Using the interpolation functions, the variables at any point in the subregions can be expressed in terms of their nodal values as

$$u = N \delta_e \qquad (10.19)$$

where u = vector of variables for which functions are assumed
 N = matrix for interpolation functions
 δ_e = nodal unknown parameters for an element.

The stress vector σ is related to the strain vector ε by

$$\sigma = D \varepsilon \qquad (10.20)$$

where the D matrix depends on the material properties like modulus of elasticity, Poisson's ratio, etc.

The strains are related to the nodal displacement by

$$\varepsilon = B \delta_e \qquad (10.21)$$

This matrix B depends on the selection of the elements, viz. triangular, quadratic, etc., curvature of the surfaces, the direction cosines of the local elements with reference to global coordination, etc.

The strain energy can be expressed as

$$\phi = \iiint_{\text{vol}} \varepsilon^T \sigma \, d_{\text{vol}} \qquad (10.22a)$$
$$= \iiint_{\text{vol}} B^T D B \delta_e \, d_{\text{vol}} \qquad (10.22b)$$

Introducing the concept of stiffness matrix K as

$$F_{\text{ext}} = K \delta_e \qquad (10.23)$$

From the principle of virtual work, Eq. (10.23) may be equated with Eq. (10.22b), and we get

$$K_e = \iiint B^T D B \, d_{\text{vol}} \qquad (10.24)$$
$$K_{\text{global}} = \iiint B^T T D T^T B \, d_{\text{vol}} \qquad (10.24a)$$
$$p = Kd \qquad (10.25)$$

From Eq. (10.25), the nodal displacement can be computed. From the nodal displacement the stresses can be computed as

$$\sigma = D\,\varepsilon$$
$$= D\,B\,d_e \tag{10.26}$$

For the overall analysis, the stiffness matrix (K) of all the elements are assembled appropriately. For the assembly of the elements, the basic data needed are the element—node relationship and nodal displacement relationship.

10.2.6 Choice of Element

Several types of elements have been developed by the researchers working in the area. The isoparametric series possesses distinct advantages over others by virtue of ease of their applicability to curved surfaces.

For analysis of tank, the plates/sheets are idealized by eight noded quadrilateral isoparametric elements, and stiffeners by three noded quadratic elements. These elements are found to be most suitable.

One of the salient features of formulation of these elements is introduction of thickness vector of the plates/shells and stiffeners and to allow the rotation of the normals, which in turn accounts for the bending and shear stresses in the plates and stiffeners. The formulation of these elements provides the simplicity of analysis of two-dimensional plates without any compromise on the accuracy of results

Another significant aspect of this formulation is the introduction of stiffener elements and its merging with the plate elements. This sophistication has been incorporated without adding any complexity in the formulation or computation.

10.2.7 Formulation of Eight-Noded Isoparametric Shell Element

In finite element method formulation, the part which is to be analyzed is subdivided into numbers of element variables, with the region of element related to the nodal unknown variables through interpolation function, as given by Eq. (10.19).

The interpolation function is known as shape-function. When the same shape-function is used for defining the geometry as well as displacements of the element, then the element is known as isoparametric element. To account for the curved surfaces in the tank structure, the curvature parameters ξ, η, ζ have been introduced along there directions and these are connected to the local X,

Y, Z coordinate system of the elements by the following matrix (Jacobian) relationship. (See Fig. 10.10).

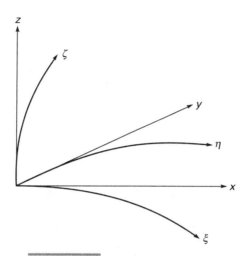

Figure 10.10 Coordinate system.

$$\left\{\begin{array}{c}\dfrac{\partial f}{\partial \xi}\\ \dfrac{\partial f}{\partial \eta}\\ \dfrac{\partial f}{\partial \zeta}\end{array}\right\} = \left[\begin{array}{ccc}\dfrac{\partial x}{\partial \xi} & \dfrac{\partial y}{\partial \xi} & \dfrac{\partial z}{\partial \xi}\\ \dfrac{\partial x}{\partial \eta} & \dfrac{\partial y}{\partial \eta} & \dfrac{\partial z}{\partial \eta}\\ \dfrac{\partial x}{\partial \zeta} & \dfrac{\partial y}{\partial \zeta} & \dfrac{\partial z}{\partial \zeta}\end{array}\right]\left\{\begin{array}{c}\dfrac{\partial f}{\partial x}\\ \dfrac{\partial f}{\partial y}\\ \dfrac{\partial f}{\partial z}\end{array}\right\} \quad (10.27)$$

or
$$\{f, \xi\eta\zeta\} = [J]\{f, xyz\} \quad (10.27a)$$

The shape functions for eight-noded quadrilateral isoparametric elements are (Fig. 10.11)

$$N_i = \frac{1}{4}(1 + \xi\xi_i)(1 + \eta\eta_i)/(\xi\xi_i + \eta\eta_i^{-1}) \quad (10.28)$$

for $\quad i = 1, 3, 5$

$$N_i = \frac{1}{2}(1 - \xi^2)(1 + \eta\eta_i) \quad (10.29)$$

for $\quad i = 2, 6$

$$N_i = \frac{1}{2}(1 + \xi\xi_i)(1 - \eta_i^2) \quad (10.30)$$

for $\quad i = 4, 8$

where ξ, η are the natural coordinates of point of interest and ξ_i, η_i are the nodal coordinates of the node.

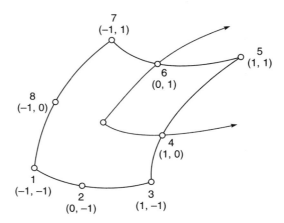

Figure 10.11 *Two-dimensional eight-noded quadrilateral elements.*

Once the shape functions are assumed, the next step is to arrive at expressions for shape fields and displacement fields for element; which are given by the following expressions in terms of thickness vector of element

$$\begin{Bmatrix} X \\ Y \\ Z \end{Bmatrix} = \sum_{i=1}^{8} N_i \begin{Bmatrix} X_{ci} \\ Y_{ci} \\ Z_{ci} \end{Bmatrix} + \frac{1}{2} \zeta \sum_{i=1}^{8} N_i \begin{Bmatrix} V_{xi} \\ V_{yi} \\ V_{zi} \end{Bmatrix} \quad (10.31)$$

where X, Y, Z, are coordinates of a point in element X_{ci}, Y_{ci}, Z_{ci}, coordinates of a center point in element

V_{xi}, V_{yi}, V_{zi}, thickness vector in X, Y, Z, direction
N, shape function
u, v, w, displacements
α, β, rotations
ζ, natural coordinates in third direction
μ_i, matrix of direction cosines in V_1 and V_2 vectors

$$\begin{Bmatrix} u \\ v \\ w \end{Bmatrix} = \sum_{i=1}^{8} N_i \begin{Bmatrix} u_{ci} \\ v_{ci} \\ w_{ci} \end{Bmatrix} + \zeta/2 \sum_{i=1}^{8} N_i t_i [\mu_i] \begin{Bmatrix} \alpha_i \\ \beta_i \end{Bmatrix} \quad (10.32)$$

10.2.8 Formation of Stiffness Matrix

The stiffness matrix is obtained by equating internal and external work done for a vertical displacement, in terms in matrix form.

Structural Design of Transformer Tank   **241**

The internal work done
$$= \iiint \varepsilon\, \sigma\, d_{\text{vol}} \quad (10.33)$$
On simplifying Eq. (10.33) the internal work done
$$= \iiint B\, \delta_e\, DB\, \delta_e\, d_{\text{vol}} \quad (10.34)$$
Similarly, external work done
$$= \delta_e^T F_{\text{ext}} \quad (10.35)$$
By equating expressions (10.34) and (10.35), and expression for F_{ext} will be
$$F_{\text{ext}} = \iiint B^T DB \delta_e\, d_{\text{vol}} \quad (10.36)$$
$$F_{\text{ext}} = k\delta_e \quad (10.37)$$
Hence,
$$K = \iiint_{\text{vol}} B^T DB\, d_{\text{vol}} \quad (10.38)$$
Expressing (D) in global coordinates and d_{vol} in curvilinear coordinates Eq. (10.38) is given as
$$K = \iiint_{\text{vol}} B^T T^T\, DTBJ/d\xi\, d\eta\, d\zeta \quad (10.39)$$

10.2.9 Nodal Forces

The nodal forces are given by
$$F = \iiint_{\text{vol}} N^T P\, d_{\text{vol}} \quad (10.40)$$
where P = external force per unit volume

10.2.10 Formulation of Three-Noded Isoparametric Stiffener Element

To match with the quadratic shell element, an appropriate quadratic stiffener element (Fig. 10.12) shall be used. It can be easily combined with the shell element after accounting for its eccentricity through the transformation matrices. The stiffener element consists of three nodes situated along its centroidal axis. Each node is assumed to have six degrees of freedom, three translation u, v, w in the directions of global axis X, Y, and Z respectively and three rotation α, β, γ about the three local axis as shown in Fig. 10.12.

The steps followed in the derivation of the element stiffness matrix for the stiffeners are similar to eight-noded shell/plate element. The shape function for the three-noded stiffener element is given by the following equations

$$\left.\begin{array}{l} N_1 = \dfrac{1}{2}\xi(\xi - 1) \\ N_2 = (1 - \xi)^2 \\ N_3 = \dfrac{1}{2}\xi(1 + \xi) \end{array}\right\} \quad (10.41)$$

where the centroidal axis of stiffener lies along ξ direction [Refer Fig. 10.12(a) and (b)].

The displacements are expressed using the above shape functions.

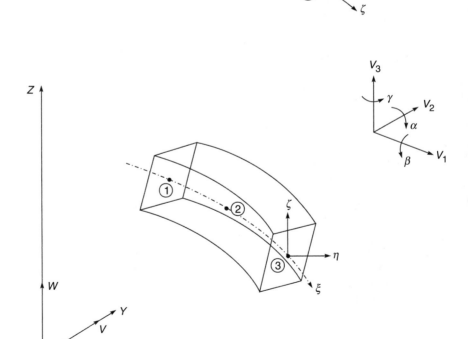

Figure 10.12 (a) One-dimensional stiffener element.
(b) Three-dimensional stiffener element.

10.2.11 Assembly and Solution

To analyze any structure (and tank in particular), the element stiffness matrix of all the elements in the structure are assembled. Where structural symmetry exists as in the case of a tank, only the symmetrical-half needs to be analyzed. Assembly of element stiffness matrix is possible by transformation of these matrices with reference to a common system of coordinates known as global system. The boundary conditions for the structure provide the fixed values of displacements in specified directions. The reactions at the supports and the loads at given points are other input data. From these sets of known values, the displacement/loads are computed at other remaining nodes of the structure. This results in a very large number of simultaneous linear equations to be solved. Unless a computer with very large core memory is available, one needs a special computer program for solution of such a large number of equations. One particular solution package using line storage on disc/tape is known as frontal and found to be suitable for a large-size structural problem.

10.2.12 Salient Features of Computer Program

A generalized computer program based on FEM (finite element method) has been developed to analyze in-plane, bending and sheer stresses at all the nodal points of transformer tanks of any kind of profile and structure in particular. This package can also analyze, in general, any plate/sheet structures with or without stiffeners. The computation also accounts for the thickness of plates/shells. It also caters for various types of loading applicable on tanks, such as oil pressure, vacuum, gravity loads, point loads, dynamic loads, etc.

The basic blocks of computation are illustrated in the flow chart given in Fig. 10.13.

10.2.13 Results

A typical tank for a large transformer is shown in Fig. 10.14. The element node relationship for side wall is shown in Fig. 10.15. Due to symmetry of the tank, only half of the tank has been taken. The results for normal deflections under vacuum pressure are shown in Fig. 10.16. The contours for major principal stresses at top under vacuum pressure are plotted in Fig. 10.17. The computer output provides the information about unsafe elements as well as weld sizes for stiffeners.

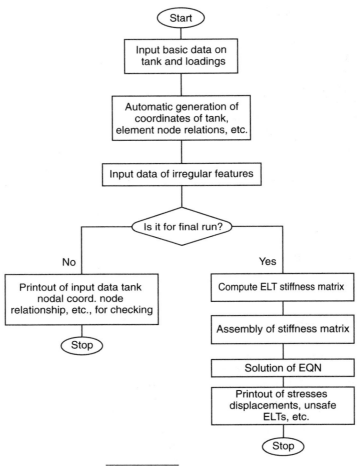

Figure 10.13 *Flow chart.*

10.3 Testing of Tanks

The oil pressure and vacuum testings are conducted on tanks to ensure against leakages and to check for strength.

10.3.1 Oil Pressure Test

The oil pressure testing on tank is done to ensure leakage proof welding joints. For this the oil connection is made at the base of tank and all the openings are blanked properly. The oil is filled up to tank

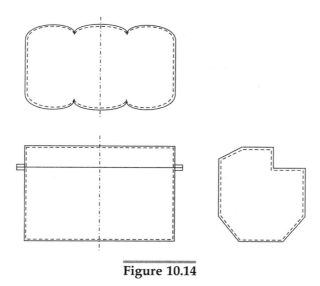

Figure 10.14

cover and the required pressure is applied by a pump. The pressure is maintained for a few hours and all the weldings are checked manually. If any leakage is found, the oil is drained and the welding is rectified. After rectification oil pressure is again applied. The tank deflection readings are measured before starting of oil pressure, at full oil pressure and after releasing the oil pressure.

10.3.2 Vacuum Test

After oil pressure test, the oil pipe connection is opened and complete oil is drained. This opening is blanked and vacuum pump connection is made at top of the tank wall and oil pressure gauge is replaced by a vacuum gauge. After ensuring all the fittings, the vacuum pump is started and the required vacuum is measured by a vacuum gauge. During vacuum testing the air leakage points are detected by an air leakage detecting instrument. Wherever the leakages are found in gasketed rims, the bolts are tightened. The temporary deflection at full vacuum is measured. If it is considerably high and consequently a rectification is needed, the vacuum is released and rectification covered on the tank. Subsequently, it is again vacuum tested. The full vacuum is maintained for one hour. After releasing vacuum the deflection readings are taken for finding out the permanent deflection. This shall be within specified allowable limits of deflection, depending on the size of the tanks.

246 *Transformers*

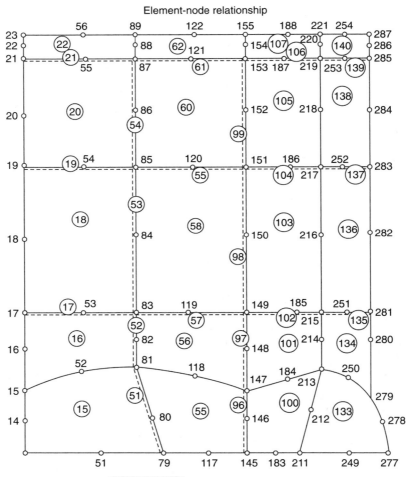

Figure 10.15 *Side wall of a tank.*

10.3.3 Measurement of Stresses

The strain measurement on tanks are normally not done, however, for comparing calculated stresses with measured stresses this can be done. The strain gauges are fixed to the tank structure with a suitable adhesive at various locations where the stresses are required to be measured and properly cured. The locations of strain gauges are shown in one of the tanks in Fig. 10.18. A gauge consists

Structural Design of Transformer Tank 247

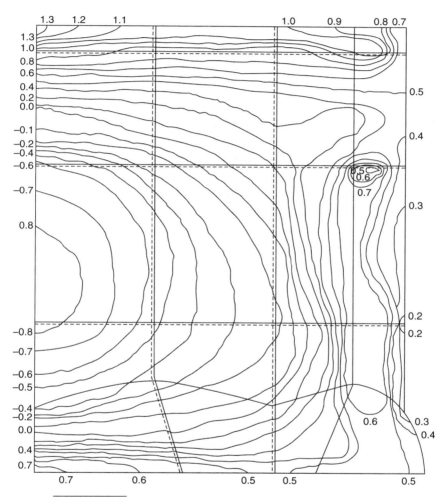

Figure 10.16 *Normal displacement under vacuum in mm.*

of a fine wire suitably fixed to the body of the structure. Underload strains are developed on the body, resulting in displacements of the points to which the ends of gauges are affixed. This results in change in the resistance of the gauge wire, which can be measured electrically by a suitable electric/electronic bridge. One typical strain gauge used in the experiment is a rossete-delta with six wires connected to the respective ends of wire gauges and brought to the bridge for measurement. The bridge is balanced for the null point

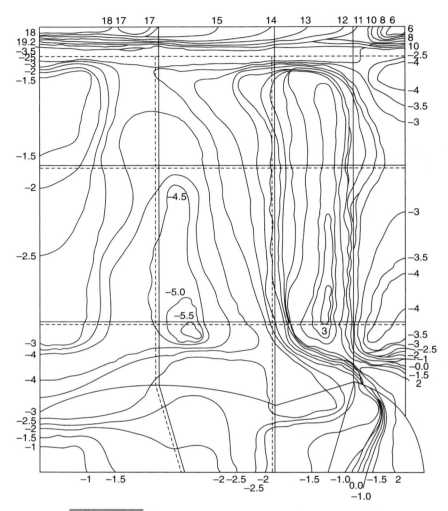

Figure 10.17 *Principal stresses under vacuum in mm.*

before commencement of measurements. The readings of strain gauges are recorded simultaneously. The tank is subjected to full vacuum and readings are taken. The strain gauges are fixed inside and outside the tank wall on large tanks to compare the top and bottom principal stresses on the surface.

The strain gauge gives values of strains in the direction of rossets. From this, the two principal stresses σ_1, σ_2 and their directions may be computed from the following expressions:

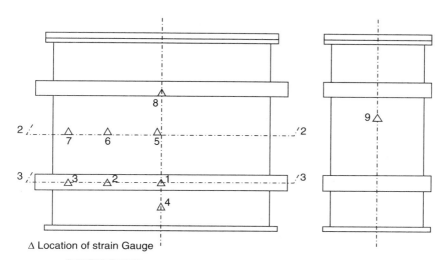

Figure 10.18 *Transformer tank showing strain gauges.*

$$\sigma_{1,2} = E \frac{\varepsilon_1 + \varepsilon_2 + \varepsilon_3}{3(1+\mu)} \pm \frac{1}{\mu} \sqrt{\left(\varepsilon_1 - \frac{\varepsilon_1 + \varepsilon_2 + \varepsilon_3}{3}\right)^2 + \left(\frac{\varepsilon_2 - \varepsilon_3}{\sqrt{3}}\right)^2} \quad (10.42)$$

$$Q = \tan^{-1} \frac{(\varepsilon_1 - \varepsilon_3)/\sqrt{3}}{\left(\varepsilon_1 - \frac{\varepsilon_1 + \varepsilon_2 + \varepsilon_3}{3}\right)} \quad (10.43)$$

where E is Young's modulus in kg/cm^2
$\varepsilon_1, \varepsilon_2, \varepsilon_3$ are strains in three directions
μ is the poisson's ratio.

Nomenclature

p = Pressure in kg/cm^2
b = Width of the plate in cm
l = Longitudinal length of plate in cm
t = Thickness of the plate in cm
E = Young's modulus in kg/cm^2
I = Moment of inertia in cm^4
Z = Section modulus in cm^3
f = Stress developed in kg/cm^2
δ = Deflection in cm
f_{ss} = Stresses for simply supported plate in kg/cm^2

f_{EB} = Stresses for fixed and plate in kg/cm^2
I_{XX} = Moment of inertia along X-X direction cm^4
I_{YY} = Moment of inertia along Y-Y direction cm^4
K_1, K_2, K_3 = Factors depending on boundary conditions
M = Bending moment, kg cm
$\varepsilon_1, \varepsilon_2, \varepsilon_3$ = Strains in three directions, mm
σ_1, σ_2 = Principal stresses in kg/cm^2
σ_x = Stress in x direction in kg/cm^2
μ = Poisson's ratio

Subscripts p for plate
s for stiffener
ss for simply supported
EB encastered beam

REFERENCES

1. Timoshenko, S., and S. Woinowsky Krieger, *"Theory of Plates and Shells,"* 2nd Edition, McGraw-Hill, New York, 1959.
2. Shakya, M.K., *"Stress Analysis of Transformer Tanks by Strain Gauges,"* M. Tech. Dissertation, M.A.C.T., Bhopal, 1980.
3. Perry, C.C., and H.R. Lissner, *"Straingauge Primer,"* McGraw-Hill Book Company, New York, 1962.
4. Cook, R.D., *Concepts and Applications of Finite Element Analysis,* John Wiley & Sons, Inc., 1974.

CHAPTER 11

Transformer Auxiliaries and Oil Preservation Systems

S.C. Verma
J.S. Kuntia

Transformer auxiliaries play a vital role in ensuring proper functioning of the main equipment. Some of the auxiliaries provide protection under fault conditions.

Transformer oil being a major insulation requires special attention against contamination by moisture and oxygen for preservation of quality.

TRANSFORMER AUXILIARIES

11.1 Gas Operated (Buchholz) Relay

The relay serves as main protection for any minor or major faults that may develop inside a transformer. Such faults always result in generation of gases which causes the operation of mercury switches giving signal for audible alarm or isolates the transformer from the network.

It comprises a cast housing which contains two pivoted buckets, each bucket being counter-balanced by a weight. Each assembly carries a mercury switch, the leads from which are taken to a moulded terminal block.

Operation

The relay is mounted in the pipe at an inclination of 3–7° as shown in Fig. 11.1. In a healthy condition of the transformer, the relay is full of

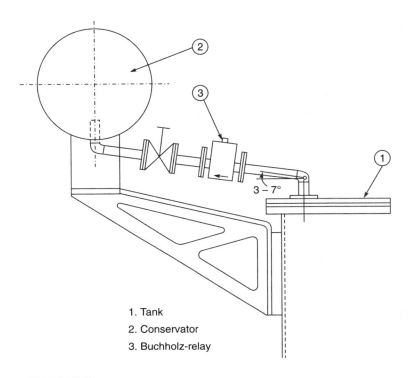

Figure 11.1 *Method of mounting Buchholz-relay on transformer.*

oil and both the mercury switches are open. In the event of a minor fault like damage to core bolt insulation, local overheating, etc., the arcing causes slow generation of gas in the oil, which passes up in the pipe and gets trapped in the relay housing. As gas accumulates, the oil level in the relay falls, leaving the top bucket full of oil.

When a sufficient volume of gas is collected in the relay, the top bucket, because of its extra weight due to contained oil, tilts, overcoming the balance weight which closes the mercury switch and initiates an audible alarm.

With a major fault like short-circuit between turns, coils or between phases, the generation of gases is rapid and the gas and the displaced oil surges through the relay and impinges on the baffle plates, causing the lower assembly to tilt and close the mercury switch. This provides a signal for tripping the circuit breaker, which disconnects the transformer from the network.

11.1.1 Relays for Earthquake Zones

Shock and vibrations acting on the tube of the mercury switch can cause the mercury within it to move and momentarily bridge the switch contact, even though it may be tilted in open-circuit position. This is considered to be a malfunction of the relay, as it is caused by external disturbances like earth tremors and not by a fault within the transformer. Consequently, relays for such situations are provided with magnet operated reed switches in place of mercury switches.

The external appearance and principle of operation of these relays is exactly similar to those with mercury switches. The reed type switches have rhodium contacts located midway along the glass tube which has an atmosphere of nitrogen.

11.2 Temperature Indicators

Temperature indicators are precision instruments, specially designed for protection of transformers and perform the following functions.

- Indicate maximum oil temperature and maximum or hottest spot temperature of winding.
- Operate an alarm or a trip circuit at a predetermined temperature.
- Switch on the cooling equipment when the winding attains a preset high temperature and switch it off when the temperature drops by an established differential (so as to avoid too frequent on and off operation of the switch).

Normally two separate instruments are used for indicating oil and winding temperatures.

11.2.1 Construction and Principle of Operation

These indicators normally work on the principle of liquid expansion, the liquid being sealed in the bellows. Figure 11.2 shows a sectional view. The indicator is provided with a sensing bulb placed in an oil filled pocket on the transformer tank cover. The bulb is connected to the instrument housing by means of flexible connect-

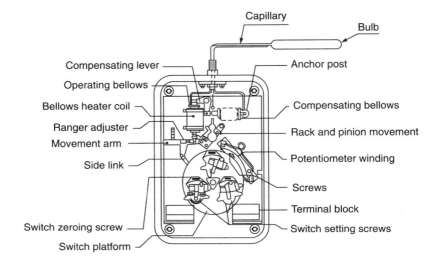

Figure 11.2 *Sectional view of a temperature indicator.*

ing tubing consisting of two capillary tubes. One capillary is connected to the operating bellow of the instrument and the other to a compensating bellow. The operating system is filled with a liquid which changes volume with varying temperature. The compensating bellow acts upon the operating bellow through a linkage compensating for variations in the ambient temperature. With change in the volume of the liquid, the bellows expand or contract, transmitting the movement through a linkage mechanism to the indicating pointer and switching disc. Up to four dry electrode mercury switches, each mounted on a steel carriage, are provided on a winding temperature indicator. The make-and-break temperature of each switch can be independently adjusted.

Oil and winding temperature indicators work on the same principle, except that winding temperature indicator is provided with an additional bellow heating element. As it is not possible to measure the winding temperature directly, it is done indirectly by means of a *thermal image* process.

The heating element is fed by a current transformer, with a current proportional to the load in the winding temperature to be indicated. The temperature increase of the heating element is thereby proportional to the temperature increase of the winding over top oil temperature. As bulb of the instrument is located in the hottest oil zone, it senses the maximum oil temperature. The operating bellow thus gets additional movement, simulating the increment of winding temperature above maximum oil temperature and thus indicates hottest spot temperature of winding.

11.3 Pressure Relief Valve

The pressure relief valve plays a significant role in the protection of power transformer systems. As mentioned before, a major fault inside the transformer causes instantaneous vaporization of the oil, leading to extremely rapid build-up of gaseous pressure. If this pressure is not relieved within a few milliseconds, the transformer tank can get ruptured, spilling oil over a wide area. The consequent damage and fire hazard possibilities are obvious. A pressure relief device provides instantaneous relieving of dangerous pressure.

11.3.1 Construction and Operation

Figure 11.3 shows a cross-sectional view of the valve. The valve is generally mounted on the tank cover above an opening. The valve has a corresponding port which is normally sealed by a stainless steel diaphragm (4). The diaphragm rests on an 'O' ring (3) and is kept pressed by two heavy duty springs (6). As the pressure inside the tank rises above a preset limit due to a major fault, the diaphragm gets lifted instantaneously and excessive pressure drops, the diaphragm then restores to its original position. The lift of the diaphragm is utilized to operate a flag indicator (10) and a micro-switch with the help of a rod (8).

Another type of device which is also used for the same purpose is called *explosion vent*. Figure 11.4 shows the general construction of an explosion vent. In the event of a serious fault, due to excessive pressure, the top diaphragm ruptures, thus releasing the pressure.

Due to certain superior features of spring loaded pressure relief valve like smaller size, elimination of equalizer pipe and provision of a switch for alarm annunciation in event of its operation, it is finding a widespread preference over the explosion vent.

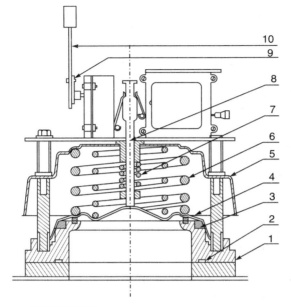

Figure 11.3 *Sectional view of a spring-loaded pressure relief valve.*

1. Base
2. Gasket ring
3. O ring
4. Diaphragm
5. Cover
6. Spring
7. Rod retaining spring
8. Switch operating rod
9. Flag carrier plate
10. Flag indicator

11.4 Oil Level Indicator

Normally all transformers are provided with an expansion vessel called conservator, to take care of expansion in the oil volume due to rise in temperature, when the load on the transformer increases or due to increase in ambient temperature. The oil level in the conservator consequently goes up. Conversely, it falls when the temperature or load reduces. It is essential that the oil level in the conservator is maintained above a pre-determined minimum level. All large transformers are, therefore, fitted with a magnetic oil level gauge which also incorporates a mercury switch. The switch closes and actuates an audible alarm in the event of oil level dropping to near empty position in the conservator.

11.4.1 Construction of Operation

Figure 11.5 shows a cross-sectional view of the inner details and a schematic diagram. A float is used as a sensor which moves with the

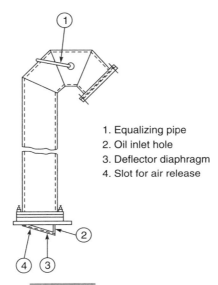

Figure 11.4 *Explosion vent.*

rise and fall of the oil bevel. Its movement gets transmitted to the switch mechanism by means of bevel gear and magnetic coupling, which ensures a complete seal between the conservator and switch compartment. The pointer is also magnetically operated and picks up the correct oil level.

11.5 Bushing and Cable Sealing Box

It is necessary to bring the low and high voltage leads out of transformer tank, to be able to make connections between transformer and generator or transmission lines, etc. This is accomplished by terminating these leads through what are known as bushings or cable box.

A bushing is a structure carrying a conductor through a partition in the tank and insulating the conductor from partition.

11.5.1 Cable Sealing Box

Generally used for termination of leads of low voltage, a cable sealing box is designed for the purpose of receiving and protecting the

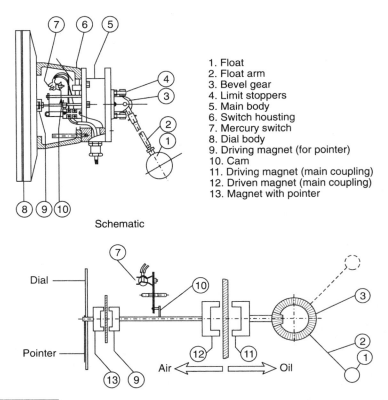

Figure 11.5 *Sectional view and schematic of a magnetic oil level indicator.*

end of a metal-sheathed cable or cables and containing a suitable insulating medium.

It is a unit complete with bushings to which the terminals of the transformer can be connected. The insulating medium in a cable box can be air or a bituminous compound. Figure 11.6 shows a typical 3-pole, 9-gland cable sealing box.

TRANSFORMER OIL PRESERVATION SYSTEMS

Transformer oil deterioration takes place due to moisture. Moisture can appear in a transformer from three sources, viz. by leakage past gasket, by absorption from air in contact with the surface, or by its formation within the transformer as a product of deterioration as

Transformer Auxiliaries and Oil Preservation Systems 259

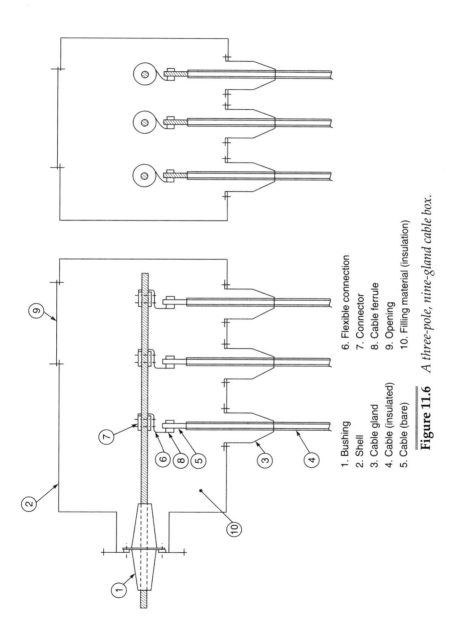

1. Bushing
2. Shell
3. Cable gland
4. Cable (insulated)
5. Cable (bare)
6. Flexible connection
7. Connector
8. Cable ferrule
9. Opening
10. Filling material (insulation)

Figure 11.6 *A three-pole, nine-gland cable box.*

insulation ages at high temperature. The effect of moisture in oil is to reduce the electric strength, especially if loose fibers or duct particles are present.

Method available to reduce oil contamination from moisture are silicagel breather, thermosyphon filter, sealed conservator tanks using gas cushion, rubber diaphragm or air-cell seals refrigerated dryers.

11.6 Silicagel Breather

A silicagel breather is most commonly employed as a means of preventing moisture ingress. It is connected to the conservator tank, which is fitted to transformer to allow for changes in volume due to temperature variations. As the load reduces, air is drawn into the conservator through a cartridge packed with silicagel dessicant, which effectively dries the air. Freshly regenerated gel is very efficient, it will dry the air down to a dew point of below −40°C, but quickly falls in efficiency. A well-maintained silicagel breather will generally operate with a dew point of −35°C, as long as a large enough quantity of gel has been used for the cycling duty. Figure 11.7 shows such a breather.

Silicagel may be reactivated by heating in a shallow pan at a temperature of 150° to 200°C for two to three hours when the crystal should have regained their blue tint.

11.7 Gas Sealed Conservators

In this method, the contact between transformer oil and atmospheric air is eliminated by providing cushion of an inert gas over oil surface in the conservator vessel. The gas pressure is always higher than atmospheric pressure to avoid ingress of air. The gas normally used for this purpose is nitrogen having high purity and dryness.

Construction and Operation

High pressure nitrogen gas at 15 MPa flows out of the cylinder and is admitted in the conservator after passing through multi-stage

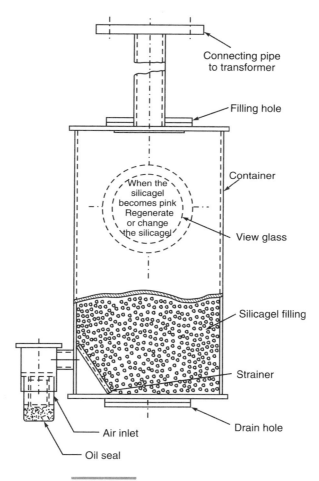

Figure 11.7 *Silicagel breather.*

pressure reducing valves. The pressure reducing valves automatically cut off the nitrogen gas supply when the pressure in the conservator reaches 3–5 kPa. Due to increase in ambient temperature and load the gas pressure builds up.

The system is designed to relieve any excessive pressure through a relief valve provided for the purpose. When the pressure drops below 3–5 kPa, the valves open to admit nitrogen from cylinder and this cycle continues until the cylinder becomes empty.

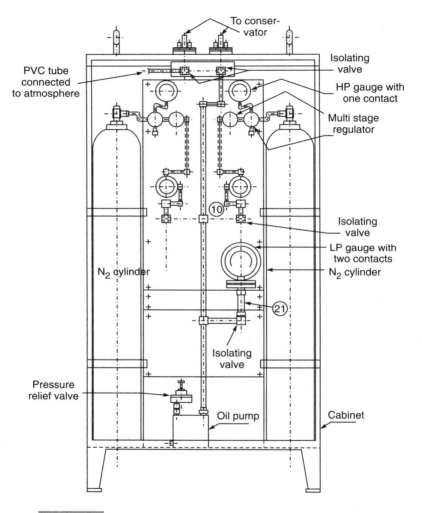

Figure 11.8 *A typical general arrangement of nitrogen system.*

The system may be provided with switches in the pressure gauges to operate the alarm under the following abnormal operational conditions:

(a) When the pressure in transformer exceeds relief over valve pre-set operating pressure.
(b) When the pressure in transformer drops below 3–5 kPa.
(c) When the cylinder pressure is reduced to around 1 MPa.

11.8 Thermosyphon Filters

Thermosyphon filters are intended for prolonging transformer oil life by extracting harmful constituents like water, acids, etc., from oil. Such filters are normally installed on ONAN or ONAF cooled transformers.

The constructional details of such a filter are shown in Fig.11.9. Normally the filter is mounted directly on the transformer tank. The filter, generally of cylindrical shape, has a number of perforated steel trays filled with an adsorbent material. The adsorbents generally employed are silicagel and active alumina, the latter being slightly more effective.

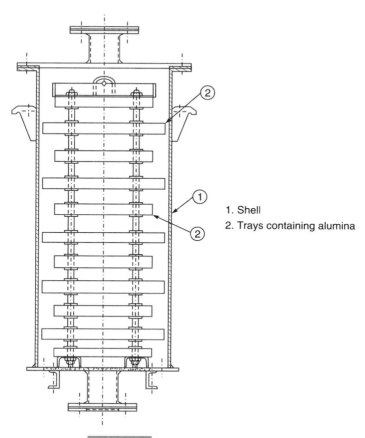

1. Shell
2. Trays containing alumina

Figure 11.9 *Thermosyphon filter.*

As a result of difference in the temperature between the upper and lower layers of oil in the tank of a transformer in operation, the oil circulates through the filter by convection currents. The water absorption capacity of alumina at relative humidity of 60% is about 15% by weight.

11.9 Bellows and Diaphragm Sealed Conservators

The contact between atmospheric air and transformer is prevented by a barrier which is made from a synthetic rubber compound. The general principle of such a system is shown in Fig. 11.10. In case of bellow type of barrier, as the oil level in the conservator vessel falls, air is sucked from the atmosphere through a silicagel breather, inflating the bellow. The bellow deflates as the oil level goes up. The conservator is also fitted with pressure vacuum bleeder to pass either oil or such in air in the event of over-filling or under-filling of conservator.

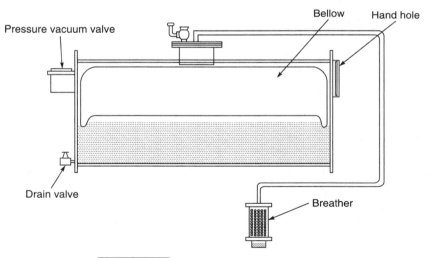

Figure 11.10 *Bellow sealed conservator.*

When diaphragm is used as a barrier between oil and atmospheric air, the conservator vessel is made in two semi-circular halves as shown in Fig. 11.11. The diaphragm is held between the two halves

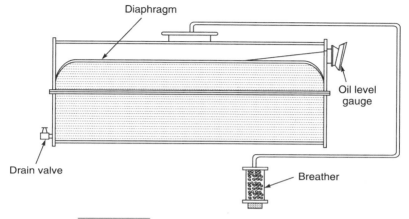

Figure 11.11 *Diaphragm sealed conservator.*

and bolted. As oil expands and moves up, it pushes the diaphragm upwards. The position of the diaphragm is indicated by the oil level indicator as the oil level indicator connecting rod is connected to the diaphragm. When the oil level falls down in the conservator, the diaphragm deflates creating vacuum, which is filled by air getting sucked through a silicagel breather.

These types of sealing systems have one advantage over the gas-sealed conservator. If gas is pressurized to a high level, it gets dissolved in oil. Over a period of time the amount of gas in oil reaches the saturation point. If at this stage, the load on transformer is suddenly dropped, or the ambient temperature falls severely, the pressure falls, oil becomes supersaturated and gas bubbles will be evolved. If there is a pump connected in the cooling circuit, it will help the generation of bubbles. These bubbles may cause insulation failure in the region of strong electric fields.

11.10 Refrigeration Breathers

In the refrigeration breather system, an air dryer is fitted to the conservator vessel. Air breathed through the unit is dried in passing down a duct cooled by a series of thermoelectric modules based on Peltier effect. Top and bottom ends of the duct are terminated in the expansion space above oil level in the conservator and air continuously circulated through the duct by thermosyphon forces.

Figure 11.12 shows general constructional feature of a refrigeration breather system. When transformer is breathing, warm moist air passes through the central duct; its cold surface which is held at about $-20°C$, collects any water vapors in air and condenses them as frost and ice. Periodically, the frost and ice on the duct is removed by reversing the current to the thermoelectric modules. The current reversal heats up the duct walls sufficiently to melt the ice, and passes out as water through a small drain tube. The current flow is again reversed in the modules and the cycle repeats automatically by a control unit.

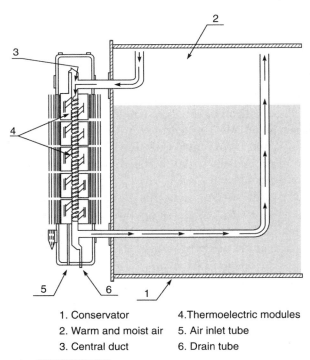

1. Conservator
2. Warm and moist air
3. Central duct
4. Thermoelectric modules
5. Air inlet tube
6. Drain tube

Figure 11.12 *Refrigerated breather system.*

Atmospheric air is breathed in through a small tube entering the unit at the top and has to pass through the central duct, ensuring that it reaches the conservator in a dry state.

CHAPTER 12

Manufacturing and Assembly

T.K. Ganguli
M.V. Prabhakar

Manufacturing and assembly of the transformer involves the following main stages and the manufacturing techniques basically depend on the design philosophy.

(a) Core building
(b) Preparation of windings
(c) Core and winding assembly
(d) Terminal gear assembly
(e) Processing
(f) Servicing and tanking
(g) Mounting of accessories for testing

12.1 Core Building

The core forms the magnetic circuit of the transformer. Core is built up from cold-rolled grain-oriented, silicon alloy sheet steel of the best magnetic properties. The material is either received in rolled coil form or in finished sizes of core sheets, ready for core building. When the material is received in coil form, it is slitted, cropped and mitered to the required dimensions based on the design requirements. CNC machines are now available which can perform the above operations with perfect control on burrs. The size, type, and construction of the core, like the number of steps, number of limbs, height, etc., depend on the number of phases and the size of the transformer.

The sequence of operations for core building and assembly are:
  (a) Slitting of core steel rolls to required width on slitting machines.
  (b) Cropping and mitering to the required dimensions.
  (c) Hole punching in the laminations where required.
  (d) Stacking of laminations of different size to the required thickness.
  (e) Laying of clamp plates and end frame and its leveling.
  (f) Assembly of insulation between clamp plate/end frame to core laminations.
  (g) Preparation of oil duct in core.
  (h) Core building.
  (i) Clamping of core after assembly of the top end frame.
  (j) Tightening of core.
  (k) Lifting of core by use of a cradle, and carrying out isolation checks after treatment of insulation items.

The core assembly is now ready for further processing and assembly.

12.2 Preparation of Windings

The windings form the electrical circuit of a transformer. Various types of windings, i.e., spiral, helical, disc, layer, etc., are used, depending upon the rating of the transformer and the design considerations. Generally, low voltage windings are either spiral or helical and high voltage windings are either layer or disc type. Various types of windings have been discussed in detail in Chapter 5.

The sequence of manufacture of windings is given below:
  (a) Loading of moulds (formers) on the winding machines.
  (b) Loading of the conductor reels on stands.
  (c) Dressing of the mould, i.e., assembly of insulation spacers and blocks on the mould.
  (d) Manufacture of the winding on horizontal/vertical winding machines depending on design, type of winding, number of conductors, to be handled at a time and the type of conductor.

(e) Preparation of the leads, etc.
(f) Dismantling of winding from the machine.
(g) Preparation of the joints between conductors, if any.
(h) After removing the winding from the winding machine, each winding is clamped between top and bottom plates through tie rods, and kept in an oven for heating. The windings are individually shrunk to the required axial dimensions by heating in the steam-heated oven and by applying the required pressure. Heating ensures removal of moisture from the insulation items. This process is called stabilization of the windings, and the windings are stabilized to such an extent that they do not shrink further during service.

12.2.1 Layer Winding Assembly

Layer winding is also clamped and shrunk as above. Winding is then placed with bottom end in top position and the bottom end insulation consisting of angle rings, angle washers, flanged paper insulation, block washers, etc., is built up. The position of winding is then reversed, bringing top of winding on top and the top insulation arrangement is completed in this position. During this process, insertion of line shield rings on both ends and filler paper, preparation of leads, placement of circular barriers, etc., are also completed. After completion of insulation arrangement, the coil is shrunk to the required axial dimensions.

Typical insulation arrangement of layer winding is shown in Fig. 12.1.

The winding is then kept in clamped condition for either composite winding assembly or assembly on core legs.

12.3 Winding Assembly

The winding assembly is carried out in clamping structure consisting of bottom and top face plates and tie rods. Steel face plates are selected on the basis of maximum diameter of the winding to be assembled. Number of windings to be assembled depends

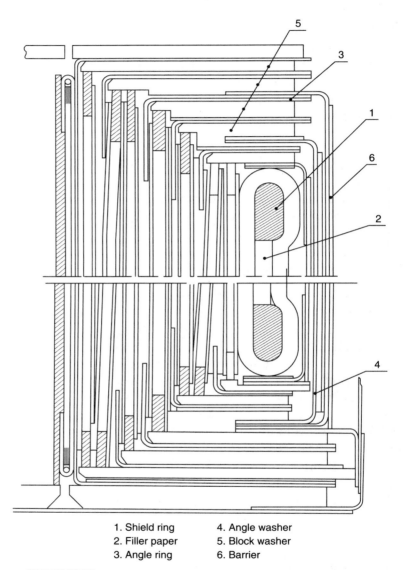

1. Shield ring
2. Filler paper
3. Angle ring
4. Angle washer
5. Block washer
6. Barrier

Figure 12.1 *Typical insulation arrangement of layer winding.*

upon the type of transformer, i.e., whether generator transformer auto-transformer or system transformer, etc. Generally, in a generator transformer low voltage, tapping and high voltage windings are involved whereas in auto-transformer tertiary, common, tapping and series windings are involved. A typical sequence of operation for winding assembly of a transformer having three windings when they are pre-assembled as follows:

(a) Winding 2 is lowered over winding 1, placing radial spacers and pressboard cylinders in between and keeping bottom end of each coil at the top. Next winding 3 is lowered over winding 2.

(b) Insulation items, i.e., block washers, angle rings, etc., are assembled. During insulation assembly, various leads, i.e., tapping, line leads, etc., are properly positioned and partly insulated at this stage.

(c) Winding assembly is reversed to bring top of the coil at the top. Insulation items of top portion are then assembled, and various leads are also prepared.

(d) Barrier arrangement is then assembled and tied with tape bands.

(e) Winding assembly is clamped in clamping structure and is shrunk to the required axial dimensions. The assembly is kept under pressure for further assembly on core legs.

The individual windings alternatively can also be assembled on the core leg instead of being pre-assembled.

12.4 Core and Winding Assembly

Core is placed on suitable levelled platform. Top-end frame and yoke laminations are removed. Bottom insulation items are placed on bottom yoke/end frame on each core leg. Various windings are either lowered one by one or composite winding assembly is lowered on each core leg. After lowering of the windings, top insulation arrangement is completed. Top-end frame and yoke laminations are placed back in position. Coils are kept under pressure by either coil clamping bolts or spring loaded hydraulic devices provided on the top-end frame.

12.5 Terminal Gear Assembly

After relacing of the top yoke, the preparation of the Terminal Gear Assembly is done as described below:

(a) Cutting of the leads as required.
(b) Crimping/brazing of the leads with cables.
(c) Brazing of bus bars.
(d) Fixing of different cleats.
(e) Crimping/brazing of cables with terminal lugs.
(f) Mounting of the tap changer/tap switch.
(g) Preparation of HV line lead.

In this stage, connections between phases to form the required vector group, tapping lead connections, line and neutral leads formation, etc., are completed. Low-voltage connections are done on one side of the winding and are designated as LV terminal gear. On the opposite side, high-voltage connections are done and are designated as HV terminal gear. Medium voltage leads (in system or auto-transformer) are taken out on LV side and tapping connections on either LV or HV side depending upon design layout. Generally in generator transformer, a three-phase on-load or off-circuit tap-changer is mounted on one end and in case of auto-transformer three single-phase tapchangers are mounted in front of the windings. Tapchangers are supported from end frame during terminal gear assembly. All leads, i.e., line and neutral leads of low-voltage, medium-voltage and high-voltage windings, tapping leads, etc., are laid out and connected using different types of joints (i.e., bolted, crimped, soldered or brazed) and insulated for the required insulation level. Leads are properly supported by cleats mounted on end frames. The clearances between various leads, coil to leads, leads to end frame and other parts are maintained and checked.

12.6 Placement of Core and Winding Assembly in Tank

After completion of terminal gear assembly, the core and winding assembly is placed in the tank. The tank may either be of conventional or bell-shaped construction depending upon requirement. In case of bell-shaped construction, the top part of the tank and for conventional construction the bottom tank is prepared for this purpose.

Preparations include mounting of shunts, barriers, etc., on tank walls as required and also laying of gaskets on the flange joints. In conventional tank, the core and winding assembly is lowered in the bottom tank.

In bell-shaped tank construction, top tank is lowered on the core and winding assembly.

After placement of core and winding assembly in tank, various electrical and mechanical clearances, viz. coils to tank, line, neutral and tapping leads to tank, etc., are checked. After the above checks are over, all the openings in tank and cover are blanked by blanking plates and gaskets. The transformer is then sent for drying and impregnation.

12.7 Processing

The quality of a transformer and, consequently its performance and life depends essentially on the factory processing. The assembled active part of the transformer after tanking or before tanking as the case may be, is heated in an oven for extraction of moisture under vacuum. Sometimes, vacuum drying is done by drawing vacuum in its own tank and keeping the whole transformer in an oven.

Besides the conventional method of drying, the most modern vapor phase drying method is also used, as explained in detail in Chapter 13.

Much importance is attached to proper dry out, as otherwise windings would become loose in service, and may get distorted under short circuit conditions.

Once the moisture is extracted to the desired level, oil is filled under vacuum and the transformer is soaked in oil.

12.8 Servicing of Transformer

The transformer active part is then taken out and all the terminal gear cleating of the leads are retightened and the coils are clamped and secured in position. The transformer tank cover is then placed in its position and the transformer is now ready for further assembly of bushings, etc.

12.9 Tanking

After servicing the transformer is assembled with all the necessary fittings. The oil is filled under vacuum and it is circulated in order to get the desired ppm and BDV levels before being offered for final testing and subjected to high voltage testing.

CHAPTER 13

Drying and Impregnation

M.P. Singh
M.V. Prabhakar

In the construction of a transformer, the insulation system is the most important feature, and hence requires maximum attention. Normally, the oil, paper, and pressboard insulation has to be freed of dirt, dust, and moisture for obtaining the optimum insulating properties.

Cellulose insulation used in power transformers and reactors has 8 to 10% of moisture by weight at ambient temperature, being a hygroscopic material. Water is injurious to transformer insulation system, since it reduces the electric strength and the resistivity of the oil and accelerate the deterioration of solid insulation. Figure 13.1 shows the effect of moisture content on impulse voltage strength of oil and paper. The decrease in impulse voltage strength is clearly evident as soon as the moisture content of the paper is greater than 0.1%. Paper with a 1.5% moisture content ages ten times faster than on with only 0.3% as shown in Fig. 13.2. Processing eliminates all types of impurities and foreign particles from

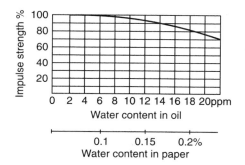

Figure 13.1 *Impulse voltage strength of paper.*

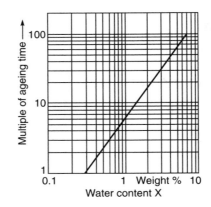

Figure 13.2 *Influence of moisture in paper on ageing time of paper: a multiple of the ageing time of oil-impregnated paper with moisture content X compared to that of a well-dried, oil-impregnated paper with 0.3% residual moisture.*

transformer, e.g., dust, dirt, fine metallic particles and fibers of diverse origins. Any impurity and particle which remains inside transformer after vacuum drying is taken away by transformer oil during circulation, which are separated at the filter. During vacuum drying and oil impregnation, the gases trapped inside solid insulation and occluded among intricate profile of insulation components are also removed, which result in low partial discharge inception voltage. Optimal utilization of the excellent properties of oil-cellulose dielectric system depends upon efficient drying, degassing and impregnation process adopted and hence the proper processing of transformers is a vital requirement.

13.1 Basic Principles of Drying

Cellulosic insulation is dried by creating conditions in which water vapor pressure (WVP) around the insulation is less than that in the insulation. The vapor pressure in the insulation is increased by heating the insulation and the vapor pressure around the insulation is decreased by removing water vapor. Figure 13.3 shows pressure plotted against humidity and temperature, from which it is seen that a 20°C temperature rise increases the internal pressure by more than 100% (a factor of two). Basically one should aim at achieving the highest processing temperature consistent with the type and

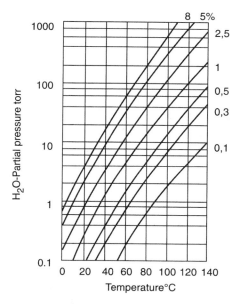

Figure 13.3 *Partial insulation water vapor pressure v. temperature, for various insulation moisture contents.*

ageing properties of the insulation. The upper temperature limit is set by the maximum permissible drying temperature for paper, i.e., about 110°C (or 130°C in an oxygen-free atmosphere). Drying efficiency also depends on diffusion coefficient of the insulation material. This coefficient is dependent on the material to be dried, its temperature, pressure and moisture content as shown in Fig. 13.4. A 20°C temperature rise increases the diffusion coefficient by about 100% and reduces drying time by half.

Two basic processes are adopted for drying of transformers.
  (a) Conventional vacuum drying
  (b) Vapor phase drying

PRINCIPLES OF DRYING

13.1.1 Drying Time

The drying time of the insulation increases considerably with the operating voltage and size of the transformer. The longer the drying time, the higher the partial discharge inception voltage.

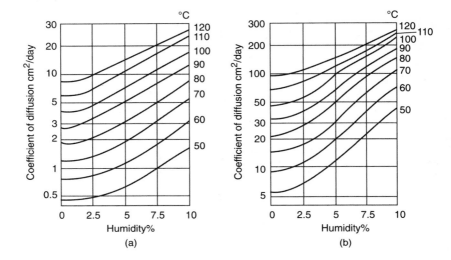

Figure 13.4 *Diffusion coefficient of low density oil-free pressboard for various drying temperatures.*
(a) under atmospheric pressure.
(b) over a 0.1 to 1.0 torr vacuum range.

The time of drying depends on the final moisture content required in the insulation pressure rings, supporting components and pre-compressed pressboards basically determine the minimum drying time.

13.2 Conventional Vacuum Drying

The moisture content in a cellulose insulation can be brought down to 2% from 8–10% by dry hot air circulation. For further reduction of moisture content, heating under vacuum is resorted to, so that the time required for drying can be reduced.

The core and windings assembly is placed in the clean transformer tank. Drying is carried out either in a vacuum vessel or in an air circulating drying oven. In the latter case, vacuum is drawn in the transformer tank.

13.2.1 Loading of Transformer

The tank containing core and windings is loaded into the vacuum vessel. For bell shaped tanks the openings on the top of the tank are kept open such that complete cellulose insulation is oil impregnated subsequently. For drying in its own tank, the tank along with core and windings is loaded into the drying oven. Tank cover is placed in position and all openings are closed.

13.2.2 Vacuum Drying

The transformer is heated initially to 100°C for about 24 hrs and a high vacuum in vessel or in its own tank is drawn. Higher the vacuum, better and quicker is the drying. Vacuum level to be achieved also depends upon voltage class of transformer. Vacuum corresponding to an absolute pressure as low as 1.33 Pa (0.01 torr) can be obtained. In order to maintain the transformer temperature at the required value, the pressure is increased to atmospheric level by the admission of hot dry air at intervals during the first few days of drying. This is necessary since winding temperature drops considerably as latent heat of transformer is utilized in conversion of insulation moisture into water vapors.

During vacuum drying, the quantity of water collected at pump is recorded at regular intervals. Insulation resistance, power factor and dispersion factor of windings are also monitored. Vacuum drying is considered to be complete, when desired condensate extraction rate is achieved and other parameters indicated above become reasonably steady.

13.2.3 Reconditioning of Insulating Oil

There is a tendency for oil to absorb water due to breathing during transport and storage. Impurities, be they solid ones like hygroscopic fibers, suspended particles or liquid ones like dissolved water, bring about a considerable reduction of the dielectric strength of oil. Oil at 20°C saturated with water (44 ppm) attains only about 25% of the original electric strength with a water content of 10 ppm. Under the same conditions, the resistivity of oil also drops considerably. Air, especially oxygen, dissolved in the oil represents a risk not only with regard to the formation of bubbles but also due to accelerated oxidation process leading to the chemical deterioration of oil. Hence, it is necessary to treat and purify the oil before impregnating the windings, so as to attain a degree of purity which meets the operational requirements. Oil is reconditioned to eliminate the following elements.

(a) Solid impurities
(b) Free and dissolved water
(c) Dissolved gases

The physical means that are used for treatment of oil include several types of filtration, centrifuging and vacuum dehydration techniques. If vacuum treatment is employed, temperatures up to 80°C can be used; if not, it is advisable to limit the temperature of oil to 60°C to prevent oxidation. The filters used should be capable of removing particles larger than one micron diameter.

Along with drying of transformer, raw oil is reconditioned in a vacuum dehydrator, till desired characteristics are obtained.

13.2.4 Oil Impregnation

After completion of vacuum drying, the reconditioned oil is admitted to the transformer without breaking vacuum in the vessel or tank, as the case may be. The transformer is allowed to cool to a temperature of about 70°C before filling with oil. Temperature of reconditioned oil is maintained at $50 \pm 10°C$. Care is taken to ensure that pressure in vessel or tank is not too low to vaporize the oil. The oil flow rate normally does not exceed 3 kl/h so that all residual gases occluded in crevices among insulation items are released and the possibility of gas bubbles being trapped at any place inside tank is prevented. Oil is filled until all cellulose insulation is covered.

13.2.5 Redrying and Oil Circulation

Oil impregnated transformer is kept under vacuum for a few hours so that solid insulation is completely soaked with oil. Winding insulation shrinks to some extent during drying as the initial water in cellulose is extracted. Hence, windings and cleats of terminal gear are retightened after taking the assembly out of tank. A little moisture reabsorbed by the insulation due to exposure to atmosphere during this period is extracted by redrying under vacuum and subsequently oil is again impregnated inside the transformer. Oil is circulated to regain its electric strength and moisture content which deteriorate due to contact with materials used in the construction of transformer. Solid particles, fibers of different origins and all other impurities are separated at filter. A settling time is given before conducting high-voltage dielectric tests on the transformer, so that suspended fine particles are settled and occluded gas is fully released.

13.3 Vapor Phase Drying

The main difference between conventional vacuum drying and vapor phase drying is that, in the latter process the heat carrier is a vapor of low viscosity solvent, more like kerosene, with a sufficiently high flash point instead of air. The vapor is condensed on the transformer and then re-evaporated in the plant. For this reason, vapor phase installations include an evaporator and condenser system in addition to the vacuum equipment and vacuum vessels associated with conventional drying equipment. Typical schematic of a vapor phase drying system is shown in Fig. 13.5, which is applicable to transformers dried in the vacuum vessel as well as in their own tank. It may be seen from Fig. 13.5 that solvent heat conveyor system consists of storage, evaporation, condensation filtration, solvent feedback and control arrangement.

13.3.1 Heat Carrier

The solvent used should possess the following properties for effective and efficient drying.
    (a) Vapor pressure must be distinctly below that of water, so that a large pressure difference assists efficient water diffusion from the beginning of the heating phase (Fig. 13.6).
    (b) Evaporation heat should be as high as possible.
    (c) The presence of small amounts of the heat-carrier in the solid or oil insulation must have no effect on their ageing or general properties.
    (d) High ageing stability, allowing practically unlimited use in the drying process. A solvent storage tank is normally required to be refilled after a few years.
    (e) Flame point should be above 55°C.

The physical properties of solvent vapor and air are given in Table 13.1.

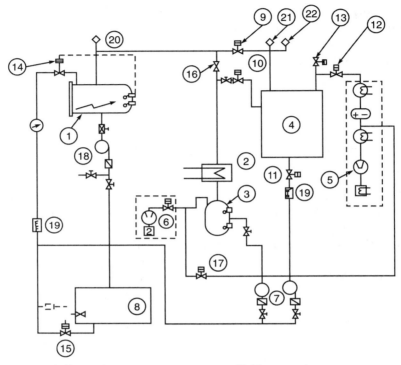

1. Evaporator
2. Condenser
3. Collecting tank
4. Vacuum tank of transformer
5. Vacuum pumping group
6. Vacuum pump
7. Conveying pumps for solvent
8. Storage tank
9. Solvent vapor return valve
10. Solvent vapor inlet valve
11. Valve for condensate
12. Vacuum valve
13. Aeration valve
14. Filling valve for evaporator
15. Stop cock
16. Bypass valve
17. Reversing valve
18. Drainage pump
19. Filters
20. Pressure switch-1
21. Pressure switch-2
22. Pressure switch-3

Figure 13.5 *Schematic of vapor drying system.*

The following solvents meet the above characteristics and are normally used in vapor phase drying systems.

Shellsol H (Shell)

Somenter T (Esso)

Varsol 60

Varsolene 60

Essovarsol 60 E

Drying and Impregnation

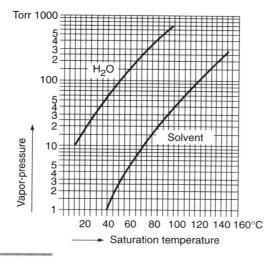

Figure 13.6 *Vapor pressure curve of solvent and water.*

➤ **Table 13.1** Physical Properties of Solvent Vapor and Air

Sl. No.	Physical properties	Hydrocarbon solvent	Air
1.	Specific density	0.785 g/cm³ (liquid)	1.25 kg/m³ (gaseous)
2.	Molecular weight	160	29
3.	Heat of Vaporization	306×10^3 W s/kg	—
4.	Specific heat	2.09×10^3 W s/kg°C (liquid)	1×10^3 Ws/kg°C (gaseous)
5.	Inlet temperature in vacuum vessel	130°C	110°C
6.	Outlet temperature from vacuum vessel at start of heating	90°C	90°C
7.	Vapor pressure at 130°C	140 torr	—
8.	Energy provided per mole	62.7×10^3 W s	581×10^3 W s
9.	Energy provided per mole	179 m³ at 130°C	31.4 m³ at 110°C
10.	Energy released per m³	351×10^3 W s	18.4×10^3 W s

13.3.2 Requirements for Plant and Transformer

As the heat carrier vapor used in the vapor phase drying system is inflammable, necessary precautions must be taken to avoid any possibility of explosion. Maximum allowed leakage rate for a vacuum vessel is 15 torr liter/s. Due to this reason, it is recommended that insulation resistance and dielectric dissipation factor of windings be measured during fine vacuum stage, only when the pressure is less than 3 torr. Explosion is likely to materialize only in the event of a sudden, massive leakage of atmospheric air and if temperature and absolute pressure inside the vessel reach to the potential explosion range. Figure 13.7 shows the potential explosion range for a mixture of heat carrier vapor and air.

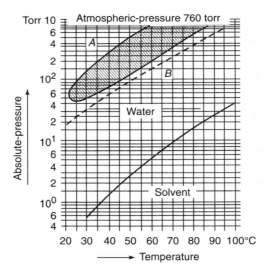

Figure 13.7 *Explosion limits for a mixture of air and solvent. (a) lower limit (b) upper limit.*

Care should be taken to ensure that paints applied on inside and outside surface of tank, end frame, clamp plate, copper bus-bar and other ferrous components are compatible with solvent vapors at 130°C. Since, it is desired that insulation be impregnated just after vacuum drying without exposing the winding to atmosphere for tanking, suitable paints which have no reaction with solvent vapors are to be applied on all parts, instead of conventional paint system. Further, transformer is required to be designed such that outlet for condensed solvent from all possible areas in the tank is available and there is no possibility of large amount of solvent being trapped in intricate insulation arrangement.

13.3.3 Drying Process

Transformer can be dried in a vacuum vessel or by evacuating its tank, provided leakage rate is within required limits specified in Sec. 13.3.2. When the former is followed, core and windings assembly is dried either alone or in its tank. In the latter case, an efficient heat insulated compartment with hot air circulation system is required to be provided to eliminate heat loss during drying and vacuum is drawn in the transformer tank. Sequence of operation for VPD in a vacuum vessel is described in the following paragraphs.

The drying process takes place in four stages indicated below, and as shown in Fig. 13.8.

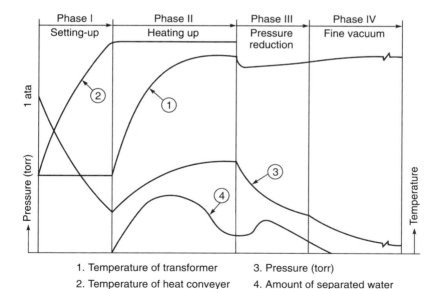

Figure 13.8 *Stages of the vapor phase drying process.*

- (a) Preparation
- (b) Heating up and drying
- (c) Pressure reduction
- (d) Fine vacuum

(a) Preparation (Setting Up)

The entire evaporator and condenser system is first evacuated to an absolute maximum pressure of 5 torr by the leakage air vacuum pump, before drawing the solvent into the evaporator and heating it to the required temperature of 130°C. The vacuum vessel valves remain closed during this operation.

In parallel with the above preparatory steps, the vacuum system evacuates the vacuum vessel containing the core and windings assembly to approximately 5 torr. For draining the condensate from the diverter switch oil compartment of on-load tapchanger, wherever it is processed along with the transformer, the drain plug in the bottom of the compartment is opened. Vessel floor is at a descending slope of 1 : 100 towards the drainage system, so that no condensed solvent remains inside the vessel. If tank containing core and windings assembly is loaded into the vessel on a horizontal trolley, tank is also kept at a slope of 1 : 100 to drain out the solvent from the tank.

(b) Heating Up and Drying

After the vessel is evacuated, vessel heating is started and this heating is continued till the end of fine vacuum phase. The vessel valves are opened at this stage, admitting vapors of heat-carrier in the vessel, most of which condense on the cold surfaces of the transformer. The condensed heat carrier is pumped back to the evaporator through a filter. Heat released by condensation gradually warms up the insulation and mass of component. Heat carrier vapor pressure in the vessel and insulation moisture vapor pressure both increase with rising temperature. As the water vapor pressure is considerably higher than that of the heat carrier, insulation moisture starts to evaporate at a comparatively low insulation temperature. This produces mixtures of water vapor, leakage air and heat carrier vapor inside the vessel, which is conveyed back to the condenser via the vapor return in the condenser, whereas the leakage air discharges to atmosphere through the vacuum pump. The condensed water and heat carrier mixture is sent into the collecting tank, in which its components settle out under gravity. The water gets collected in the bottom of tank due to higher specific weight, which is measured periodically and drained off. Final insulation drying temperature of 120–125°C is maintained for the time required, to ensure full moisture evaporation from the deeper insulation layers. The longer the heating-up phase, the shorter is the fine vacuum phase.

(c) *Pressure Reduction*

The vapor supply remains closed during this stage in which most of the heat carrier absorbed by the insulation re-evaporate, condense out in the condenser and is finally returned to the evaporator. This phase is terminated when an absolute pressure of 15 to 20 torr is reached in the vessel.

(d) *Fine Vacuum*

This is the final drying stage, which comes immediately after the pressure reduction phase. It is the same as conventional vacuum drying (see Sec. 13.2.2). The vessel is evacuated by the main vacuum system to a pressure not exceeding 0.1 torr. This phase is terminated when water extraction rate is below the desired level and insulation resistance and dissipation factor of windings become constant.

After drying of insulation by solvent vapors, other activities like oil impregnation, soaking, draining, retanking, redrying, completion of fittings and oil circulation follow in the same manner as for conventional drying (see Sec. 13.2) with the following exceptions:

(i) Since the transformer is at a temperature of 125°C at the end of V.P.D., it is cooled to a temperature depending upon the pressure in the vessel, such that oil is neither vaporized nor oxidized during oil impregnation.

(ii) If only core and windings assembly is dried without tank, the assembly is taken out from the vessel and is immediately loaded into the tank. The tank is again kept in vessel and oil is filled after evacuating vessel to the desired level and removing any moisture absorbed by the insulation due to exposure to the atmosphere.

13.3.4 Drying of Oil-impregnated Insulation

Vapor phase drying (V.P.D.) is ideal for drying transformers from site after repairs, whose windings are exposed to atmosphere for a longer period, since heat carrier washes the deteriorated insulating oil out of the insulating materials. The high diffusion coefficient of non-impregnated insulation is restored which is 20 to 30 times that oil-impregnated insulation. Hence, total drying cycle is considerably reduced and windings are cleaned of all sediment, fibers and other impurities.

13.3.5 Paper Depolymerization

Detailed comparative measurements on paper specimens under process conditions of V.P.D. and conventional drying have proved that depolymerization of paper is of the same order in these two processing methods, as shown in Figs. 13.9–13.11. Short processing times seem to be the major factor in reducing overall polymerization in V.P.D. even though drying temperature is 130°C.

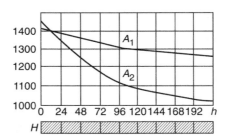

Figure 13.9 *Depolymerization of a paper for hot air drying at atmospheric pressure.*
A_1 air temperature 90°C
A_2 air temperature 110°C
H heating

13.3.6 Advantages of Vapor Phase Drying

Some of the advantages in the use of vapor phase drying of transformers over the conventional system are as under:

(a) Uniform heating of the entire mass by penetration of the hot solvent vapor ensures moisture extraction from the innermost parts of the insulation, e.g., from precompressed laminated pressboard clamping rings and crepe paper insulation over HV leads, etc.

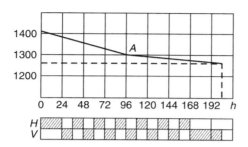

Figure 13.10 *Depolymerization of paper for hot air vacuum drying vacuum tank. H heating V vacuum A air temperature 95°C.*

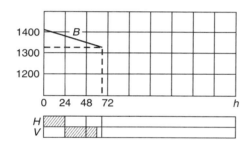

Figure 13.11 *Depolymerization of paper for vapor phase drying.*

(b) Washing action resulting from the condensation of the heat carrier vapor helps in removal of dust, dirt and fibers from the innermost spaces in a transformer, and as such oil is not contaminated and oil circulation time is considerably reduced.

(c) Ageing of insulation due to depolymerization is comparable or even less than that which occurs in the conventional drying.

(d) V.P.D. is more effective for drying transformers from site after repairs.

(e) Drying time cycle before oil impregnation is reduced to the extent of 30–40% of the usual system, thereby reducing the transformer manufacturing cycle.

(f) Rapid buildup of temperature without consequent damage to insulation due to more uniform distribution of temperature in various parts of the transformer, thereby reducing the heat up time.

(g) Effective moisture extraction due to the deeper penetration of hot solvent vapors and also higher water vapor pressure differential (Refer to Fig. 13.6).

REFERENCES

1. Mosser, H.P. et al., *Transformer Board Special Print of Scientia Electrica*, Translated into English by H. Weidmann, EHV. Weidmann Limited, St. Johnsbury, Vermont, U.S.A., 1979.
2. *Micafil News*, Vapor–phase drying of EHV Transformer, MNV 46/le November 1977, Micafil Limited 8048, Zurich, Switzerland.
3. D.P. Gupta et al., Vapour Phase Drying System for Transformers, *BHEL Journal*, Vol. 3, No. 1, 1978.
4. *IEC Publication*, 422–1973, "Maintenance and Supervision Guide for Insulating Oils in Service."

CHAPTER **14**

Testing of Transformers and Reactors

P.C. Mahajan
M.L. Jain
R.K. Tiwari

Testing is an important activity in the manufacture of any equipment. While certain preliminary tests carried out at different stages of manufacture provide an effective tool which assures quality and conformation to design calculations, the final tests on fully assembled equipment guarantee the suitability of the equipment for satisfactory performance in service. The basic testing requirements and testing codes are set out in the national and international standards. This chapter, however, is intended to cover the purpose and the methodology of performing the tests.

With a view to cover detailed information about impulse and partial discharge tests, which are of great importance, separate sections have been devoted to these tests. A separate section has been specifically devoted for short-circuit testing of transformers which is a special test having significance in the transformer's reliable service. Also, the specific requirements of reactor testing have been dealt with in a separate section. All other tests and temperature-rise test on power transformer are described in the first section. Thus, the chapter has been divided into following five sections:

    Section I : Testing of power transformers
    Section II : Impulse testing
    Section III : Partial discharge testing
    Section IV : Testing of reactors
    Section V : Short-circuit testing of transformers.

SECTION I

14.1 Testing of Power Transformers

Preliminary tests are carried out on the transformer before it is put into the tank. Final tests are carried out on completely assembled transformer.

14.1.1 Preliminary Tests

Following tests are carried out in the works at different stages, before the core and coil assembly of the transformer is placed in its tank. These checks help in detecting any fault at an early stage.

(a) Core Insulation

After the core is assembled, a 2 kV test is done to ensure that the insulation between clamp plates, core bolts and core is adequate.

(b) Core Loss Test

This is conducted on the core assembly to ensure its soundness. Some turns are wound over the core and it is energized at normal flux density. Core loss and magnetizing current are noted and compared with design value.

(c) Check of Ratio, Polarity, Vector Relationship and Winding Resistance of Transformer Assembly

Ratio test is conducted to ensure the correctness of voltage ratio between different windings on each tapping. The tolerance allowed for ratio is ±0.5% of the declared ratio or ±10% of the percentage impedance voltage, whichever is smaller. The latter tolerance limit is not applicable for auto-transformers and booster transformers, where impedance value is small. In order to get accurate ratio, a ratiometer is employed. It also indicates the polarity of transformer windings.

For a three-phase transformer, it is more usual to carry out a vector relationship test, in which one of the high-voltage and low-voltage line terminals are jointed together as shown in Fig. 14.1. Three-phase 400 V supply is connected across high voltage line

Testing of Transformers and Reactors

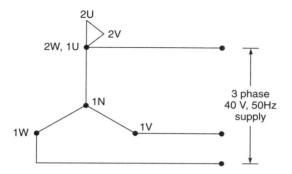

Figure 14.1 *Vector relationship test for star-delta (YN, d11) connected step-down transformer.*

terminals. Voltages between terminals 2U–1N, 2V–1N, 2W–1N, 2V–1V, 2W–1W and 2V–1W are measured.

For YN, d11 vector relationship

    2U–1N > 2V–1N > 1U–1N

and   2V–1W > 2V–1V or 2W–1W

The vector relationship for any other group can be checked in a similar manner.

The dc resistance of each winding is measured by Kelvin's double bridge to check that there is no faulty joint.

(d) Preliminary Load Loss and Impedance Voltage Measurement

Preliminary load loss and impedance measurements at reduced current are carried out to ensure that these are within guaranteed limits.

14.1.2 Final Tests

The completely assembled transformer is tested in accordance with the International Standards. The tests comprise the following:

(a) Routine tests

  (i) Measurement of winding resistance
  (ii) Measurement of voltage ratio and check of voltage vector relationship
  (iii) Measurement of impedance voltage (principal tapping) short circuit impedance and load loss

(iv) Measurement of no load loss and current
(v) Measurement of insulation resistance
(vi) Dielectric tests
(vii) Tests on-load tapchangers (where appropriate)

(b) *Type tests*
All the tests listed as above and the temperature rise test and Dielectric type tests.

(c) *Special tests*
(i) Dielectric special tests.
(ii) Measurement of zero-sequence impedance of three-phase transformers
(iii) Short-circuit test
(iv) Measurement of acoustic noise level
(v) Measurement of harmonics of the no load current
(vi) Measurement of power taken by the fans and oil pumps
(vii) Measurement of capacitances between windings to earth and between windings.
(viii) Measurement of transferred surge voltage on low voltage windings.
(ix) Measurement of insulation resistance to earth of the windings, or measurement of dissipation factor (tan delta) of the insulation system capacitances.

(a) Routine Tests

(i) Measurement of winding resistance. For calculation of I^2R-losses in the winding, it is necessary to measure dc resistance of each winding. The resistance measurement should be done after the direct current circulating in the winding has reached a steady state. In some cases this may take several minutes depending upon the winding inductance.

Temperature of the winding must be stable and for this reason, this test is carried out usually before load loss measurement. The average oil temperature is determined as the mean of top and bottom oil temperatures and is taken as average winding temperature.

(ii) Measurement of voltage ratio and check of voltage vector group. These tests are conducted as described in para. 14.1.1 (c) to check that all connections to the bushings, tapchangers, etc., have been made correctly during final assembly.

(iii) Measurement of impedance voltage and load loss. The load loss comprises the sum of the I^2R-losses in the windings and the stray losses due to eddy currents in conductors, clamps and tank. The stray losses vary with frequency. It is thus important to give supply to the transformer at the rated frequency. The load loss and impedance voltage are guaranteed at 75°C but are measured at the ambient temperature of the test room. Measured load loss is corrected to reference temperature (75°C). It is known that I^2R-losses are proportional directly to resistance, which varies with the temperature. The stray losses vary inversely with the temperature.

The test is carried out by short circuiting, usually the LV winding and by supplying the impedance voltage to HV winding. The measured power will also include small core-loss. Since the supply voltage during the test is a small fraction of normal voltage, this loss can be ignored. However, for a high impedance transformer (i.e., $Z >$ 15%) The core-loss may become appreciable and can be deducted after separately measuring it at impedance voltage.

As per IS : 2026 (Part I)—1977, the measurements can be made at any current between 25 to 100%, but preferably not less than 50% of the rated current. Load loss and impedance voltage can be corrected for rated MVA as below:

Computed loss at rated current

$$= \text{Measured loss at test current} \times \left(\frac{\text{Rated current}}{\text{Test current}}\right)^2$$

$$\% Z = \frac{\text{Test voltage}}{\text{Rated voltage}} \times \frac{\text{Rated current}}{\text{Test current}} \times 100$$

While measuring load loss and impedance at different tap positions, readings should be taken quickly, and the interval between the measurements at different taps should be adequate to avoid significant errors due to momentary temperature rise of the windings. The difference in temperature between the top oil and bottom oil should preferably be small enough to enable the average temperature to be determined accurately.

Three-wattmeter method should be used instead of two wattmeter method to avoid large value of wattmeter multiplier constant.

The power factor during load loss test can be less than 0.1 and wattmeters suitable for such low power factor should be used.

Following tolerances (Table 14.1) are applicable as per IS : 2026, for specified values of losses and impedance voltage.

(iv) Measurements of no-load loss and current. The measurement of no-load loss and current is important not only for the purpose of assessing the efficiency of the transformer, but also as a check that the high-voltage tests have not caused any damage to winding insulation. For large transformers, therefore, no-load loss measurement is carried out before and after completion of the dielectric tests.

> **Table 14.1** Tolerances

S. No.	Item	Tolerance
(a)	Total losses	+10% of the total losses
	(ii) Component losses	+15% of each component loss, provided that the tolerance for total losses is not exceeded
(b)	Impedance voltage at rated current	
	(i) Principal tapping	
	(1) two-winding transformers	±10% of the declared impedance voltage
	(2) Multi-winding transformers	±10% of the declared impedance voltage for one specified pair of windings
		±15% of the declared impedance for a second specified pair of windings
	(ii) For tappings other than the principal tapping	Tolerance shall be increased by a percentage equal to half the difference in tapping factor (percentage) between the principal tapping and the actual tapping

Measurement of no-load loss is carried out at rated frequency feeding usually LV winding. Since the no-load current is very small, the I^2R-losses in the windings will be negligible. The power factor, while measuring no-load loss for medium power transformer, is generally around 0.3. The power factor for large units, especially when working at higher flux densities is about 0.1.

The core-loss consists of hysteresis and eddy current losses. The hysteresis loss is dependent on average value used and the eddy current loss on rms value of supply voltage. Two voltmeters are used during the test, a bridge rectifier type to indicate average voltage, and a dynamometer type to indicate rms voltage. The supply voltage is set so that the specified value is indicated on the average voltmeter. With this, hysteresis component of no-load loss will be measured correctly, while the eddy current loss will be either lower or higher than the true value, depending upon the form factor of the supply voltage. Since the ratio between the two components is known for any particular quality of core steel, losses are corrected by using the following formula:

$$P = \frac{P_m}{P_1 + KP_2}$$

where P = no-load loss for sinusoidal voltage
P_m = the measured no-load loss
P_1 = ratio of hysteresis losses to total iron loss
P_2 = ratio of eddy current loss to total iron loss, and

$$K = \left(\frac{\text{rms voltage}}{1.11 \times \text{average voltage}}\right)^2$$

For normal flux densities and frequencies of 50 Hz and 60 Hz, P_1 and P_2 are each 0.5 for grain-oriented steel, and 0.7 and 0.3 respectively, for non-grain-oriented steel.

(v) Measurement of insulation resistance. Insulation resistance is measured between all windings and the tank with a megger. The insulation resistance varies inversely with the temperature. Thus the oil temperature is also recorded. Sometimes insulation resistance values at 15th and 60th second are noted to determine polarization index of the insulation system.

(vi) Dielectric tests.
(1) *Separate source voltage withstand test.* This test is intended to check the adequacy of main insulation to earth and between windings.

The line terminals of the windings under test are connected together and the appropriate test voltage is applied to them while the other windings and tank are connected together to the earth. The value of test voltage for fully-insulated winding is indicated in Table 14.2. Windings with graded insulation, which have neutral intended for direct earthing, are tested at 38 kV.

The supply voltage should be nearly sinusoidal. The peak value of the voltage is measured and for this a digital peak voltmeter associated with capacitive voltage divider is employed. The peak value divided by $\sqrt{2}$ shall be equal to the test value. The duration of application of test voltage is 60 s.

(2) *Induced over-voltage withstand test.* The test is intended to check the inter-turn and line end insulation as well as main insulation to earth and between windings.

For transformers with uniformly insulated windings, the test voltage is twice the corresponding rated voltage.

For transformers with non-uniformly insulated HV windings, which are usually designed for systems with highest voltage $U_m \geq 72.5$ kV, the values of test voltage are indicated in Table 14.2 (below the dotted line).

For test voltage up to 66 kV, three-phase transformers are generally supplied direct from a three-phase source.

For higher values, it is the usual practice to raise each HV terminal in turn, to specified test voltage by applying single-phase voltage to LV winding. The neutral terminal may be raised to any appropriate voltage, so as voltage per turn of the phase under test shall be at least twice the normal value. In order to avoid core saturation at the test voltage, it is necessary to use a supply frequency higher than the normal. Figure 14.2 indicates a typical induced over-voltage test circuit for a generator transformer.

When the frequency is chosen in the range of 150 to 240 Hz, the capacitive reactance of transformer is appreciably reduced and it draws significant capacitive current at test voltage, which causes heavy loading on the generator set. The loading on the generator can be reduced by connecting a variable reactor across the generator terminals.

Self-excitation of generator and series resonance between the capacitive load and any series inductance in the test circuit must be avoided by proper selection of test frequency. As a precaution, a sphere gap can be connected across the high-voltage winding, so that it will spark-over if any excessive voltage appears on the terminal.

Test duration is determined by the following formula:

$$\text{Test duration in seconds} = \frac{120 \times \text{Rated frequency}}{\text{Test frequency}}$$

> **Table 14.2** Rated Withstand Voltages for Transformer Windings with Highest Voltage for Equipment U_m up to and Including 420 kV

Highest voltage for equipment kV (rms)	Rated short-duration power frequency withstand voltage kV (rms)	
1.1	3	
3.6	10	
7.2	20	
12	28	
17.5	38	
24	50	
36	70	
52	95	
72.5	140	
123	185	230
145	230	275
245	360	395
420	570	630

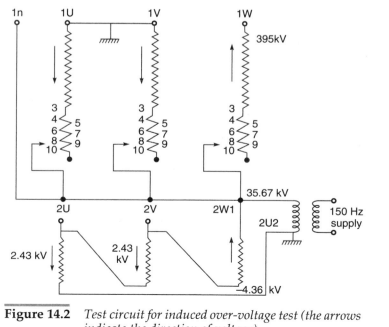

Figure 14.2 Test circuit for induced over-voltage test (the arrows indicate the direction of voltage).

The high voltage is normally measured by means of a digital peak voltmeter associated with capacitive voltage divider.

(b) Type Tests

(i) Temperature rise test. The test is conducted to confirm that under normal conditions, the temperature rise of the windings and the oil will not exceed the specified limit. Temperature rise limits for oil-immersed type transformers are indicated in Table 14.3.

➤ **Table 14.3** Temperature Rise Limits for Oil-Immersed Type Transformers

S.No.	Part	Temperature rise (°C)		Condition
		External cooling medium		
		Air	Water	
1.	Windings	55	60	When the oil circulation is natural or forced non-directed
		60	65	When the oil circulation is forced directed
2.	Top oil	50	55	When the transformer is sealed or equipped with a conservator
		45	50	When transformer is neither equipped with a conservator nor sealed

The temperature rises are measured above the temperature of the cooling air for all types of transformers except those water cooled. In the latter case, the temperature rise is measured above the inlet water temperature.

(ii) Temperature rise for top oil. Normally, LV windings of the transformer under test is short circuited and a voltage of such a value is applied to HV winding that power input is equal to no-load loss plus load loss corrected to a reference temperature of 75°C. For multi-winding transformers, the temperature of the top oil refers to the specified loading combination for which the total losses are the highest. The total losses are measured by the three-wattmeter method in the circuit and maintained constant until the top oil rise has reached a steady value.

If the total losses cannot be supplied due to plant limitations, losses not less than 80% of the total loses are supplied and the following correction factor is applied to top oil temperature rise

$$\left(\frac{\text{Total losses}}{\text{Test losses}}\right)^x$$

The value of x for

  Natural air circulation: 0.8
  Forced air circulation and water cooling: 1.0

During the temperature rise test on an oil-immersed type transformer, hourly readings of the top oil temperature are taken by means of a thermometer placed in a pocket in the transformer top cover. The temperature of oil at inlet and outlet of the cooler bank is also taken hourly and mean oil temperature is determined. Ambient temperature is measured by means of thermometers placed around the transformer at three to four points situated at a distance of 1 to 2 m from and half-way up the cooling surface of the transformer. The thermometers are inserted in oil cups, which are filled with transformer oil.

(iii) Duration of temperature rise test. The test is continued until the requirement of one of the following methods have been met:

Method (a)

The test should not be regarded as complete until the temperature rise increment of top oil is less than 3°C in 1 hour. The method shown in Fig. 14.3 shall be employed for the determination of the final oil temperature rise.

Method (b)

It should be demonstrated that the top oil temperature rise does not vary more than 1°C per hour during four consecutive hourly readings. The last reading is taken for determination of top oil rise.

(iv) Winding temperature rise. When top oil temperature rise is established, the current is reduced to its rated value and is maintained for 1 hour to allow the winding to attain normal temperature.

If the rated current cannot be supplied, the tests can be performed with a current not less than 90% of the rated current. Following correction factor is applied to determine the winding gradient corresponding to rated current.

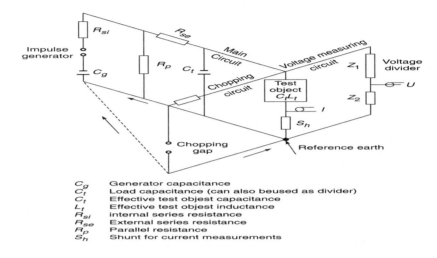

Figure 14.3 *Method for determining the final temperature rise of oil.*

$$\text{Winding gradient} = \text{Test gradient} \times \left(\frac{\text{Rated current}}{\text{Test current}}\right)^y$$

The value of y for

        natural and forced non-directed oil circulation   : 1.6
        forced-directed oil circulation   : 2.0

At the end of the test, the supply is switched off. The cooling fans or water pumps should be stopped but the oil pumps should remain running. The short-circuit connection is removed. The value of hot resistance of the winding is measured by Kelvin's double bridge or by Tettex resistance measuring equipment. A certain time, about 3 to 4 minutes, usually elapses between switching off the power supply and taking the first reading, during which the resistance of winding will be decreasing. In order to determine the temperature of winding at the instant of power switch-off, the resistances are measured at intervals over a period of about 15 min. Graph of hot resistance versus time is plotted, from which winding resistance (R_2) at the instant of shutdown can be extrapolated in the manner shown in Fig. 14.4.

From this value, θ_2, the winding temperature at the instant of shutdown can be determined as below:

$$\theta_2 = \frac{R_2}{R_1}(235 + t_1) - 235$$

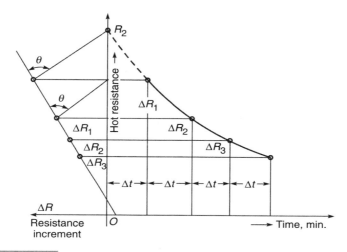

Figure 14.4 *Method for determining the winding resistance at the instant of switching off the supply.*

where, R_1 is the cold resistance of winding at temperature t_1, 235 being reciprocal of the temperature coefficient for copper. Winding temperature rise can be determined by subtracting mean ambient temperature from θ_2 and adding the drop in mean oil temperature rise, if any, from steady-state condition (that is, before current is reduced to rated value) to shut-down condition.

(v) Lightning impulse test. It is dealt with in Sec. II.

(vi) Switching impulse test. It is dealt with in Sec. II.

(vii) Partial discharge test. It is dealt with in Sec. III.

(c) Special Tests

(i) Measurement of zero-sequence impedance of three-phase transformer. The zero-sequence impedance is measured on star-connected windings, which have an earthed neutral, in order to determine the current which will flow in the event of a line-to-earth fault. Reluctance path for zero sequence flux is different in three-phase three-limb core and three-phase five-limb core and hence the value of zero sequence impedance depends upon type of core used in a transformer. Usually the value of zero sequence impedance lies between 80–90% of positive sequence impedance for transformer having three-limb core, whereas it is between 90–100% of positive sequence impedance for transformer having five-limb core construction.

Figure 14.5 shows the connection for carrying out this test on a delta-star connected transformer. The line terminals on the star connected windings are joined together and single-phase supply is applied between these and the neutral point, the delta terminals being left floating during this test.

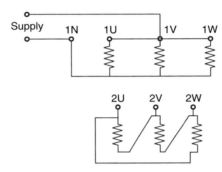

Figure 14.5 *Test circuit for the measurement of zero sequence impedance on a star-delta transformer.*

For transformers with more than one star-connected windings and neutral terminal, additional measurements of zero-sequence impedances are made, in which the line terminals and the neutral terminal of the other star-connected winding are connected together.

Auto-transformers with a neutral terminal are treated as normal transformers with two star-connected windings.

Zero-sequence impedance in ohms can be obtained as below

$Z_0 = 3$ V/I where V is the test voltage and I is the test current.

(ii) Short-circuit test. It is dealt with in Section V.

(iii) Measurement of the harmonics of no-load current. The harmonics of the no-load current in all the phases are measured by means of harmonic analyzer at rated voltage and the magnitude of the harmonics is expressed as a percentage of the fundamental component.

(iv) Measurement of acoustic sound level. Acoustic sound test measures the average sound level generated by the transformer, when energized at rated voltage and frequency at no-load with cooling fans and pumps in operation. The main sound comes from the core vibration, since there are no moving parts in the transformer. Noise level measurements are taken in accordance with NEMA-TR1. The measurement details are covered in Sec. IV.

SECTION II

14.2 Impulse Testing

Lightning is probably the most common cause of flashover on overhead transmission lines. Two mechanisms can be distinguished. In the first, the lightning stroke makes a direct contact with a phase conductor producing a voltage on the line in excess of the impulse voltage level and in the other, the stroke makes contact with an earth wire or tower and the combination of tower current and tower impedance produces a voltage near the tower top sufficient of precipitating a so-called back flashover. The terminal equipments of high-voltage transmission lines experience lightning impulses in service.

Insulation is one of the most important constituents of a transformer. Any weakness of insulation may result in failure of a transformer. A measure of the effectiveness with which insulation performs is the dielectric strength. The power frequency tests alone are not adequate to demonstrate the dielectric strength of transformers. The distribution of impulse voltage stress along the transformer winding is different from the power-frequency voltage distribution. The power-frequency voltage distributes itself throughout the winding on a uniform volts per turn basis. Impulse voltage is initially distributed on the basis of the winding capacitance and finally the inductance. Thus, it is necessary to ensure adequate dielectric strength of transformers for impulse conditions.

Switching impulses occur during all kinds of switching operations in the system, for example by switching a transformer "on" or "off" a system, switching of distribution line, etc. The magnitude and form of impulses produced differ from case to case. The magnitudes of switching impulses occurring in the network are proportional to the network voltage. The maximum voltage can be about 3.5 times the service voltage (the ratio of the maximum switching impulse voltage to the maximum peak phase-to-earth service voltage). This ratio is as low as 2.1 on some EHV systems. The need for switching impulse tests is based on the possibility that switching impulses in service can cause insulation damage to a transformer designed for a greatly reduced insulation level. The switching impulse tests are worth- while for transformers with reduced insulation level. Studies and researches on insulation arrangements have shown that a transformer designed without taking switching impulses into account can withstand switching impulse voltages equivalent to 83% of the corresponding basic impulse level (BIL).

14.2.1 Lightning Impulse

Impulse tests are made with waveshapes which simulate conditions that are encountered in service. From the data compiled about natural lightning, it has been concluded that system disturbances from lightning can be represented by three basic wave shapes—full waves, chopped waves and front of waves. It is also recognized that lightning disturbances do not always have these basic wave shapes. However, by defining the amplitude and shape of these waves, it is possible to establish a minimum impulse dielectric strength that a transformer should meet.

If a lightning disturbance travels some distance along the line before it reaches a transformer, its waveshape approaches that of the full wave (Fig. 14.6). It is generally referred to as 1.2/50 wave. A wave travelling along the line might cause flashover across an insulator after the peak of the wave has been reached. This wave is simulated by a chopped wave which is of the same magnitude as the full wave, as shown in Fig. 14.6. If a lightning stroke hits directly at or very near to a transformer terminal, the impulse voltage may rise steeply until it is relieved by a flashover, causing sudden, very steep collapse in voltage. This condition is represented by the front of wave (Fig. 14.6).

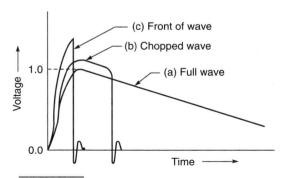

Figure 14.6 *Lightning impulse wave shapes.*

These three waves are quite different in duration and in rate of rise and decay of voltage. They produce different reactions within the transformer winding. The full wave, because of its relatively long duration, causes major oscillations to develop in the winding and consequently produces stresses not only in the turn-to turn and section-to-section insulation throughout the winding, but also develops relatively high voltages compared to power frequency stresses across the large portions of the winding and between the windings and the ground.

The chopped wave, because of its shorter duration, does not allow major oscillations to develop fully and generally does not produce as high voltages across the large portions of the windings. However, because of the rapid change of voltage following flashover, it produces higher turn-to-turn and section-to-section stresses.

The front of wave is still shorter in duration and produces still lower winding-to-ground voltages within the winding. The rapid change of voltage on the front followed by a flashover produces high turn-to-turn and section-to-section voltages very near the line-end of the winding. The front-of-wave test is not carried out on transformers but is a very important test for lightning arresters.

14.2.2 Switching Impulse

Studies on network models and practical measurements in the network regarding the form and magnitude of unipolar voltages due to switching operations reveal that these over-voltages have front times of several hundred microseconds. The form of over-voltage may be a periodically damped or oscillating one.

The IEC-60060 have adopted, for switching impulse test, a long wave having front time 250 µs and time to half value 2500 µs with tolerances. The American Standard recommends the duration of voltage wave at 90% of the crest value of at least 200 µs. The actual time to crest shall be greater than 100 µs and the time to first zero-voltage on the wave tail shall not be less than 1000 µs, except where core saturation causes the tail to become shorter. In view of the difficulties in generating these wave-shapes in laboratories and more to minimize the core saturation during testing of transformers, IEC-60076 Pt. III has specified a switching impulse wave having a virtual front time of at least 20 µs, duration above 90% of the specified amplitude of at least 200 µs and a total duration to the first zero passage of at least 500 µs. The time to crest has been chosen so that the voltage distribution within the winding will be essentially uniform.

The dielectric strength of air, particularly for a positive unipolar impulse is dependent on the front time of these unipolar impulses, and feature a pronounced minimum in the range of 100–300 *µs*. Therefore, to avoid flashover on the external insulation of the transformer, it is preferred to conduct the test with negative polarity impulse.

The switching impulse voltages are of long duration and generally do not cause any non-uniform voltage distribution along the winding, but they affect the insulation to ground, between windings and phases and the external insulation of the bushings. During the switching impulse test, the open circuited low-voltage winding on the same core limb, as the impulsed HV winding, is automatically subjected to impulse voltage stresses. These voltages are induced proportional to the transformation ratio and the flux distribution. Further tests on low-voltage windings are thus not essential.

14.2.3 R.S.O. Techniques

It is not possible to measure the voltage in different parts of a coil during the transformer impulse test. The behavior of windings can be analytically calculated with the use of a computer; alternatively, the voltage distribution in windings can be obtained with recurrent surge oscillograph techniques. The recurrent surge generator is a low-voltage equivalent of a high-voltage impulse generator with lumped parameters. Repetitive pulses of low voltage are applied to the winding terminal from a recurrent surge generator. The voltages induced and their shape in different parts of the winding and other windings can be measured and recorded with an oscilloscope. The recurring impulse wave gives a completely still picture on the oscillograph screen. The applied waveform to give desired front and tail duration can be adjusted. It is thus possible to measure the voltage distribution in the winding and the voltage differential between the windings, thus allowing a suitable choice of the insulation. The R.S.O. test, being a low-voltage test, allows the study of behavior of windings exposed to atmospheric surges. The R.S.O. test is also done on completely assembled transformer to find out circuit parameters for generation of desired impulse waveshapes.

14.2.4 Generation of Impulses and Test Circuit

Lightning and switching impulse test is conducted by means of an impulse generator by direct voltage application on the winding under test.

Impulse waves are generated by an arrangement that charges a group of capacitors in parallel and then discharges them in series. The magnitude of the voltage is determined by initial charging voltage, the number of capacitors in series at discharge and the regulation of the circuit. The waveshape is determined largely by the constants of the generator and the impedance of the load. The impulse generators built according to well known Marx multiplier circuit are capable of generating different unipolar over-voltages from several microseconds to a few hundred microseconds. For inter-phase testing of EHV transmission towers and external insulations of EHV substations with switching impulses of very long wave front from 1000 to 5000 µs, the conventional impulse generation technique becomes less effective due to problems in triggering all generator stages, arc extinction in generator spark gaps, etc. These waves can also be generated using cascade connected testing transformers by either discharging a capacitor bank through the primary winding of the transformer or by energizing the transformer during a short time period, typically less than one-half cycle of the power frequency.

Principles of Waveshape Control

The impulse waveshape is influenced by the generator capacitance (C_g), internal and external series resistances (R_{si} and R_{se}), parallel resistance (R_p), loaded capacitance (C_i) (including capacitance of the voltage divider, the chopping equipment, if used) and the effective capacitance and inductance of test object (C_t, L_i). The simplified circuits for the control of the wave front and wave tail durations for impulse test are given in Fig. 14.7.

The front time will be approximately

$$T_1 \simeq 3 \cdot R_s \cdot \frac{C_g \cdot C_i}{C_g + C_i} \tag{1}$$

For $C_g \gg C_i$,

$$T_1 \simeq 3 \cdot R_s \cdot C_i \tag{2}$$

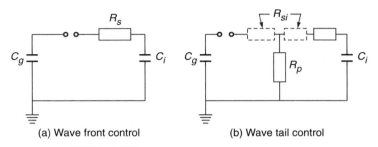

Figure 14.7 *Wave shape control circuits.*

The time to half value will be approximately

$$T_2 \simeq 0.7 (R_s + R_p)(C_g + C_i) \quad (3)$$

for $R_s \ll R_p$ and $C_g \gg C_i$,

$$T_2 \simeq 0.7 R_p \cdot C_g \quad (4)$$

For purely capacitive loads, the front and tail durations can be adjusted according to the above equations. However, in the case of transformer, the effective transformer capacitance C_t included in the total load capacitance is a different physical quantity for front and tail considerations. For the front time, C_t is defined as $C_t = \sqrt{C_s \cdot C_{g'}}$, where C_s is the winding series capacitance and $C_{g'}$ is the winding earth capacitance. For the wave tail, C_t is defined as $C_t \simeq C_{g'}$.

14.2.5 Lightning Impulse Test Circuit

For testing of transformers, the basic arrangement of test circuit has been shown in Fig. 14.8. The physical arrangement can be sub-divided into three major circuits:

— The min circuit, which includes the impulse generator, wave-shaping components and the test object,
— The voltage measuring circuit, and
— The chopping circuit, where applicable.

The standard lightning impulse is a full lightning impulse having a virtual front time of 1.2 μs and a virtual time to half-value of 50 μs with tolerances of ±3% on peak value, ±30% on front time and ±20% on time to half-value. While testing high voltage windings of large transformers, viz. 400 kV, 250 MVA, it becomes difficult to obtain the wave

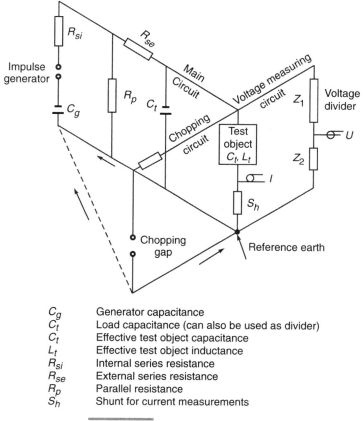

C_g	Generator capacitance
C_t	Load capacitance (can also be used as divider)
C_t	Effective test object capacitance
L_t	Effective test object inductance
R_{si}	Internal series resistance
R_{se}	External series resistance
R_p	Parallel resistance
S_h	Shunt for current measurements

Figure 14.8 *Impulse test circuit.*

front within the specified limits. This is due to the large capacitance of the transformer, the inherent self-inductance of the generator and the leads connecting the generator and transformer. On low-voltage winding of these transformers, it becomes difficult to obtain the specified wave tail due to extremely low impedance of these windings.

The effective impedance of transformer controls the wave tail duration. The effective impedance can be varied by different terminal impedances at the non-impulsed winding terminals. If they are isolated, the effective impedance will be maximum. However, they cannot be left isolated because high voltages are induced in these windings. It is desirable that the voltage on terminals that are not being tested is limited to 75% of its BIL. To ensure this, non-impulsed terminals of the windings are earthed directly through resistors (maximum up to 400 Ω).

The full-wave test sequence consists of application of a reduced full-wave and three full waves. The dielectric stress distribution in a transformer under impulse test depends on the tapping connection and the design in general. Thus, for lightning impulse test, different tappings are selected for the tests on the three phases of a three-phase transformer or the three single-phase transformers of a three-phase group, for example, the two extreme tappings and the principal tapping.

14.2.6 Test with Impulse Chopped on the Tail

Stresses set up in windings by chopped waves are quite different from those set up by full waves. The chopped wave produce greater stresses to the line portion of the winding, when compared to a full wave due to the fast rate of collapse of voltage during chopping.

Different times to chopping will result in varying stresses in different parts of the winding, depending on the winding construction and arrangement. It is thus not possible to state a time to chopping, which is most onerous either in general, or for any particular transformer, or reactor. The time to chopping between 2 and 6 μs is considered to provide severe line-end stresses in the transformer winding and has been specified for withstand test.

When measuring chopped impulses, the measuring system should be accurate to record rate and duration of the voltage collapse and the amplitude of reversed polarity. During the chopping, both the rate of collapse and the amplitude of opposite polarity are largely dependent on the geometrical arrangement of the chopping circuit and on the impedance of the circuit. The IEC 60076–3 and other standards specify the amount of overswing to opposite polarity to be not more than 30% of the amplitude of the chopped impulse. The triggered type of chopping gap provides better consistency of the time to chopping. The detection of failure during chopped wave tests is more difficult than with full waves. Chopped wave impulses do not give identical oscillations after chopping, unless the time to chop are identical. The deviations on the wave shape before the chopping give more positive indication of abnormality. However, the failure during the chopped wave test are reflected on application of full waves, immediately after the chopped wave and these current and voltage oscillograms are compared with full wave oscillograms before chopped wave application.

The standard sequence of application is:

— one reduced full impulse,
— one 100% full impulse,
— one reduced chopped impulse,
— two 100% chopped impulses,
— two 100% full impulses.

14.2.7 Switching Impulse Test of Transformer

The test circuit of direct application of switching impulses on the HV winding is shown in Fig. 14.9. Unlike in the case of lightning impulse test where the windings not being tested are grounded, for switching impulse test these windings are left open circuited as the loading impedance at short-circuit is low and prevents the application of switching impulse voltages of the required duration and magnitude. The iron core becomes magnetized due to long duration switching impulses. The application of successive impulses causes a progressive increase in the residual flux in the core. Sounds, which, under normal impulse tests would be related to defects in transformers may occur during switching impulse tests due to magnetostriction.

The front of the wave depends mainly on the value of series resistance R_{se}, the effective capacitance to ground of tested winding and the divider capacitance. As long as the core is unsaturated, the current flowing through the winding remains relatively low. After the peak of the wave is reached, the electric charge is practically conducted only through the parallel resistance R_p. The test voltage remains practically constant. Thus,

$$U = \frac{d\phi}{dt} \quad (14.1)$$

where, U = test voltage
 ϕ = magnetic flux
 t = time
 for $\phi < \phi_s$ (saturating flux)

$$\phi = U \cdot t \quad (14.2)$$

where, U = peak value of the test voltage.

314 *Transformers*

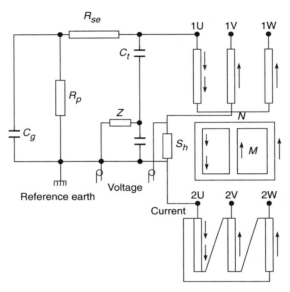

C_g = Generator capacitance	S_h	= Shunt for current measurement
C_p = Parallel resistance	U, V, W	= High voltage terminals
C_{se} = Series resistance	N	= Transformer neutral
C_t = Capacitive voltage divider	1U, 1V, 1W	= Low voltage terminals
Z = Surge impedance	M	= Iron core

The arrows indicate both direction and magnitude (one arrow is equivalent to half to half the value) of the magnetic flux and the respective voltage

Figure 14.9 *Circuit for switching impulse test on a transformer.*

The flux ϕ increases linearly and reaches after a certain time t_s, the value ϕ_s, the saturation flux for the corresponding test voltage. The impedance of the winding and consequently the test voltage collapses at saturation. The current in the winding increases rapidly and reaches the peak value when voltage reaches zero. The higher the test voltage, the faster is the saturation point reached and consequently shorter becomes the wave duration. Core saturation is, therefore, the limiting factor for the wave duration. Even a large source of energy (i.e., high energy rated impulse generator) cannot prolong the duration of the wave. Biasing magnetically the core in the opposite direction helps in increasing the wave duration considerably.

From Eq. (14.1)

$$\int U \, dt = \int_{\phi_1}^{\phi_2} d\phi = \phi_2 - \phi_1$$

Taking ϕ_2 as saturation flux and ϕ_1 as residual flux, the difference is the factor which determines the area enclosed in the total wave between the voltage trace and the time axis. For any particular test voltage, the wave duration is proportional to the difference between the saturation flux and residual flux. Thus, a residual flux of opposite polarity can increase the wave duration of that particular test voltage. This is accomplished by application of opposite polarity switching impulse of reduced amplitude (Say 50 to 60 %).

One reduced full-wave (75%) and two full-waves of rated voltage are applied on the winding under test. For fault detection, the recording of voltage wave is done, neutral current can also be recorded. Exact comparison of oscillograms is not possible as both the voltage and current waves change in magnitude and duration.

14.2.8 Measurement and Recording of Impulses

To measure the amplitude and shape of the applied impulses which have values ranging from a few tens to over thousands of kV and duration 0.2 to 250 μs for the peak, special measuring equipments are used. Oscillographs with high writing speeds and good accuracy and voltage dividers with response time suitable for extremely fast transients are required. Digital readout impulse peak voltmeters are used for amplitude measurements. There are three basic types of dividers that are suitable for impulse testing. The resistance divider utilizes the principle that the voltages across a resistor varies directly as the resistance, in the capacitance divider the voltage varies inversely as the capacitance. The compensated dividers are the combination of the two. The capacitance and compensated dividers also serve as the load capacitor while generating impulses.

For switching impulse test, capacitive types of voltage dividers are preferred. Resistive voltage dividers have an influence on the efficiency of the circuit and waveshape. They may also be thermally overloaded.

The chopping of impulse wave can be obtained with a rod gap, triggered sphere gap or with a multiple chopping gap. The rod gap and triggered sphere gap lack consistency in chopping duration, which can be obtained with a multiple chopping gap.

Oscillographic Recording

(i) Lightning impulse test. The applied voltage wave and one other parameter, whose choice depends on the selection of method of failure detection, are recorded. For best comparison, oscillograms taken at reduced and full test levels should be recorded to give equal amplitude by the use of attenuators at the oscilloscope.

Recording of Voltage

(a) *Wave shape recording.* The preferred sweep time for the wave front record is 5–10 μs (longer sweep time may be required when testing transformer neutrals) and for wave tail 50–100 μs.

(b) *Test wave recording.* For full waves, the sweep time should not be less than 50 μs and the chopped waves should be recorded at 10–25 μs sweep.

Recording of Current

The impulse currents are normally the most sensitive parameters in failure detection and record of current waves are the main criteria of the test result. The use of more than one recording channel at different sweeps gives better resolution. The criteria for selection of sweep is

(a) to obtain a clear representation of the oscillations including the high frequency components near the front of wave;

(b) to be able to detect any discrepancies occurring late in time.

Recording of current at 10 μs and 100 μs sweep covers the above requirements in general.

(ii) Switching impulse test. During switching impulse test, only the recording of applied voltage is required. The voltage record will indicate any fault developed on winding under test or other non-tested windings.

Recording of Voltage

(a) *Waveshape recording.* For the wave front record, a sweep time 100–200 μs is used. For the wave tail record, by which the time above 90% is determined, a sweep time of 1000–2000 μs is adequate.

(b) *Test wave recording.* The sweep time for test wave recording should be long enough to encompass the first zero passage, generally a sweep time of 1000–3000 μs is used. Logarithmic time sweep is also used.

Recording of Current

A switching impulse current comprises three parts:

— an initial current pulse
— a low and gradually rising value of current coincident with the tail of applied voltage
— a peak of current coincident with any saturation

It is usual to employ the same sweep time as used for voltage record.

14.2.9 Fault Detection

The detection of faults is the most important phase of impulse testing. The detection of failure with CRO is most effective and sensitive. It is based on the fact that an insulation failure will change the impedance of the transformer, causing a variation in the impulse current and in the voltage. Both the voltage and current oscillograms are taken during impulse test. Reduced full-wave oscillogram represents the characteristic waveshape of the transformer winding and circuit at a voltage stress, when transformer is considered to be healthy. Subsequent full-wave oscillograms are compared with reduced wave oscillograms taken as reference. Full-wave oscillograms prior to and after the chopped wave application match when winding is sound.

For failure detection during lightning impulse test, three different currents can be measured and used separately, or in combination as shown in Fig. 14.10.

(a) The neutral current
(b) The capacitively transferred current
(c) The tank current

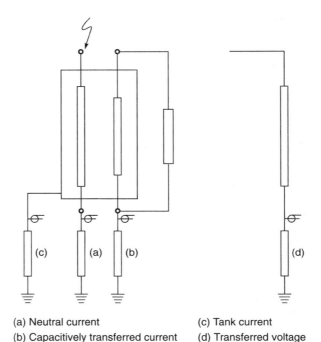

(a) Neutral current
(b) Capacitively transferred current
(c) Tank current
(d) Transferred voltage

Figure 14.10 *Methods of failure detection.*

In addition, the transferred voltages to non-tested windings can also be used for detection of failure.

For failure detection in switching impulse test, normally the measurement of applied voltage is adequate, but the current flowing through the winding to earth can also be used. For recording of current, a non-inductive resistor is used as the impedance for producing the voltage drop.

14.2.10 Interpretation Oscillograms

(a) *Lightning Impulse*

Interpretation of oscillograms is a skilled task and it is often difficult to decide the significance of discrepancies, because of the large number of possible disturbance sources, while investigating discrepancies. First, it should be checked that the test circuit and earthing are not causing disturbances. Sometimes the recording oscillograph generate small disturbances which can be mistaken as a fault inside the transformer.

Secondly, a check should be made of any nonlinear elements within the test object, e.g., tapping protection or core earthing as a source of the disturbance.

Voltage oscillograms

Applied voltage oscillograms are relatively insensitive and indicate major faults in the insulation. However, the indication of the fault in the following cases can be obtained from voltage oscillograms.

- Faults in a major insulation at or near the line end will result in a rapid and total collapse of the voltage.
- A total flashover across the winding under test will result in a slower voltage collapse, normally occurring in steps.
- A flashover across the part of the winding reduces the impedance of the winding, thus resulting in a decrease of the time to half value. Oscillations may also occur on the voltage wave.
- Minor faults such as breakdown of coil-to-coil or even turn-to-turn insulation may sometimes be detected as high frequency oscillations on the voltage oscillogram.

Transferred voltage oscillograms also indicate these faults and the sensitivity of measurement is higher than that of applied voltage.

Current oscillograms

Current oscillograms are the most sensitive means of failure detection. However, this sensitivity is accompanied by the possibility of indicating apparently a number of defects, which may not directly be associated with failure like sparking in the external earth circuit, poor earth connection within the tank, core sparking, etc. Thus, judgment and discrimination are required in the interpretation of current oscillograms to avoid the possibility of a healthy transformer being judged unsound.

Current oscillograms almost always verify and magnify the small disturbances found on the voltage oscillograms. Major changes in current oscillograms indicate probable breakdown within the windings and earth. Small, localized high frequency oscillations spread over two or three microseconds are a possible indication of severe discharges or partial breakdown in the insulation between turns or coils or coil connections. An indication of even a short-circuited turn is obtained in a current trace, as a change in shape or amplitude, which may not be revealed in a voltage oscillogram. A short-circuit means a decrease of the inductance, which results in a change in the current through neutral shunt.

To have better judgment, current oscillograms are recorded simultaneously on two time sweeps, a faster sweep and a slower sweep.

(b) *Switching Impulse*

Voltage oscillograms
All types of faults normally cause a significant change in the voltage wave, either as a temporary dip or more often as a complete collapse of the wave or a shortening of the tail. The voltage records for a switching impulse tests are a sufficiently sensitive means for detection of faults.

The wave tail shortening due to fault in general, is clearly distinguishable from the variation in the wave tail length due to magnetization of the core.

Current oscillograms
The current oscillograms taken on successive impulses may not be comparable, since the voltage waves themselves may not be identical. Current oscillograms should be looked for burst of oscillations, which would occur at approximately the same time as any distortion in voltage wave.

REFERENCES

(For Section II)

1. IEC Standard 60076–3. Bureau Central de la Commission Electrotechnique Internationale, Geneva, Switzerland.
2. IEC Standard 60060–2 and 60060–3. Bureau Central de la Commission Electrotechnique, Internationale, Geneva, Switzerland.
3. IEEE Guide for Transformer Impulse Tests C57.98 IEEE No. 93, June 1968.
4. Feser, K., *Problems Related to Switching Impulse Generation at High Voltages in the Test Plant*—Haefely Publication.
5. Sie, T.H., Switching Surge Tests of Power Transformers—*Bulletin Oerlikon* No. 383/384.
6. Switching Surge Tests for Oil Insulated Power Transformers—IEEE Committee Report, *IEEE Trans.*, PAS 87 No. 2, 1968.

SECTION III

14.3 Partial Discharge Testing

Partial discharge test has recently been introduced as a routine test for transformers and reactors designed for system voltages ≥ 300 kV. This is basically a long duration test intended to check the insulation with regard to voltage under normal operating conditions and temporary over-voltages, generally originating from switching operations and faults (e.g., load rejection, single-phase faults). The magnitude of these over-voltages do not exceed 1.5 p.u. The test proposed in the standards is based on the assumption that a voltage of 1.5 times the rated voltage applied for 30 min is suitable for checking the ability of the insulation to withstand the stresses mentioned above by judging its electrical discharge behavior.

14.3.1 Definition of P.D.

A partial discharge can be defined as localized electrical discharge in insulating media which only partially bridges the insulation between conductors.

Some of the conditions that can initiate partial discharges in transformers are:

— Improper processing or drying of the insulation.
— Over stressed insulation due to a lack of proper recognition of the voltage limitation of the insulation.
— High stress areas in conducting parts, which can be caused by sharp edges on either the conducting part or the ground plane.

These factors may cause ionization in cavities within solid insulation, in gas bubbles in insulating liquids, or along dielectric surfaces. Although, involving small amounts of energy, the partial discharges may lead to progressive deterioration of the dielectric properties of insulating materials.

14.3.2 Theory of P.D. Measurement

Consider Fig. 14.11, the ideal case where a single cavity exists within the dielectric. The cavity can be compared to an air condenser C_2. The

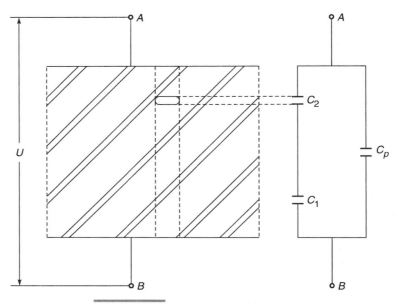

Figure 14.11 *Dielectric with a cavity.*

remainder of the column of the dielectric having same diameter as the cavity is represented by a perfect condenser C_1 in series with C_2. The remainder of the dielectric forming a perfect condenser is represented by C_p.

When a test voltage U is applied, distribution within the insulation will assume its steady-state value $u_1(t)$ and $u_2(t)$ across C_1 and C_2 respectively, according to the relative capacitive coupling conditions. If $u_1(t)$ across C_1 approaches the break down voltage U_z, then the dissipated charge within C_1 will be

$$q_{1E} = C_1 \cdot U_z$$

While C_1 is bridged by the discharge of q_{1E}, C_p will transfer a charge into C_2 through C_1 until charge equilibrium is attained between C_2 and C_p. This charge transfer will cause a voltage drop across the terminals A–B. The charge dissipated within C_1 will then be

$$q = C_p \cdot \Delta U = C_2 (U_z - \Delta U)$$

for $C_p \gg C_1$, ΔU will be $\ll U_z$, hence

$$q = C_2 \times U_z$$

The total charge dissipated within C_1 is then given by

$$q_1 = q + q_{1E}$$

This charge, however, cannot be measured, because it does not pass through connections external to the insulation and further, the values of C_1 and C_2 are not known. The charge, accessible in this condition, is the apparent transferred charge of the insulation, which is determinable by measurement across the terminals A–B. According to the IEC Publication 270, the apparent charge is defined as follows.

"The apparent charge 'q' of partial discharge is that, which, if injected instantaneously between the terminals of the test object, would momentarily change the voltage between its terminals by the same amount as the partial discharge itself."

14.3.3 Principle of Measurement

The absolute value $|q|$ of the apparent charge is referred to as the discharge magnitude and is expressed in coulombs. Figure 14.12 shows a basic partial discharge test circuit.

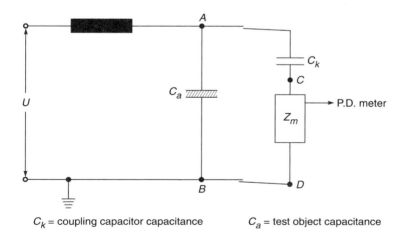

C_k = coupling capacitor capacitance C_a = test object capacitance

Figure 14.12 *Basic partial discharge test circuit.*

On the inception of P.D. in the test object, the voltage across the terminals $A = B$ will be $U - \Delta U$ at the instant $t = 0$, because of the voltage drop ΔU caused by the charge transfer. However, across the capacitor, the voltage at the instant $t = 0$ will still be U, due to the time constant introduced by the series resistance Z_m (measuring impedance).

The charge transferred from C_k via Z_m into C_a, i.e., the measured charge will be

$$q_m = \int_0^\infty i(t)\,dt = \frac{\Delta U}{Z_m} \int_0^\infty e^{-t/\tau} \cdot dt = \frac{\Delta U}{Z_m} \cdot \tau$$

with

$$\tau = Z_m \left(\frac{C_k \cdot C_a}{C_k + C_a} \right)$$

the above equation becomes

$$q_m = \Delta U \cdot C_a \frac{C_k}{C_k + C_a}$$

$$= q \cdot \frac{1}{\left(1 + \dfrac{C_a}{C_k}\right)}$$

∴ The apparent charge

$$q = q_m \left(1 + \frac{C_a}{C_k}\right) = q_m \cdot K$$

The factor K, by which the measured charge has to be multiplied to obtain the apparent charge, is called the correction factor. It is observed that this factor is affected by the circuit characteristics, especially by the ratio of the test object capacitance to that of the coupling capacitor.

For each new test object, the correction factor is determined by calibration of the P.D. instrument with the test object connected in the circuit. A calibration charge of known magnitude q_0 is injected between the terminals A–B and the calibration charge q_{0m} indicated on the P.D. measuring instrument is registered.

The correction factor is then calculated as $K = \dfrac{q_0}{q_{0m}}$.

14.3.4 Test Circuits

The three basic types of test circuits for measurement of partial discharges are described in IEC 60070(2). Each of these circuits mainly consist of

— a test object, which in general can be regraded as a capacitor, C_a
— a coupling capacitor, C_k
— the measuring circuit consisting of a measuring impedance, connecting leads and the measuring instrument (generally some wide band instruments like ERA Model III or V and Nonius differential detector according to Dr. Kreuger).

For high voltage transformers, reactors, etc., usually, the measuring circuit is connected between the bushing tap (when available), and the tank (Figs. 14.13 and 14.14) so that the bushing capacitance acts as the coupling capacitance to the HV terminal. It is sometimes neces-

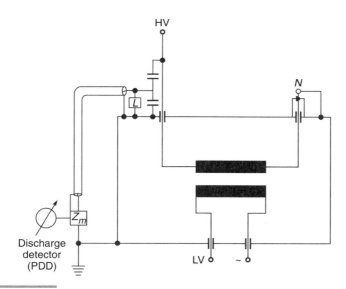

Figure 14.13 *Test circuit for P.D. measurement on transformer at bushing test tap.*

sary to connect an adjustable inductance across the measuring terminals or a matching reactance network, giving maximum response over the measuring frequency band.

14.3.5 P.D. Measurement

Typical circuit arrangement for P.D. test is shown in Fig. 14.15. The connections shown are for the HV line terminal and the neutral

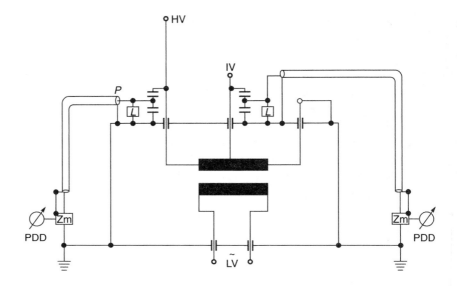

Figure 14.14 *Test circuit for P.D. measurement at two terminals.*

terminal, either during three-phase of single-phase excitation. Figure 14.15(a) and (b) show three-phase and single-phase supply arrangements respectively. The preferred arrangement is to test three-phase transformers by single-phase excitation, so that the discharges are confined to the phase under measurement.

Measurement by multi-terminal method consists in calibrating the complete measuring circuit and measuring the discharges during excitation of the transformer at the required test voltage. The measurements are made at neutral M_1 and at the bushing tap M_2 of the line terminal as shown in Fig. 14.15(c). Measurements at other terminals of the transformer are also done similarly.

14.3.6 Duration of Test and P.D. Values

Figure 14.16 indicates the time sequence of the test voltage, as per IS : 2026.

The test voltages between line and neutral terminals are expressed in P.U. of $U_m/\sqrt{3}$ as follows:

$$U_1 = \sqrt{3} \ U_m/\sqrt{3} = U_m, \text{ the highest voltage for equipment}$$
$$U_2 = 1.5 \ U_m/\sqrt{3} \text{ or } 1.3 \ U_m/\sqrt{3}$$

Testing of Transformers and Reactors

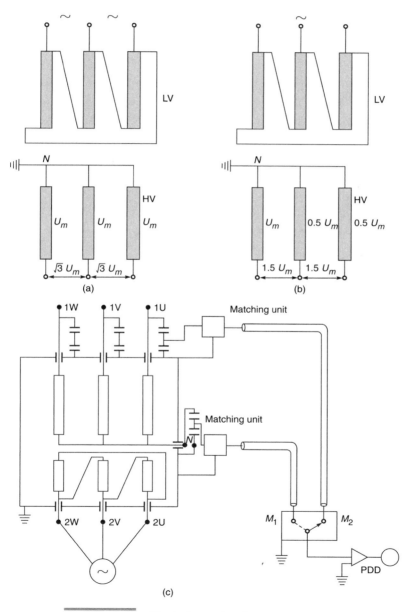

Figure 14.15 *Typical test circuit arrangements.*

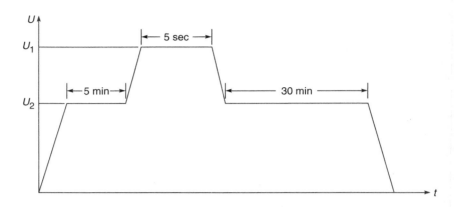

Figure 14.16 *Time sequence for the application of test voltage.*

Throughout the application of test voltage, the value of apparent charge should not exceed the values specified in standards, i.e., at 1.3 $U_m/\sqrt{3}$: 300 pC and at 1.5 $U_m/\sqrt{3}$: 500 pC.

14.3.7 Disturbances and Measurement Precautions

Disturbances are the electrical discharges which do not come from the test object.

The principal forms of disturbances are as follows:

— Interference from the supply mains.
— Interference from the earthing system.
— Pick-up from other high-voltage test or electromagnetic radiation.
— Discharges in the test circuit.
— Contact noise caused by bad contacts in the circuit.

For the correct interpretation of the partial discharge behavior of the test object, it is necessary to ensure that the origin of the measured partial discharges lies in the test object and not in the measuring circuit or the surroundings. To know the disturbances from the measuring circuit, the measurement is taken before applying the voltage. If necessary, proper corrective action is taken to limit the discharge level well below the maximum permissible value for the object under test. The instrument deflection is considered only when the discharge from the test object exceeds the ambient level. When the voltage is raised, the unearthed metallic objects lying in the surrounding area can also cause disturbances, as they behave as floating potential points. It is, therefore, desirable that as far as possible, the surrounding area should be free from loosely-lying metallic parts which are not necessary for the test.

The disturbances caused by electromagnetic fields of rotary machines, rectifiers, radio transmitters, etc., can be effectively checked by proper shielding of the laboratory, which acts as a Faraday cage. This also proves to be a very effective measure for carrying out P.D. measurements, without the need for often very lengthy and troublesome search for causes of disturbances.

Above all, the high voltage connections also should be made discharge free, and the sharp points should be properly shielded using corona control rings or shields.

14.3.8 Location of P.D.

Precise location of P.D. requires great skill and experience. Several methods are in vogue for the location of partial discharges. Some of them are described here briefly.

(a) *Acoustic Detection*

For the location of corona discharges in air, directionally selective microphones with high sensitivity above the audible frequency range are used.

(b) *Visual Detection*

Visual observations are carried out in a dark room with or without the aid of field-glasses of large aperture, Alternatively, a photographic record can be made, but fairly long exposure time is usually necessary. For special purposes, photomultipliers are sometimes used.

(c) *Electrical Location*

The method consists in multi-terminal calibration by injecting known charge between each pair of terminals available (e.g., HV terminal and tank, neutral terminal and tank, HV and neutral terminal, HV and bushing tap). This procedure makes it possible to know the instrument response at different measuring terminals for each simulated P.D. position. A correct evaluation of the P.D. source location can be done by comparing the ratio between two simultaneous readings at different terminals during the HV test with the corresponding ratio obtained during circuit calibration. The closest pair of ratios give the approximate electrical location of the P.D. source.

(d) *Ultrasonic Method*

The methods described above do not enable geometrical location of the source of P.D. For example, a P.D. source, electrically located close to the middle terminal of an auto-connected transformer may be at any place along the connecting leads, between the series and common windings, or at the adjacent winding ends. It, therefore, becomes difficult to ascertain whether the discharges are harmless or they need to be eliminated.

Ultrasonic method facilitates location of the space coordinates of the P.D. source with respect to the transformer tank.

The principle of ultrasonic method of location is very simple. Partial discharges produce high frequency electrical pulses that spread through the windings and can be picked up at the bushing tap almost instantaneously.

Partial discharges, being very similar to small explosions, also produce pressure waves that are propagated in all directions and can be collected by the transducers placed on the outer surface of the tank as shown in Fig. 14.17.

Each transducer picks up the ultrasonic signal with delays, depending on its distance from the P.D. source. The time of propagation of the pressure wave may be assumed as equal to the delay between electric pulse at the terminals and the signal detected by the transducer.

The delays obtained with a few transducers make it possible to trace the origin of the partial discharge by a trigonometric procedure. In principle, the delay of the ultrasonic signal can be observed with an oscilloscope, triggered by the terminal electric pulses. But in practice, such a procedure is difficult, since the ultrasonic signal is mixed with random and spurious signals due to a high partial discharge repetition rate, background noise and transducer vibrations. The electrical pulses appear at time intervals shorter than those necessary for ultra-

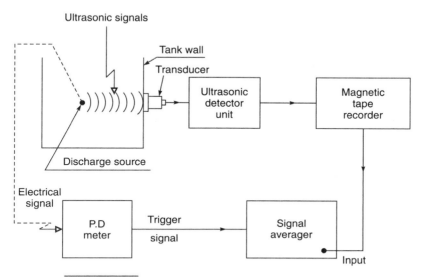

Figure 14.17 *Location technique of discharge source.*

sonic wave propagation and transducer damping. Thus the ultrasonic signal at the transducers is practically continuous (Fig. 14.18).

Moreover, if the discharges have a low amplitude (i.e., low signal-to-noise ratio), the ultrasonic signal may be masked by background noise.

This problem can be overcome by using a signal averager instrument for recording the ultrasonic signals. As the noise is generally of random nature, the summation averaging procedure of signal averager results in signal-to-noise ratio improvement, proportional to the square-root of the number of summations. However, high frequency disturbances may interfere with the electric pulse used for triggering.

14.3.9 Characteristics of Partial Discharges

The characteristics of partial discharges are shown in oscillograms in Fig. 14.19, according to their origin, i.e., partial discharges of a metallic point, partial discharge caused by several metal parts with free potential against each other and partial discharges caused by bad contacts.

332 *Transformers*

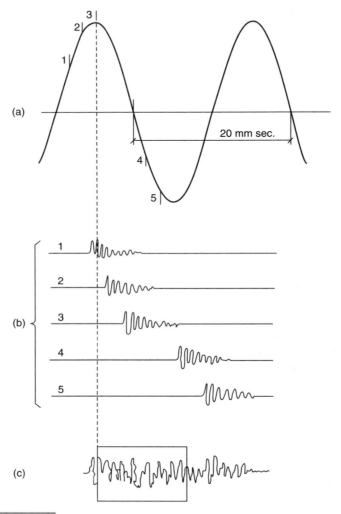

Figure 14.18 *Ultrasonic signal masking due to PD high repetition rate.*
  (a) *test voltage and high frequency pulses*
  (b) *ultrasonic signal produced by single P.D. and relevant delays*
  (c) *their super position at the transducer.*

Testing of Transformers and Reactors

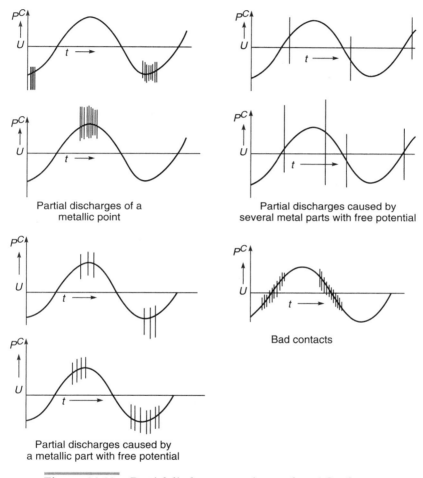

Figure 14.19 *Partial discharge superimposed on AC voltage.*

REFERENCES

(For Section III)

1. *Specification for Power Transformers* IS : 2026 (Part III)—1977.
2. *Partial Discharge Measurements*, IEC Publication 60270.
3. *Methods for PD Measurement*, IS : 6209—1998.
4. Nieschwietz, H., and W. Stein, *Partial Discharge Measurements as Means of Quality Control of High-Voltage Power Transformers*, Conference on Partial Discharge in Electrical Insulation, April 1976, Bangalore.
5. The Prachauser: *Locating Partial Discharges in High Voltage Equipment*, ibid.
6. "Measurement of Partial Discharges in Transformers," *Electra*, No. 19, Nov. 1971, pp 13–65.
7. Recognition of Discharges—*Electra*, No. 11, pp 61–98.
8. Mora, P., and J. Poittevin, *Location of Partial Discharges in the Transformers to be Detected with the Help of Ultrasonic Waves*.

SECTION IV

14.4 Testing of Reactors

The fundamental difference between a transformer and a shunt reactor is that the latter cannot be tested at no-load, like a transformer. The reactor will draw rated power when rated voltage is applied.

Large and expensive test plant is needed for carrying out most of the tests on shunt reactors, especially on large EHV units. Power capacities of the factory test laboratories have in the past, normally been decided by the requirements for power transformer tests, viz. load-loss test and heat-run test. When EHV shunt reactors of large size were designed and manufactured for the first time two decades ago, the test sources were insufficient even for testing the reactors up to rated voltage for short time. This was the situation for large low-voltage reactors also.

Manufacturers of large EHV transformers have, during recent years, increased their test facilities for transformers. This has at the same time increased the resources for reactor testing and special facilities for such tests have also been added.

Recommendations of IEC Publication 60289 *Reactors*, apply to the following types of reactors

— series reactors, such as current limiting reactors and load sharing reactors;
— shunt reactors;
— single-phase neutral earthing reactors and arc-suppression coils;
— three-phase neutral electromagnetic couplers and earthing transformers.

Because of the similarity between power transformers and the reactors, many of the provisions of IEC: 60076 "*Power Transformers*" apply also to these reactors. Tests on reactors comprise the following:

(a) *Routine Tests*

(i) Measurement of winding resistance
(ii) Measurement of reactance

(iii) Measurement of loss
(iv) Inter-turn over voltage withstand test
(v) Separate source voltage withstand test
(vi) Measurement of insulation resistance of windings
(vii) Partial discharge measurement (as applicable)
(viii) Lightning impulse voltage withstand test (routine and type test)
(ix) Switching impulse voltage withstand test (as applicable)

(b) *Type Tests*

In addition to the routine tests, the following type tests are conducted, if specified by the purchaser

(i) Temperature rise test
(ii) Lightning impulse voltage withstand test

(c) *Special Tests*

The following special tests are conducted on the reactors where applicable

(i) Measurement of zero-sequence impedance
(ii) Measurement of vibration and acoustic sound level
(iii) Measurement of third harmonics in phase currents
(iv) Lightning impulse voltage withstand test on neutral
(v) Short-circuit current withstand test—applicable to series reactors and earthing transformers only

While the majority of the tests applicable to different types of reactors are similar in details, a few tests are the particular requirements for particular reactors.

As large EHV shunt reactors pose some problems in testing, the tests generally applicable to shunt reactors have been discussed in detail. Wherever necessary, special notes have been added to bring out important differences in requirements for other types of reactors.

14.4.1 Measurement of Winding Resistance (Routine Test)

This is similar to testing of power transformers and covered in Sec. I.

14.4.2 Measurement of Loss (Routine Test)

The measurement of loss is carried out preferably at rated voltage and frequency. Should this be impracticable, the measurement method is subject to agreement between manufacturer and purchaser on the understanding that the test voltage chosen is as high as possible, and that tests at different voltages are conducted for the purpose of extrapolation.

In case of series reactors, impedance is a guaranteed figure. It is measured at rated frequency by applying a voltage, which results in the circulation of the continuous rated current. The ratio of the applied voltage to the rated current gives the impedance value, e.g., in case of three-phase series reactor, the impedance value is taken as

$$\frac{\text{Phase-to-phase applied voltage}}{\text{Continuous rated current} \times \sqrt{3}}$$

Unlike transformers, the loss in the various parts of the reactor cannot be separated by measurement. It is thus preferred that in order to avoid correction to reference temperature, measurement is performed when the average temperature of the windings is practically equal to the reference temperature. If this is impracticable, the loss may be broken down into core-loss, I^2R-loss and stray loss and the individual components corrected to rated voltage and to the reference temperature in accordance with the requirements of standards. The core-loss is taken to be the calculated core-loss and is deemed to be independent of temperature. The I^2R component of the measured loss is calculated from the measured resistance of the winding. The remaining loss is considered to be the stray loss.

If from the results of type tests on a similar reactor, the losses both at reference temperature and in the cold state are known, the loss on the reactor under routine test can be corrected to reference temperature as follows:

If W_r is the cold-state loss of routine tested reactor at temperature t_1,
  W_t is the cold-state loss of type tested reactor at temperature t_2,
  W_{ft} loss of type tested reactor at reference temperature,
  W_{fr} loss of routine tested reactor at reference temperature (to be determined),

then

$$W_{fr} = W_r \frac{W_{ft}}{W_t} \times \frac{235 + t_2}{235 + t_1}$$

14.4.3 Inter-turn Over-voltage Withstand Test (Routine Test)

Because of large power requirements, this test is generally impracticable even with increased frequency for large reactors, This test, when impracticable, is replaced by an impulse-voltage test. Details of impulse test are similar to those applicable for transformers and are covered in Sec. II.

14.4.4 Separate Source Voltage Withstand Test (Routine Test)

The reactor is tested at the specified test level. The test voltage is applied between the line and neutral terminals joined together and the earth for 60 s.

14.4.5 Measurement of Insulation Resistance (Routine Test)

The insulation resistance of the winding with respect to core and endframe and tank connected together to earth is measured. The oil temperature is measured immediately prior to the test and recorded.

14.4.6 Partial Discharge Measurement

In order to check the level of internal partial discharges at specified over-voltage, partial discharge measurement is performed on all EHV reactors of 300 kV and above class, as a routine test. The methodology of conducting the test is similar to that applicable for power transformers and is covered under Sec. III.

14.4.7 Lightning Impulse Voltage Withstand Test (Routine and Type Tests)

Impulse test is intended to check the ability of a reactor to withstand steep-wave impulse-voltages between turns of the windings. Details of the test with full wave and chopped wave are covered under Sec. II.

14.4.8 Switching Impulse Voltage Withstand Test

The test is intended to check the internal main insulation to earth of the reactor against switching transient over-voltages caused by switching operations in the systems.

This is a routine test for windings with highest voltage ≥ 300 kV. The method of testing is similar as applicable for power transformers and is covered under Sec II.

14.4.9 Temperature Rise Test (Type Test)

This test is performed at rated voltage and frequency to demonstrate the thermal performance of the reactor. When the capacity of test plant does not permit the test to be performed at rated voltage, it can be done, subject to agreement between manufacturer and purchaser, at a voltage as close as prescribed to the rated voltage and the results are corrected to rated conditions in accordance with IEC : 60076.

14.4.10 Measurement of Zero-Sequence Impedance

This test is not applicable to series reactors. For star-connected three-phase shunt reactors, the zero-sequence impedance (in ohms per phase at rated frequency) is measured by applying rated phase-to-neutral voltage between the line terminals connected together and the neutral.

14.4.11 Measurement of Vibration and Noise Level

(a) *Vibration Measurement*

The vibration measurements are taken to check the safety of the reactor tank against any impending mechanical damage.

The vibrations are generated due to the magnetostrictive strain in the core and can magnify due to resonance of the core itself. Local resonance of auxiliary components may also magnify the vibrations.

For conducting vibration measurements, the reactor is energized at the specified voltage and at the rated frequency. An accelerometer is fixed at the location where measurement is intended. The vibration signal from accelerometer is transmitted to a vibration meter which indicates the vibration level in microns.

(b) Acoustic Sound Level Measurement

In order to ascertain the noise intensity, sound level measurement is made during factory tests. The main purpose of this measurement is to verify that the design has not hit mechanical resonance, either for the active parts or for the different parts of the tank.

The vibration of the iron circuit is the major cause of noise generated in reactors. The noise level depends upon the flux density and the magnetostrictive strain.

Sound level measurements are taken in accordance with NEMA-TR1 1971. The reactor is energized at the specified voltage at rated frequency. The measurements are taken at approximately the center height of the reactor tank, if its height is less that 8 ft and at approximately one-third and two-third heights of the tank, if its height is 8 ft or more. The measurements are taken around the reactor at intervals of 3 ft. The microphone is kept at a distance of 1 ft from the main noise emitting surface. In case the radiator fans are in operation, the microphone is kept at a distance of 6 ft from any portion of the radiators or coolers. The correction for background noise, if any, is applied. The noise level of a reactor is the arithmetic average of all the readings obtained at different locations with appropriate correction for background, if applicable.

14.4.12 Measurement of Third Harmonics

The third harmonics in the phase currents add together at the earthed neutral. Consequently the earth current can be high, leading to disturbances. It is a normal requirement that the total harmonic content in the phase current of a star-connected HV reactor, and particularly the zero-sequence component appearing in the neutral shall stay below a specified limit, which is of the order of 1–3% of rated current.

In order to ascertain the magnitude of third harmonic current, the reactor is energized at rated voltage and frequency with the neutral solidly grounded and measurements done. The verification by measurements is sometimes difficult because the harmonics in the energizing voltage are of comparable order, to the contribution from reactor reactance non-linearity.

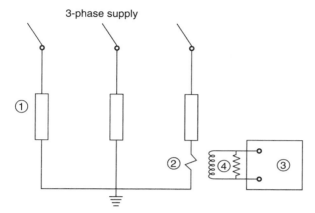

Figure 14.20 *Scheme for measurement of current harmonics in a shunt reactor.*

14.4.13 Lightning Impulse Voltage Withstand Test on a Neutral Terminal

When the neutral terminal of winding has been designed for impulse withstand voltage, it is verified by applying an impulse voltage through any one of the line terminals or through all three line terminals of a three-phase winding connected together. The neutral terminal is connected to earth through an impedance and the voltage amplitude developed across this impedance, when the impulse applied to the line terminal is equal to the rated withstand voltage of the neutral terminal.

Alternatively, an impulse test corresponding to the rated withstand voltage of the neutral is applied directly to the neutral, with all line terminals earthed. In this case, a longer duration of the front time is allowed, up to 13 microseconds.

14.4.14 Short-circuit Current Withstand Test (Special Test on Series Reactors)

This test is designed to prove the mechanical capability at rated short-time current, and in the case of a reactor with a magnetic core or shield, to measure the impedance corresponding to the rated short-time current.

The test consists of two applications of rated short-time current, each of a duration of ten cycles, the first peak having a crest value of $1.8 \times \sqrt{2}$ times the rms value. (Certain service conditions may result in asymmetry factors less than $1.8 \times \sqrt{2}$.)

Three-phase reactors are subjected to two three-phase tests—one with maximum asymmetry of the short-circuit current in the center phase, the other with maximum asymmetry in an outer phase.

The proof of the reactor having withstood the test satisfactorily is determined by remeasuring the impedance by visual inspection and by repetition of the routine dielectric tests at 75% of the initial value.

14.4.15 Specialities of Reactor Testing

Requirement of large reactive power is sometimes a constraint for testing of large EHV reactors at rated voltage. If the test facilities permit, the reactor should be energized up to the rated voltage and routine tests be conducted.

The difficulties in testing can mainly be due to

— Inadequate power capacity of test plants:
— Lack of a reliable and straightforward method of loss measurement. In the absence of International consensus, IEC recommends two alternative methods apart from the conventional wattmeter method, which is subjected to some errors.
— Limitations imposed by factory conditions affecting measurement accuracies.

The power frequency over-voltage tests and the loss tests have been described hereafter with special reference to the specialities involved in testing.

14.4.16 Over-voltage Tests

A check of level of internal partial discharges at specified test voltage is becoming a generally agreed condition for acceptance of EHV transformers. Application of an over-voltage test for a large shunt reactor, corresponding to the induced voltage test for a transformer, is a major difficulty, because of the prohibitive reactive power demand, even for short duration testing. Testing at elevated frequency is the only practical possibility (as for transformers) and is examined here by elementary calculations.

Symbols used:

U_N — rated voltage $\quad U_1$ —
f_N — rated frequency $\quad f_1$ — Corresponding test
X_N — rated impedance $\quad X_1$ — quantities
Q_N — rated reactive load $\quad Q_1$ —

K — ratio of test quantity and rated quantity:

Subscript defines the quantity under consideration.

The voltage and frequency used for the test are:

$$U_1 = K_u \cdot U_N ; K_u > 1 \qquad (1)$$
$$f_1 = K_f \cdot f_N ; K_f > 1$$

Reactor impedance

$$X_1 = X_N \cdot K_f$$

The current and flux density at any point:

$$I_1 = I_N \cdot K_a \cdot (K_f)^{-1} \qquad (2)$$
$$B_1 = B_N \cdot K_u \cdot (K_f)^{-1} \qquad (3)$$

The reactive load:

$$Q_1 = Q_N \cdot (K_u)^2 (K_f)^{-1} \qquad (4)$$

I^2R loss vary with the square of the current. Eddy losses in the winding and in metal parts are proportional to the square of frequency and to a varying power of the flux density. In the winding with stranded conductors, it is assumed proportional to B^2, but for solid parts with pronounced skin effect, the power tends toward $B^{0.5}$. Losses in the winding dominate, however, eddy losses in the winding are reduced to small proportion of the I^2R losses. The total winding losses are

$$P_1 = (I^2R)_N \cdot (K_u)^2 \cdot (K_f)^{-2} + P_{eN} \cdot (K_u)^2 \qquad (5)$$

It is to be noted that eddy losses P_e in the winding are independent of frequency as long as the test voltage in constant.

Example 14.1   For inter-turn over-voltage withstand test on EHV reactors with

$$K_u = 2.0, K_f = 4, P_{eN} = 0.12 \, (I^2R)_N$$

Winding losses at rated conditions, and during the test will be:

$$P_N = (I^2R)_N \cdot 1.12$$

$$P_1 = (I^2R)_N \cdot (2.0)^2 \cdot (4)^{-2} + 0.12\,(I^2R)_N \cdot (2.0)^2$$
$$= 0.73 \cdot (I^2R)_N$$

The reactive power demand will be
$$Q_1 = Q_N \cdot (2.0)^2 \cdot (4)^{-1}$$
$$= Q_N \cdot 1$$

Thus it is observed that the reactive power consumption at 200 Hz is equal to the rated power.

Example 14.2 For partial discharge test with
$$K_u = 1.5,\ K_f = 4,\ P_{eN} = 0.12\,(I^2R)_N$$

Winding losses at rated conditions, and during the test will be:
$$P_N = (I^2R)_N \cdot 1.12;$$
$$P_1 = (I^2R)_N \cdot (1.5)^2 \cdot (4)^{-2} + 0.12(I^2R)_N \cdot (1.5)^2$$
$$= (I^2R)_N\,0.41 = 0.37 \cdot P_N$$

The reactive power demand is
$$Q_1 = Q_N \cdot (1.5)^2 \cdot (4)^{-1} = 0.56 \cdot Q_N$$

It is observed that there is reduction in active and reactive power consumption. However, the effect of these conditions on local hot-spots should be examined.

In a shell-form winding, first, the local flux pattern is very regular, varying from a maximum value at the inner edge to zero at the outer edge. This variation means that local eddy losses at the inner edge are three times as high as that indicated by the average for the whole winding. A comparison between rated conditions and test conditions analogous to Eq. (5) above, thus gives:
$$P_N(\text{hot-spot}) \sim (I^2R)_N + 3 \times 0.12\,(I^2R)_N \simeq 1.36 \cdot (I^2R)_N$$
$$P_1(\text{hot-spot}) \sim (I^2R)_N \cdot (1.5)^2 \cdot (4)^{-2} + 0.36\,(I^2R)_N \cdot (1.5)^2$$
$$\sim 0.95 \cdot (I^2R)_N$$

with $K_u = 2.0$
$$P_1(\text{hot-spot}) \sim 1.69 \cdot (I^2R)_N$$

It remains to be checked, how the static shields, if used, would behave. Their losses would go up by the square of test voltage, independent of frequency.

In a gapped core reactor, the same basic relations apply, but the distribution of fringing flux around the air gaps and consequently, the local winding eddy losses are greatly dependent on design.

It may be concluded that

- The inter-turn over-voltage test, the test level for which is double the rated voltage, is practicable at an elevated frequency of 200 Hz, but may cause thermal injury to insulation.
- The alternative test recommended to replace this test is the full-wave lightning impulse test. Although the lightning impulse test is more onerous, because of high voltage gradients in a few initial turns resulting from non-linear impulse voltage distribution along the winding, yet this test is the most appropriate substitute test.

The partial discharge test, which is recommended at 1.5 times the rated voltage, can be conducted with advantage at an elevated frequency. This precludes the possibility of any risk of thermal injury to the reactor under test.

14.4.17 Loss Measurement

It is well known that the difficulty associated with loss measurement on reactors lies in determining these with sufficient accuracies at power factors less than 0.004. The common methods of loss determination in practical use are:

(a) Calorimetric measurement with oil-to-water heat exchanger.
(b) Bridge method giving the loss angle of reactor impedance.
(c) Direct wattmeter measurement.

It is generally agreed that the calorimetric test can be considered as the reference test method, which gives good accuracy when carried out with care. It is virtually unjustifiable as a routine test due to considerable cost and time involved in measurements. Normally Bridge method is used for loss measurement of reactors however periodic calibration of the bridge is recommended with calorimetric method. Wattmeter method is the most straightforward method when accuracy requirement is not extremely high.

(a) *Calorimetric Loss Measurement*

The principle of calorimetric test is to determine the power carried away in the cooling medium by measuring its rate of flow and its temperature rise. In water-cooled apparatus with closed cooling systems, measurements can be performed either in air or oil or in the water circuit. In view of the difficulties in making measurements, where air is the cooling medium, and uncertainty as regards the specific heat of oil, the calorimetric measurements are carried out predominately on the water circuit. To the power carried away by the cooling medium, the heat lost by virtue of radiation and convection, or by other means not assessed in the cooler, is added.

The reactor is energized at rated voltage and frequency. The external surface of the reactor, including the pipe work, is thermally insulated by glass-wool to minimize the surface radiation. Oil is circulated in the winding by a pump. The losses generated in the reactor, after thermal equilibrium has been reached, are transferred from the reactor to the exterior, either in the cooling water or by convection and radiation from the surfaces of the tank, piping and cooling equipment. Temperatures of oil and water are measured at the inlet and outlet of the cooler. Water flow is measured by a water-flow meter. When steady-state is reached, winding temperature is determined by resistance cooling curve, and power consumption by the pump is measured. Losses of the reactor at the test temperature at steady-state condition comprise:

— Losses evacuated by the water circuit.
— Losses through the lagging insulation.
— Losses of the pump motor.

Figure 14.21 shows a general circuit arrangement for loss measurement on a three-phase shunt reactor by calorimetric method.

Cooling arrangement. As per the standards, the loss guarantees are at an average winding temperature of 75°C. The calorimetric test must therefore be so controlled that thermal equilibrium is reached when the winding is at or near this temperature. In the case of a water-cooled reactor, the reactor's normal cooler is used for the calorimetric test even though difficulty can sometimes be encountered in reaching 75°C in the windings, because of low cooling-water temperature. If the reactor is normally air-cooled, a separate oil water-cooler is fitted in place of the air-coolers.

Accuracy of the method. A rigorous analysis of the expected probable errors of calorimetric test, in general, is hardly possible in view of many unexpected errors during the test. However, when measurements are made with all precautions, the accuracy of the calorimetric method can be better than that achieved by a conventional method,

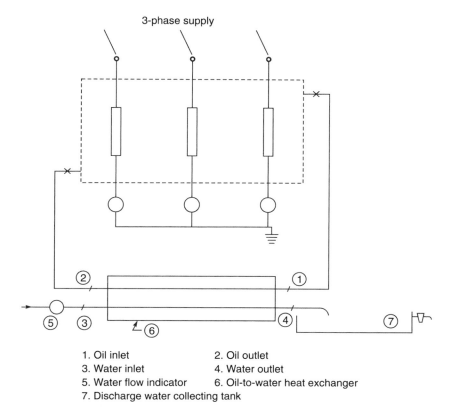

1. Oil inlet 2. Oil outlet
3. Water inlet 4. Water outlet
5. Water flow indicator 6. Oil-to-water heat exchanger
7. Discharge water collecting tank

Figure 14.21 *Arrangement for loss measurement on a three-phase shunt reactor by calorimetric method.*

viz. the wattmeter method, because this measurement is not affected by the reactive power.

Though calorimetric method is more expensive and takes more time to perform than conventional methods, yet in certain cases, e.g., for the determination of losses of a large high voltage shunt reactor, this method is the only reference method, which gives good accuracy and can be employed to calibrate the bridge.

(b) *Bridge Measurement*

With bridge circuit, a voltage proportional to the reactor current but exactly 90° out-of-phase with respect to this current, is induced in the mutual inductance M. This proportional but out-of-phase voltage is measured magnitude-wise and phase-wise relative to the terminal voltage of the reactor by means of Schering bridge by comparing the test object with an almost loss-free standard capacitor, with accurately known capacity and dissipation factor. The bridge is balanced by adjusting resistance and capacitance until perfect balance is achieved.

Figure 14.22 shows the scheme of the bridge method for loss measurement on a shunt reactor.

Factors which affect accuracy. The bridge circuit, incorporating a mutual inductor in series with the high-voltage reactor, is a precision component, which can be calibrated for loss angle with high accuracy, the compressed gas capacitor used for dielectric loss measurement is also accurate. However, in practice, the connection is complicated by the introduction of a current transformer in cascade with the mutual inductor.

In a circuit without current transformer, the mutual inductor primary constitutes an addition to the impedance of the specimen, and the modification of the reactor current is negligible. With a

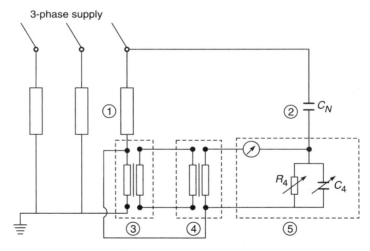

1. Shunt reactor under test
2. Compressed gas capacitor
3. Precision current transformer
4. Mutual inductor
5. Schering bridge

Figure 14.22 *Bridge circuit for power loss measurement.*

current transformer added, however, the mutual inductor primary is a loading on the current transformer, and the resulting phase shift represents the greatest individual term of uncertainty in the whole set-up. In order to bring this bridge circuit up to a satisfactory level, it is necessary to have a mutual inductor designed for the full reactor current.

Further, the bridge circuit arrangements, particularly the ones used for the measurements on objects which are producing harmonic waves, e.g., high voltage inductances with closed iron-cores or power transformers under no-load are frequency sensitive and small frequency fluctuations in the applied voltage tend to reduce measurement precision. Therefore, a measuring voltage free from frequency fluctuations is necessary.

The bridge method is very sensitive of magnetic stray fields and the current transformer is especially sensitive.

However, the bridge method has its own merits over the calorimetric method.

— Compared to the calorimetric method, the use of bridge affords considerable economy in time.
— Use of a high-voltage compressed gas capacitor as a standard capacitor is an advantage because it is generally available in most high-voltage testing laboratories.

REFERENCES

(For Section IV)
1. *IEC Publication* 289 (1989); Reactors.
2. IS : 5553 Part II—1989; Shunt reactors.
3. BS : 4944: 1973—*Specification for Reactors, Arc-suppression Coils and Earthing Transformers for Electric Power Systems.*
4. *NEMA Standard Publication* No. TR 1—1971: Transformers, regulators and reactors.
5. Sollergren, B., *Special Report for Group 12 (Transfomers)*; CIGRE, Vol. 1970.
6. Grundmark, B., *High Voltage Shunt Reactors—Trends in Design and Testing.* Report 12—03 Ibid.
7. De Bourg, H. and others, *Calorimetric Loss Measurement On Alternators and Reactors*, Report No. 119; CIGRE, Vol. 1964.

SECTION V

14.5 Short-Circuit Testing of Power Transformers

14.5.1 Title the Short-Circuit Test Requirement:

It is given in IEC 60076-5: 2000 and IS2026 (Part I). For large power transformers, the margins are quite high for thermal effect whereas dynamic effect is quite complex and requires tedious calculations. This section elucidates the testing requirements of short-circuit tests and short-circuit duty of power transformers.

This test is a special test and the facilities required for the same are generally not available with the manufacturers as a very large amount of power is required. There are only a few laboratories in the world that are equipped with such large rating test plants. Depending upon the voltage class, size (MVA rating) and the impedance voltage of the transformer, specific confirmation is to be obtained from the reputed laboratories for conducting this test.

As per IEC 60076-5, power transformers have been categorized as under:

    Category I   — Up to 2500 kVA
    Category II — From 2501 kVA to 100 MVA
    Category III — Above 100 MVA

The symmetrical short-circuit current is calculated using the measured short-circuit impedance of the transformer plus the system impedance. The short-circuit apparent power of the system is generally specified by the utility.

14.5.2 Requirements Prior to Short-Circuit Test

The transformer should pass all routine tests as specified in IEC 60076-1. However, the Lightning Impulse test is not required to be conducted prior to the short-circuit test. The routine test report must be available prior to the short-circuit test. At the beginning of short-circuit test, the average winding temperature shall be preferably between 10° C to 40° C.

14.5.3 Peak Value of Short-Circuit Current

The first peak of asymmetrical short-circuit current is given by

$$\hat{I} = I \times \sqrt{2} \times k$$

where I = symmetrical short-circuit current
  k = accounts for initial offset of test current
$\sqrt{2}$ = conversion factor for rms to peak value;
$k \times \sqrt{2}$ is called peak factor and depends on X/R ratio
where X = sum of reactances of the transformer and system $(X_t + X_s)$ in ohms
  R = sum of resistances of the transformer and system $(R_t + R_s)$ in ohms

If system impedance is taken into account and X/R ratio of the system is not known, then X/R ratio of transformer is considered.

Value of factor $k \times \sqrt{2}$ for different X/R ratios as given in standards are I:

X/R	1	1.5	2	3	4	5	6	8	10	14
$k \times \sqrt{2}$	1.51	1.64	1.76	1.95	2.09	2.19	2.27	2.38	2.46	2.55

Normally, X/R ratio is >14 hence
  (1) For Category I transformers, $k \times \sqrt{2} = 1.8 \times \sqrt{2}$, i.e., 2.55
  (2) For Category II and Category III transformers, $k \times \sqrt{2} = 1.9 \times \sqrt{2}$ i.e., 2.69

14.5.4 Tolerance on Asymmetrical Peak and Symmetrical RMS Value of Short-Circuit Current and duration of Short-Circuit Test

As per IEC 60076-5 (2000), the tolerances allowed are:
On peak asymmetrical short-circuit current = ±5% of specified.
On symmetrical short-circuit current = ±10% of specified.
Duration for short-circuit test = 0.5 sec for category I with ±10% Tol.
    = 0.25 sec for category II and III with ±10% Tol.

14.5.5 Short-Circuit Testing Procedure of Two Winding Transformers

Two types of short-circuit test procedures can be followed:
  (1) Pre-set Method
  (2) Post-set Method

(1) Pre-set Method

In this method, one winding is short-circuited prior to application of the voltage to the other winding in order to obtain the desired short-circuit current in both windings. In a power transformer, since in most cases LV is near to core, hence normally HV is supplied with power and LV is pre-shorted. This is essential in order to avoid saturation of magnetic core which could lead to excessive magnetising current superimposed on short-circuit current during the first few cycles.

(2) Post-set Method

One of the windings is applied with AC supply of desired value, and on the other winding, post short-circuit is done through a synchronous switch. Thus, there is very low magnetisation inrush current. This method, however, requires a power three times as high as the pre-set method.

14.5.6 Short-Circuit Test Shots and Connections

Short-circuit test requires 3 shots on each phase. In case of 1-phase transformer, one shot each shall be on the maximum tap, normal tap, and minimum tap respectively, while in the case of 3-phase transformers, a total of 9 shots are applied, 3 on each phase. Thus, 3 shots shall be applied at the maximum tap on one of the outer phases, 3 shots at normal tap on middle phase, and 3 shots at minimum tap on the other outer phase.

Further, the short-circuit test connections are given below in the preferential order:

(a) *3-phase Connection*: For a 3-phase transformer, a 3-phase supply is to be used as far as possible. (See Fig. 14.23)

Zs = system impedance
S = synchronous switch for post-set method or a rigid connection for pre-set method.

(b) *1½ Phase Connection:* In case it is not possible to meet requirements of cl. 14.5.6 (a) for 3-phase (especially for category II and III transformers), then 1½ phase connection may be used with a single phase supply on HV side. The connection is given in Fig. 14.24

(c) *Single Phase Connection:* In case of a single phase transformer, this connection is used. Similarly, if power supply is

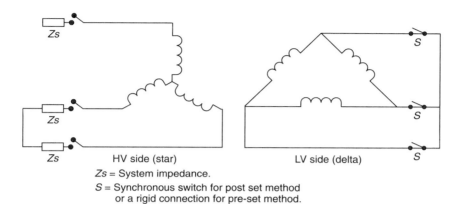

Figure 14.23 *3-Phase method for S.C. test of transformer.*

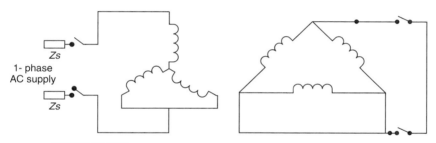

Figure 14.24 *1½-Phase method for S.C. test of transformer.*

insufficient for conducting 3-phase or 1½ phase method in a 3-phase transformer, then single phase method is used (in case star connected winding and neutral is available). The connection is given in Fig. 14.25.

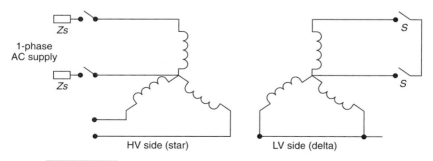

Figure 14.25 *Single phase method for S.C. test of transformer.*

14.5.7 Testing Procedure

— Calibration shots up to maximum 70% are applied to adjust and check the proper function of test setup with respect to moment of switching on, current setting.
— The routine test should be complete and the test report should be available.
— Short-circuit is applied at 100% level.
— Oscillographic record of applied voltage, current, as well as of secondary side should also be recorded for each shot.
— Reactance shall be measured after each shot, and reactance variation shall be within ± 2% for core type transformers with concentric coils of category I and II, and within ± 1% for category III transformers.
— Condition of passing of short-circuit test:
  (a) Results of short-circuit test and measurements and checks performed do not reveal any condition of faults.
  (b) Dielectric tests successfully conducted/repeated.
  (c) Out-of-tank inspection does not reveal any defects like displacement, shift of laminations, deformation of windings, connection or support structure etc.
  (d) No trace of electrical discharges are found.
  (e) Reactance variation shall be measured after each shot, and reactance variation shall be within ±2% for category I and II transformers, and within ±1% for category III transformers. In case of category III transformers, if the reactance variation is in the range of 1 to 2%, the acceptance is subject to mutual agreement between the manufacturer and the utility. For this, a more detailed inspection of the windings after disassembly needs to be carried out and analyzed to establish the cause of the deviation.

14.5.8 Optional Diagnostic Measures

One way to detect changes in the transformer due to forces during short-circuit of transformer is to measure the Frequency Response Analysis (FRA) before and after the short-circuit test. This is done by measurement of frequency response (voltage transfer function or admittance), which can be used to compare the frequency of the windings response before and after the short-circuit test. Mechanical change and voltage independent defects, like winding shorts are reflected in the finger print.

Addendum to Chapter 14

Test Results of a Few Typical Transformers

	315 MVA, 3-phase 50 Hz auto-transformer ONAN/ ONAF/OFAF cooled	250 MVA, 3-phase 50 Hz generator transformer ONAN/ ONAF/OFAF cooled	200 MVA, 1-phase 50 Hz generator transformer OFWF cooled	100 MVA, 3-phase 50 Hz auto-transformer ONAN/ONAF cooled
1. No-Load Ratio at Principal Tap	400/220/33 kV	15.75/420 kV	21/400 kV $\sqrt{3}$	220/110/11 kV
2. Connection Symbol	YN, a0, d11	YN, d11	YN, d11 when connected in 3-phase bank	YN, a0, d11
3. Lightning Impulse				
HV	1300 kVp	1300 kVp	1300 kVp	900 kVp
IV	950 kVp	—	—	450 kVp
LV	250 kVp	125 kVp	125 kVp	170 kVp
4. Switching Impulse				
HV	1050 kVp	1050 kVp	1050 kVp	—
5. Partial Discharge at $1.5 \dfrac{U_m}{\sqrt{3}}$	20 pC	35 pC	65 pC	—

(Contd.)

(Contd.) Addendum to Chapter 14

Test Results of a Few Typical Transformers

	315 MVA, 3-phase 50 Hz auto-transformer ONAN/ ONAF/OFAF cooled	250 MVA, 3-phase 50 Hz generator transformer ONAN/ ONAF/OFAF cooled	200 MVA, 1-phase 50 Hz generator transformer OFWF cooled	100 MVA, 3-phase 50 Hz auto-transformer ONAN/ONAF cooled
6. Separate Source Voltage withstand				
HV/Earth	38 kV rms	38 kV rms	38 kV rms	38 kV rms
LV/Earth	95 kV rms	50 kV rms	50 kV rms	28 kV rms
7. No-load Loss	101.4 kW	128.3 kW	92.4 kW	30.69 kW
8. % No-Load Current	0.06	0.07	0.14	0.05
9. Load Loss	274.1 kW	620.5 kW	476.7 kW	187.7 kW
10. % Impedance				
HV–IV	13.34	—	—	—
HV–LV	51.38	14.51	14.66	8.36
IV–LV	33.10	—	—	—
11. Top Oil Rise	28.3°C	36.3°C	13°C	30.5°C
12. Winding Rise	48.6°C	45.0°C	31°C	44.4°C

CHAPTER 15

Standards on Power Transformers

V.K. Lakhiani
S.K. Mahajan

Standards are evolved to meet a generally recognized demand, taking into account the interest of manufacturers and users and fulfilling the needs of economy. Today, a product cannot even be visualized without a standard. A standard is a useful guide in all facets of a product—conception, design, manufacture, testing, installation, operation, maintenance, etc.

Indian Standard 2026 "Specification for Power Transformers" is the governing standard on power transformers. This has been revised in 1977 with a view to bring it in line with the 1976 revision of IEC Publication 76 "Power Transformers." It is now in five parts and is an exhaustive standard on power transformers.

Besides IS: 2026, other related standards, in extensive use, are concerning electrotechnical vocabulary, fittings and accessories, on-load tapchanger and its application guide, loading guide, new oil, maintenance of oil, selection, installation and maintenance of transformers, etc. The scope of these standards is briefly touched upon in this chapter.

New standards in the offing are those connected with bushing air clearance, guide to testing and application guide on power transformers. An introduction to these new standards has also been made.

Efforts in different forums have been made to come out with a standard specification of the transformer. "Central Board of Irrigation and Power (CBIP) Specification" and "Central Electricity Authority (CEA) Report" are the offshoots of such concerted efforts. A brief mention has been made about these specifications. More and more adaptations of these standard specifications, which purport to supplement IS: 2026 are in the interest of the transformer industry.

15.1 First Revision of IS: 2026

15.1.1 State-of-the-Art

IS: 2026 was first published in 1962 and covered initially naturally cooled oil-immersed transformers. By subsequent amendments (a total of 8 in number) forced cooled transformers were included, use of synthetic liquids as coding medium was permitted and requirements for minimum windings were incorporated.

The first revision was brought out in 1977 with a view to bringing it in line with the revision of IEC Publication 76—1967, which was revised in 1976.

At present the requirements for power transformers are covered in five parts as follows:

PART I : 1977 General (First Revision)
PART II : 1977 Temperature Rise (First Revision)
PART III : 1981 Insulation levels and dielectric tests (Second Revision)
PART IV : 1977 Terminal markings, tappings and connections (First Revision)
PART V : 1994 Transformer/Reactor bushing minimum electrical clearances in air.

The second revision to Part III published in 1981. This revision was exactly in line with IEC 76-3 (1980). However since then IEC 76-3 has undergone changes and now available as IEC 60078-3 (2000) : Insulation levels, dielectric tests and external clearances in air. In this version of IEC 60076-3 following new major points have been introduced compared to IEC 76-3 (1980).

— Long duration AC (ACLD) test as a special test for $72.5 <$ Um ≤ 170 kV and Routine test for Um > 170 kV
— LI is routine test for Um > 72.5 kV
— SI is routine test for Um > 170 kV
— Short duration AC (ACSD) as routine test for Um ≤ 170 kV and special test for um > 170 kV
— Short duration AC (ACSD) test with P.D. for Um > 72.5 kV
— Induced over voltage test/P.D. test are not called and instead ACLD and ACSD tests have been introduced

Part 4 of IEC 76 tappings and connections has been clubbed with IEC 76-1 revised in 1993.

15.1.2 IS : 2026-1977 vis-à-vis IS : 2026-1962—Major Points of Difference

A few major differences partwise are enumerated below:

Part I

 (a) The kVA ratings have been extended to an unlimited range in the same series.
 (b) Dry-type transformers have been covered.
 (c) Terminology is now covered by IS : 1885 (Part xxxviii)–1977 "Electrotechnical Vocabulary : Transformers" (first revision).
 (d) The reference ambient temperatures have been revised. Maximum ambient air temperature and maximum daily average ambient air temperature have been increased by 5°C to 50°C and 40°C, respectively. Concept of maximum yearly weighted average ambient temperature has been introduced and a hot-spot temperature of 98°C over this weighted temperature of 32°C has been recognized, thereby giving a 66°C hot-spot rise.
 (e) The requirements with regard to ability to withstand short-circuit have been revised. A test under the category of "Special Test" for assessing the dynamic ability of the transformer to withstand short-circuit has been included. The thermal ability to withstand short-circuit has to be demonstrated by calculation and the duration of the symmetrical short-circuit current to be used for the calculation of the thermal ability to withstand short-circuits has been specified as 2 s.

   In fact, IEC 60076 has a separate Part 5 on this subject, but in case of IS : 2026 (1977) the same has been clubbed with Part I.
 (f) Tolerances on losses, impedance voltage, etc., have been revised. No-load current is also made a subject of guarantee with +20% tolerance. The component losses have +15% tolerance, while the total losses have +10% tolerance. Impedances at taps other than those at principal tapping have more tolerance than applicable for principal tapping by a

percentage equal to half the difference in tapping factor (percentage) between the principal tapping and the actual tapping.
(g) Operation test on on-load tapchangers and auxiliary circuits insulation tests have been included in the routine tests.
(h) Supplementary tests now termed "Special Tests" include besides zero sequence impedance; short-circuit test, measurement of acoustic noise level, measurement of the harmonics of the no-load current and measurement of the power taken by the fans and oil pumps.
(i) Requirement for overfluxing for generator transformers has been added. A generator transformer subjected to load-rejection conditions shall be able to withstand 1.4 times the rated voltage for 5 s at the transformer terminals to which the generator is to connected.

Part II

(a) Identification of transformers according to cooling methods has been lined up with the international practice of specifying by 4 letters, e.g., ONAN, ONAF, OFAF, ODAF, OFWF, etc.
(b) Directed flow type cooling has been recognized and has 5°C more permissible winding temperature rises.
(c) The temperature limits are the same as those in IEC 60076-2 (1993) viz. 100°C for top oil and 105/110°C for winding, depending upon whether non-directed oil circulation or directed oil circulation is adopted. As deviation from IEC, 5°C more rises have been permitted for water cooled trans-formers.
(d) Temperature rise limits are now valid for all tappings. For multi-winding transformers, the temperature rise of the top oil refers to the specified loading combination, for which the total losses are the highest. Individual winding temperature rises shall be considered relative to that specified loading combination, which is the most severe for the particular winding under consideration.
(e) Total losses to be fed for temperature-rise test should not be less than 80%. Similarly, rated current should also be not less than 90%. Values of indices 'x' and 'y' for the correction of oil rise and gradient have also been revised.

(f) Winding resistance at the instant of switching off the supply is now to be determined only by extrapolation of the curve resistance versus time. The earlier method of applying correction factors based on watts per kilogram of copper is discontinued. The graphical method, though consumes time, is more reliable, because average oil temperature can be obtained on the graph which can serve as a cross-check. Further, the trend of the curve shows whether the resistance readings were stabilized.

Part III

(a) Below 300 kV, insulation requirements and dielectric withstand tests are the same as before. The lightning impulse test continues to be the type test. The other dielectric tests are separate source power-frequency voltage withstand test (for line terminals in case of uniform insulation and for neutral terminal in case of non-uniform insulation) and induced over-voltage withstand test, both to be performed as routine tests. During the induced over-voltage test, at least twice the rated voltage should appear across the winding, having uniform insulation but in case of winding with graded insulation, the line terminals must receive rated voltage according to related table and excessive voltage between line terminals is avoided. Neutral raising is allowed if it is so designed. An impulse test for the neutral terminal is a special test if a rated impulse withstand voltage for the neutral terminal has been specified. A longer duration of the front time up to 13 μs is allowed for the impulse test on a neutral terminal.

(b) For voltage class ≥ 300 kV, there are two alternative methods for specifying and testing the dielectric requirements, viz. method 1 and method 2.

Method 1 is the same as applicable to voltage class < 300 kV, non-uniform insulation, except that lightning impulse test is a routine test.

Method 2 introduces newer tests like switching impulse and partial discharge measurement. According to this method, the verifying dielectric tests are:

(i) Lightning impulse test for the line terminals (routine tests)

(ii) Switching impulse test for the line terminals (routine test)
(iii) Separate source power-frequency voltage withstand test for the neutral (routine test)
(iv) Induced power-frequency voltage withstand test with partial discharge measurements (routine test)
(v) Impulse test on the neutral, if specified (special test)

(c) Test levels for method 1 are available up to 420 kV class and for method 2 they are given up to 765 kV class. It is expected that beyond 420 kV class, method 2 will be normally chosen.
(d) In case of switching impulse test which is intended to verify the switching impulse voltage withstand of the line terminals to earth and between line terminals in case of three-phase transformers, 1.5 times rated test voltage appears between line terminals according to prescribed test circuit. This test is much more severe regarding higher phase-to-phase oil and air clearances.
(e) Impulse test whether routine test or type test, is to be conducted on all the three phases of a three-phase trans-former. In fact, each phase is to be tested such that one phase is in the maximum, another in the principal and the third in the minimum voltage-tap position.
(f) Chopped wave impulse test is now a special test. Application of shots in case of chopped wave test is more stringent now, since two chopped waves are to be followed by two full waves, as against one full wave as before.
(g) The rules for highest voltage winding apply to the trans-former in toto. Transformer is to be designed and tested according to one method alone, either by method 1 or method 2 and not by combination of the two methods for different windings. Method of test should be known right at the enquiry stage.

Part IV
(a) The phase marking ABC have been replaced by UVW following the international practice.
(b) The corresponding IEC Pub 76-4 (1976) does not contain the information concerning terminal markings and the same is now included in IEC 60076-1.

(c) Information about neutral and connection has been included in the connection symbols.
(d) Line terminal is now terminal No. 1 and neutral is invariably 2. In the pre-revised system, line terminal was subscripted by 2. Marking of tapping terminals now start from the line end, instead of neutral, in the ascending order.
(e) Principal tapping, is the mean tapping position and is essentially the full power tapping. All guaranteed parameters usually refer to this tapping.

15.2 Other Related Standards

To supplement IS : 2026, there are a number of related standards. A list of these is furnished in Appendix A. A few important ones are described below:

15.2.1 IS : 1885 (Part 38)–1993 (Second Revision) Electrotechnical Vocabulary : Power Transformer and Reactors

This standard covers definitions of terms applicable to transformers. The first revision of this standard published in 1977 aligns it with the first revision of IS : 2026.

15.2.2 IS : 3639-1966 Fittings and Accessories for Power Transformers

This standard attempts to standardize accessories and fittings for transformers to facilitate interchangeability and rationalize use of fittings, such as would be commonly acceptable to manufacturers and users. Presently, it covers accessories like valves and valve flanges, flanges for Buchholz relay, breather pipe, thermometer pockets, level indicator, rollers and earthing terminals.

This standard is being revised to make the description elaborate and most up to date and include other fittings like radiators, fans, pumps, etc.

15.2.3 IS : 2099-1986 Specification for Bushings for Alternating Voltages above 1000 V (Second Revision)

This standard covers rated values, performance requirements and tests for bushings for three-phase alternating current systems, having rated voltages above 1000 V and frequencies between 15 and 6000 Hz. It also provides information for ordering of bushings. However, it does not give dimensions of bushings, for which a separate Indian Standard IS : 3347 "Dimensions for porcelain transformer bushings" which gives the dimensions of the porcelain bushings up to 123 kV ratings exists. The scope of this standard is being extended to include condenser bushings.

The standard is based on IEC Pub 137 and British Standard BS 223.

15.2.4 IS : 8468-1977 "Specification for On-load Tapchangers"

This standard covers on-load tapchangers and their motor-drive mechanisms. This is based on the first revision of IEC Pub 214 "on-load tapchangers." Besides rating and service conditions, the standard describes at length the type the routine tests to be performed on tapchangers.

15.2.5 IS : 8478-1977 Application Guide for On-load Tapchangers

This application guide is intended to assist in the selection of suitable on-load tapchangers for use in conjunction with the tapped windings of transformers.

Since the on-load tapchanger represents only a small part of the total cost of the equipment in a power transformer, it should be freely chosen to suit the equipment. The guiding factors for selection of a suitable tapchanger are insulation level, current, breaking capacity, tapping positions, discharge problems, etc. The guide also gives the information required with enquiry or order.

15.2.6 IS : 335-1993 Specificationfor New Insulating Oils (Fourth Revision)

This standard is based on IEC 60296 (1986), which covers two classes of oil, class I and class II. Two classes have been included in the IEC recommendation, because of the difference in natural requirements largely dictated by climatic conditions. In this standard, however, only one class, mostly based on class I is covered, which is considered to be suitable for the conditions prevailing in this country.

This standard specifies the requirements for unused insulating oil as delivered in bulk, such as tank wagons and road tankers, or drums suitable for immersion or filling of transformers. The oil covered by this standard are uninhibited oils free from antioxidant additives.

The standard gives schedule of characteristics, both chemical and electrical, which are acceptance norms for the new oil. Consequent on the tendency for water absorption and deterioration, the limits for power factor, resistivity and electrical strength apply to tests made within two weeks of delivery.

If tests are not carried out within two weeks, the oil may be treated and tested. It shall then meet the requirements.

Unused oil complying with the requirements of this standard are considered to be compatible with one another and may be mixed in any proportion. (Please refer to Ch. 3 "Materials Used in Transformers.")

15.2.7 IS : 1866-2000 Code of Practice for Electrical Maintenance and Supervision of Mineral Insulating Oil in Equipment (Third Revision)

This standard is based on IEC 60422 (1989), "Maintenance and supervision guide for insulating oil in service" and is a companion standard to IS : 335–1993.

Insulating oil, complying with IS : 335, from the time the oil is received, is subject to deterioration or contamination in storage, or in handling or in service. Accordingly, a periodic treatment to maintaining it in fit condition, is required and eventually it may have to be replaced by new oil. This standard covers the causes of oil deterioration and recommendations are made for the various tests and their interpretation, whether reconditioning, reclaiming replacement is required, also giving permissible limits for satisfactory use. The methods used for reconditioning and reclaiming are also briefly given.

Unfortunately, no IS gives characteristics of oil to be filled in and just after filling in the transformer, for works testing.

15.2.8 IS : 10028 Code of Practice for Selection, Installation and Maintenance

This standard supercedes the earlier standard IS 1886–1967 which covered information about installation and maintenance of power and distribution transformers.

New standard IS 10028 also includes selection of transformers in addition of installation and maintenance. It is in three parts:

Part 1–1985 Selection
Part 2–1981 Installation
Part 3–1981 Maintenance

15.2.9 IS : 6600-1972 "Guide for Loading of Oil Immersed Transformers"

This guide in the present form is applicable to oil immersed transformers of type ONAN and ONAF. The guide indicates how oil immersed transformers may be operated for different ambients and duties without exceeding the acceptable limit to deterioration of insulation through thermal effects.

Basically, the cooling medium is at 32°C (maximum weighted average ambient) but, deviations from this are provided for in such a way that the increased use of life when operating with a cooling medium temperature above 32°C in summer, is balanced by the reduced use of life, when it is below 32°C in winter.

This standard, based on IEC 60354 "Loading guide for oil-immersed transformers" is a very useful tool for the user.

15.3 New Standards

New standards being formulated are on air clearances of bushings, application guide and testing guide on power transformer. A brief introduction of these follows:

15.3.1 Minimum Bushing Air-Clearance

IS 2026 (Part 5) 1994 titled "Transformer/Reactor Bushings-minimum external clearances in air-specification" was published in May 1994.

This standard specifics clearances in air between live parts of bushings of oil immersed power transformers and reactors and to object at earth potential.

15.3.2 Applications Guide for Power Transformers

A new standard on the above subject has been published in 1983 as IS : 10561-1983. This standard which is based on IEC. 606 (1978) "Application guide for power transformers" is intended to assist in the determination and selection of transformer characteristics. Subjects like specification of tapping quantities, selection of winding connections for transformers for three-phase systems, parallel operation of transformers, loading capability of the neutral point of windings in star or zigzag connection and calculation of voltage drop (or rise) for a specified load condition have been discussed at length.

This standard is a useful guide for purchasers at the time of purchase and during subsequent usage.

IEC 606 is withdrawn and now available as IEC 60076 Part 8 (1997).

15.3.3 Guide to Testing of Transformers

A draft "Indian Standard" is under formulation to cover allied subjects to testing of transformers, viz. class of accuracy and frequency of calibrations of the measuring instruments, sequence of testing, purpose and significance of the various tests and necessary guidance on testing, etc. This standard will be a coordinating link between various testing codes and a useful guide for the test engineers.

15.4 Standard Specification of a Power Transformer

15.4.1 Minimum Specification of a Power Transformer

The following information should be definitely available with enquiry and order for the design and drafting of the transformer.
- (i) Type of transformer, for example separate winding transformer, auto-transformer or booster transformer.
- (ii) Number of phases, single or polyphase.
- (iii) Frequency.

(iv) Rated power (in kVA) and for tapping ranges exceeding ±5% required power on extreme tappings.
(v) Rated voltages for each winding.
(vi) Connection symbol.
(vii) Requirement of on-load/off-circuit tapchangers or links— number of tappings, tapping range, location of tappings, particular voltage required to be varied and whether constant flux/variable flux/combined voltage variation.
(viii) Impedance voltage at rated current and principal tapping for different pairs of windings and at least on the extreme tappings in case of parallel operation if required.
(ix) Indoor or outdoor type.
(x) Type of cooling and if different types of cooling involved, rated power for each type of cooling.
(xi) Temperature rises and ambient temperature conditions including altitude and in case of water cooling chemical analysis of water.
(xii) Number of cooling banks, spare capacity if any and number of standby cooling pumps/fans.
(xiii) Highest system voltage for each winding.
(xiv) Method of system earthing for each winding.
(xv) Whether windings shall have uniform or non-uniform insulation, and in case of non-uniform insulation, power frequency withstand voltage of neutral and impulse withstand level if an impulse test on the neutral is required.
(xvi) For windings having system highest voltage greater than 300 kV, the method of dielectric testing, whether method 1 or method 2.
(xvii) Withstand voltage values constituting insulation level of line terminals, viz. test levels of lightning impulse, switching impulse, one minute power frequency, long duration power frequency with partial discharge measurement, wherever applicable.
(xviii) Limitation of transport weight and moving dimensions and special requirements, if any, of installation, assembly and handling.
(xix) Whether a stabilizing winding is required.

(xx) Overfluxing conditions or any other exceptional service conditions.
(xxi) Loading combinations in case of multi-winding transformers and when necessary active and reactive outputs separately, especially in case of multi-winding auto-transformers.
(xxii) Details of auxiliary supply voltage (for fans, pumps, OLTC motor, alarm and control).
(xxiii) Controls of tapchangers.
(xxiv) Short-circuit levels of the systems.
(xxv) Vacuum and pressure withstand values of the transformer tank.
(xxvi) Noise level requirement.
(xxvii) Number of rails and rail gauge for movement along shorter and longer axes.
(xxviii) Fittings required with their vivid description.
(xxix) Any other appropriate information including special tests if any and capitalization formula for the losses.

15.4.2 Need for Standardization of Specification

Transformer is a tailor-made product and requires a total effort in the design and drafting, even if a single parameter is changed. Standardization of the specification and design parameters of this vital equipment of energy transport will not only help in ensuring optimal deployment of available resources but also go a long way in economizing the capital costs. Standard design would mean quicker deliveries, lesser outages and a big leap towards achieving national grid.

Efforts to standardize transformer specification have been made in different forums. The CEA report and the CBIP specifications are the outcome of a concerted effort over the years.

15.4.3 CEA Report of the Committee for Standardization of the Parameters and Specifications of Major Items of 400 kV Equipment

The Government of India, in the Ministry of Energy (Department of Power) constituted a committee with active participation of Central Electricity Authority (CEA) in March 77 (BHEL was also represented) to make recommendations for standardization of the parameters and specifications of 400 kV equipment. The committee submitted its report in April 1978.

Chapter 1 of the report gives the specification of 400 kV transformers. Based on the characteristics of the available lightning arresters, a lightning impulse withstand level of 1300 kVp and a switching impulse withstand level of 1050 kVp have been recommended for 400 kV transformers. Method 2 as per IS : 2026 (Part III)–1981 has been adopted for specifying the insulation levels and dielectric tests.

250 MVA rating has been standardized for the matching generator transformers for 210 MW sets and 600 MVA generator transformer rating is proposed for the 500 MW sets. The interconnecting 400/220 kV auto-transformers are recommended to have 250, 315, 500 and 630 MVA standard ratings. The 400/132 kV auto-transformers shall have 100 and 200 MVA ratings. All auto-transformers have loaded 33 kV tertiary windings with one-third rating. The generator transformer shall have off-circuit tap switch and the auto-transformer shall be equipped with on-load tapchanger. All other important design parameters, viz. tapping range, impedances, operating conditions, etc., and fittings and accessories have also been standardized. Standardization is a continuous process and must go on. With this view in mind, the reconstituted Working Group No. 1 responsible for transformers has reviewed and updated the recommendations of Chapter 1. This revision is much more elaborate and also included standard dimensions of bushings from interchangeability point of view.

15.4.4 CBIP Specifications

Central Board of Irrigation and Power (CBIP) brought out a publication of "Specification for Power and Distribution Transformers" in 1968 which was revised in September 1976 to make it up to date and more comprehensive. Chapters on testing, commissioning, protection, etc., have also been added. It is revised again in October. 1999 and available as "Manual on Transformers," Publication No. 275.

Appendix I

I. Indian Standards Related to Power Transformers

(a) Specifications

1. IS : 1885 (Part: 38)–1993 Electrotechnical vocabulary, Transformers and Reactors (Second revision)
2. IS : 2026 (Part I)–1977 Power transformers: Part I General (first revision)
3. IS : 2026 (Part II)–1977 Power transformers : Part II Tempera-ture rise (first revision)
4. IS : 2026 (Part III)–1981 Power transformers: Part III Insulation level and dielectric tests (second revision)
5. IS : 2026 (Part IV)–1977 Power transformers: Part IV Terminal marking, tappings and connections (first revision)
6. IS : 11171-1985 Dry type power transformers
7. IS : 2026 (Part V)–1994 Power transformers : Part V Transformer/Reactor Bushing-minimum external clearances in air.

(b) Fittings and Accessories

1. IS : 778-1984 Gunmetal gate, globe and check valves for waterwork purposes
2. IS : 2099-1986 Bushings for alternating voltages above 1000 V (second revision)
3. IS : 3347 (Part I-Part VII–Section 1) Dimensions for porcelain transformer bushings, Section 1–Porcelain parts, Voltage 3.6-123 kV
4. IS : 3347 (Part I-Part VII–Section 2) Dimensions for porcelain transformer bushings, Section 2–Metal parts, Voltage 3.6-123 kV
5. IS : 2312-1997 Propeller type AC ventilating fans
6. IS : 3024-1965 Grain oriented electrical steel sheets and strips
7. IS : 3151-1982 Earthing transformers
8. IS : 3231-1965 Electrical relays for power systems protection
9. IS : 3401-1992 Silicagel (third revision)
10. IS : 3588-1987 Electric axial flow fans
11. IS : 3624-1987 Specification for pressure and vacuum gauges

12. IS : 3637-1966 Gas operated relays
13. IS : 3639-1966 Fittings and accessories for power transformers
14. IS : 4253 (Part II)-1980 Specification for cork composition sheets
15. IS : 6088-1988 Oil to water heat exchangers for transformers
16. IS : 7421-1988 Porcelain bushings for alternating voltage up to and including 1000 V
17. IS : 8468-1977 On-load tapchangers
18. IS : 8603-1977 (Part I to III) Dimensions for porcelain transformer bushings for use in heavily polluted atmosphere
19. IS : 8765-1978 Ceramic insulating materials for electrical purposes
20. IS : 9147-1979 Cable sealing boxes for oil immersed transformers
21. IS : 9700-1991 Specification for activated alumina
22. IS : 11333-1985 Flame proof dry type transformers for use in mines

(c) Selection, Installation, Operation and Maintenance

1. IS : 104-1979 Readymixed paint, brushing, zinc chrome, priming
2. IS : 900-1992 Code of practice for installation and maintenance of induction motor.
3. IS : 1255-1983 Code of practice for installation and maintenance of paper insulated power cables
4. IS : 1554-(Part I)–1988 PVC insulated heavy duty electric cables (for working voltages up to and including 1100 V)
5. IS : 1866-2000 Code of practice for electrical maintenance and supervision of mineral insulating oil in equipments
6. IS : 10028 Code of practice for selection, installation, maintenance of transformers
7. IS : 2266-1989 Steel wire ropes for general engineering purposes
8. IS : 2932-1993 Enamel, synthetic, exterior undercoating and finishing
9. IS : 3043-1987 Code of practice for earthing
10. IS : 3638-1966 Application guide for gas operated relays
11. IS : 3832-1986 Hand operated chain pulley blocks
12. IS : 3842 (Part I)–1967 Application guide for electrical relays for AC systems. Part I, Over current relays for feeders and transformers
13. IS : 3842 (Part XII)–1976 Application guide for electrical relays for AC systems. Part XII, Differential relays for transformers.

14. IS : 5216-1962 Guide for safety procedures and practices in electrical work
15. IS : 5561-1970 Electric power connectors
16. IS : 13234-1992 Guide for short-circuit current calculations in 3-phase AC system
17. IS : 6034-1989 Insulating oil conditioning plant
18. IS : 6132 Shackles
    Part 1–1971 General requirements
    Part 2–1972 Dimensions of D-shackles
    Part 3–1972 Dimensions of bow shackles
19. IS : 6600-1972 Guide for overloading of power transformers
20. IS : 7689-1989 Guide for the control of undesirable static electricity
21. IS : 8270 (Part I-V) Guide for preparation of diagrams, charts and tables for electrotechnology
22. IS : 8478-1977 Application guide for on-load tapchangers
23. IS : 8923-1978 Warming symbol for dangerous voltages
24. IS : 9434-1992 Guide for sampling and analysis of free and dissolved gases and oil from oil filled electrical equipment
26. IS : 9615-1980 Guide on general aspects of electromagnetic interference suppression
27. IS : 10561-1983 Application guide for power transformers

(d) Transformer Oil and Oil Testing

1. IS : 335-1993–New insulating oils (Fourth revision)
2. IS : 1448 (P : 10)–1970 Methods of test for petroleum and its products, cloud point and power point
3. IS : 1448 (P-16) 1990 Methods of test for petroleum and its products, cloud point and power point–Density of crude petroleum and liquid petroleum products by hydrometer method
4. IS : 1448 (P-21)–1992 Methods of flash point by Pensky Martens apparatus
5. IS : 1548 (P-25)-1976-Determination of kinematic and dynamic viscosity
6. IS : 1783-1993–Drums, large fixed ends
7. IS : 1866-2000–Code of practice for Electrical, Maintenance and Supervision of Mineral Insulating Oil in Equipment. (Third Revision)
8. IS : 2362-1973–Determination of water by the Karl Fischer Method

9. IS : 6103-1971 Method of test of specific resistance (Resistivity) of electrical insulating liquids
10. IS : 6104-1971 Method of test for interfacial tension of oil against water by ring method.
11. IS : 6262-1971 Method of test for power factor and dielectric constant of electrical insulating liquids
12. IS : 6792-1992 Method of determination of electric strength insulating oils
13. IS : 6855-1973 Methods of sampling for liquid dielectrics
14. IS : 9434-1992 Guide for sampling and analysis of free and dissolved gases and oil from oil filled electrical equipment

(e) Insulation Co-ordination and High Voltage Testing

1. IS : 1876-1961 Method for voltage measurement by means of sphere gaps
2. IS : 2071 – Methods of high voltage testing
   Part 1 – 1993 General definitions and test requirements
   Part 2 – 1974 Test procedures
   Part 3 – 1976 Measurement devices
3. IS : 2165-1977 Insulation coordination
4. IS : 3716-1978 Application guide for insulation coordination
5. IS : 4004-1985 Application guide for nonlinear resistor type surge arresters for AC system
6. IS : 4850-1968 Application guide for expulsion type lightning arresters
7. IS : 6209-1982 Methods for partial discharge measurements
8. IS : 8690-1977 Application guide for measuring devices for high voltage testing

II. IEC Publications

1. 60076-1 (2000) "Power transformers"
   Part 1 : General
2. 60076-2 (1993) Part 2 : Temperature rise
3. 60076-3 (2000) Part 3 : Insulation levels and dielectric tests
4. 60076-5(2000) Part 5 : Ability to withstand short-circuit
5. 60214 (1989) "On-load tapchangers"

6. 60542 (1976) "Application guide for on-load tapchangers"
7. 60354 (1991) "Loading guide for oil-immersed transformers"
8. 60551 (1995) "Measurement of transformer and reactor sound levels"
9. 60296 (1986) "Specification for new insulating oil for transformers and switchgear"
10. 60076-8 (1997) "Application guide for power transformers."

CHAPTER 16

Loading and Life of Transformers

D.P. Gupta

In any electric power system, right from generation to utilization, there are several points where power transformation takes place. The total capacity of transformers used for this purpose in a system would be around six to eight times the installed generating capacity. Inspite of transformers having the highest loss-efficiency among the various electrical equipment, the cumulative energy losses are substantial. Their construction also requires large quantities of vital material like copper, aluminium, core and structural steels, cellulose insulation, oil, etc. Considering depletion of world resources, there is an ever increasing emphasis on conserving materials and energy on one hand and capital on the other.

This chapter aims at bringing about a better understanding of the behavior of transformer under varying conditions of ambient air temperature, load and winding temperatures and how they affect the life of the transformer, so that it would help in conserving the vital materials, energy and capital resources by maximizing their use without unduly sacrificing safety, reliability and normal expected life.

Materials can be put to optimum use by the manufacturer of the transformer through an in-depth understanding of the properties and the stresses that are likely to occur in service. But it is the purchaser who can optimize the utilization of the existing transformers or funds invested in them, through maximizing their use by better understanding of the inbuilt capabilities.

It is necessary to establish a common base of understanding between user and manufacturer concerning general consequences of loading beyond name plate rating and particular degree of loading capability required for a specific transformer.

16.1 Life of a Transformer

A transformer has practically no moving parts, except tapchangers or cooling fan or pump motors. Therefore, it cannot wear out like rotating machinery. With adequate protection against corrosion, the copper windings, laminated cores and fabricated parts will last indefinitely. But insulating materials, mostly made from cellulose materials, deteriorate from the effects of temperature, moisture and oxygen. Out of these factors, it is the temperature which must be kept within known limits to prevent rapid deterioration of insulation. The time in which insulation deteriorates under normal usage to the point of failure, may be very long by human standards. The life of a transformer is normally dependent upon the life of the insulation. When insulation fails, the transformer life has ended. The term *transformer life* gives an impression as if it was quite definite, but in fact a transformer hardly ever *dies*. It is usually *killed*, by some unusual stresses breaking down a weakened part leading to the end of the transformer. The two factors which normally contribute to the eventual failure are:

(a) Deterioration of insulation over a span of time with temperature, moisture and oxygen.
(b) Operating stresses, mechanical, electromagnetic, thermal, beyond the strength of those parts which have considerably weakened over a period of time.

The life of a transformer has ended when probability of its failure becomes too high. It is practically impossible to determine the probability that any transformer will fail within a few months or years; but this probability actually exists. In other words, it is not possible to predict precisely the life of a transformer working under given temperature and load conditions. Presently, a scientific basis for evaluating degree of risk is not available. Dielectric strength of insulation, immersed in oil, does not deteriorate until the material has become brittle and cracked and it is, therefore, possible for a transformer to continue to operate long after the mechanical life of its insulation has been virtually used up, unless it is subjected to excessive mechanical stresses like short-circuit, handling or other mechanical shocks. It is known to users of transformers that many times a transformer which has been giving satisfactory service for years, when shifted to other location has failed due to damage to its weakened insulation. It does not follow, therefore, that a transformer will fail when its insulation becomes brittle. However, if it is subjected to some excessive stresses, it may fail even before the mechanical strength is completely used up.

Insulation is deteriorating and losing its life all the time, depending upon the temperature at which it is operating. Therefore, it becomes necessary to understand the various factors which contribute to the deterioration of insulation; especially the temperature and laws relating to ageing of insulating materials, so that a transformer can be put to optimum use without too rapid a loss of life, resulting in premature failure.

16.2 Ageing of Insulation

In oil immersed transformers, paper or cellulose material along with oil forms the major insulation. Therefore, the insulation must maintain adequate dielectric strength against voltage surges and adequate mechanical strength against short-circuit forces.

Paper and pressboard insulation when heated under oil for long periods of time, lose mechanical strength but dielectric strength is hardly affected until the paper is charred to the point where free carbon becomes conducting or too brittle to withstand mechanical shocks. Deterioration to this extent is complicated by free water liberated by decomposition, leading to lowering of dielectric strength. Depolymerization of insulation takes place when deterioration sets in.

Long before this occurs, mechanical strength will be reduced considerably. There are several ways of defining mechanical strength of paper, but the most meaningful and easily measured quantity is tensile strength. Although there is no simple relationship between loss of tensile strength and loss of effective transformer life, it seems reasonably logical to use loss of tensile strength as a measure of life.

16.2.1 Causes of Loss of Tensile Strength of Paper in Oil

The combination of factors to which paper insulation is subjected to in service are several, including temperature, time and presence of moisture, oxygen and various other reaction products in oil. Many studies have been made in the past which give reasonable closeness of relationship between tensile strength, temperature and time. However, it has been found that very small amounts of moisture accelerate the loss of tensile strength (Fig. 16.1). Presence of oxygen with moisture can lead to various kinds of reactions depending upon temperature and presence of catalytic materials.

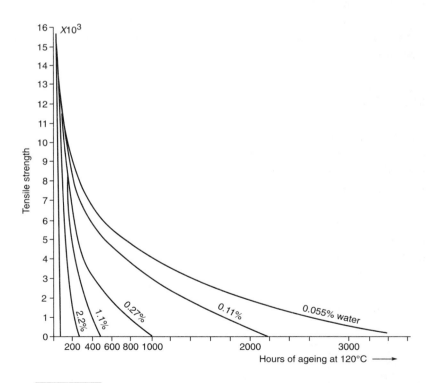

Figure 16.1 *Effect of moisture on mechanical strength of oil-immersed paper at 120°C in sealed container.*

16.3 Law of Insulation Ageing

The insulation of a transformer tends to age and deteriorate when heated. The higher the temperature, the faster is the insulation deterioration. During periods of subnormal operating temperature, the loss of life of the insulation will be less than normal. But when the operating temperatures are greater than normal, the loss of life will be higher than normal. Consequently, transformer may be safely operated for a time at above normal temperatures, provided the loss of insulation life during this period is adequately compensated for by operation for a sufficiently long time at temperatures lower than normal.

The law of insulation ageing is not linear. It is now generally accepted that the life of insulation deteriorates with temperature and is governed by the law of Arrhenius:

$$\text{Life} = e^{(A + B/T)} \quad (1)$$

where A and B are constants, (derived by experiment for a given insulating material), T is the absolute temperature and e is the Napierian base. In the range of 80° to 140°C winding hot-spot, this law can be expressed in a more convenient form called Montsinger relation:

$$\text{Life} = e^{-p\theta} \quad (2)$$

where p is a constant and θ is the temperature in °C. Various investigators have not always agreed on the length of life at any given temperature. However, they do agree, that between 80° to 140°C, the rate of loss of life due to ageing of transformer insulation is doubled for every 6°C rise in temperature. Until a few years back, the insulation ageing was based on 8°C rule, which has been replaced by 6°C rule, considered more near the actual behavior of cellulose insulation under varying temperatures.

The Montsinger relation can be used to obtain the relative rate of using life at any hot-spot temperature over the normal hot-spot temperature (98°C) at rated load and reference ambient and temperature rise conditions dealt in Chapter 5, paragraphs 5.1 and 5.2. Relative rate of using life,

$$V = \frac{\text{Rate of using life at any hot-spot } \theta_c}{\text{Rate of using life at rated hot-spot } \theta_{cr}}$$

$$= 2(\theta_c - \theta_{cr})/6$$

or
$$= 2(\theta_c - 98)/6$$
$$\theta_c = 98 + 19.93 \log_{10} V \tag{3}$$

If a transformer operating at hot-spot θ_c uses up one day's life (based on concept of normal life at 98°C hot spot) in t hours:

$$\begin{aligned} t &= 24/V \\ &= 24 \times 10^{(98-\theta_c)/19.93} \end{aligned} \tag{4}$$

Values of θ_c for various values of t in hours are given in Table 16.1:

> **Table 16.1** Values of θ_c for Various Values of t

t, hr/day	θ_c
24	98
16	101.5
12	104
8	107.5
6	110
4	113.5
3	116
2	119.5
1.5	122
1.0	125.5
0.75	128
0.5	131.5

16.3.1 Concept of Weighted Ambient

As the law of insulation ageing is not linear (Fig. 16.2), an equal variation of ambient over the mean in a given period of time and constant load will result in corresponding variations in the winding hot-spot temperatures and a net loss of insulation life over the normal. Therefore, the temperature which, if maintained continuously during the period of time under consideration, would result in the same ageing of insulation as that occurring under the actual condition of ambient temperature variations, is called *weighted ambient temperature*. The weighted ambient temperature is, therefore, not the arithmetic mean value of the ambient temperatures over a given period of time, and has essentially the element of life built in it.

Loading and Life of Transformers

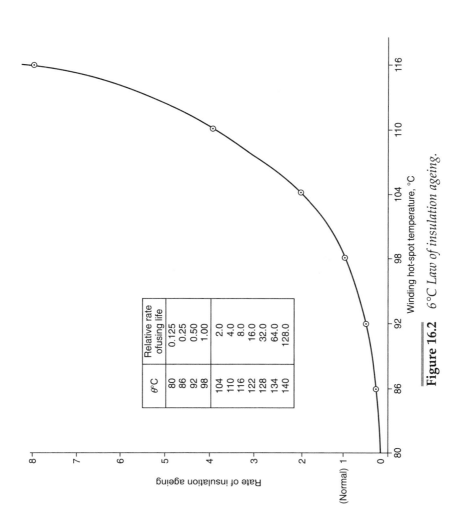

Figure 16.2 6°C Law of insulation ageing.

16.4 Significance of Weighted Value of Ambient Temperature

16.4.1 Ambient Temperature

Any location on the surface of the earth sees variations in the ambient temperatures over the day, month and full year due to the rotation of earth on its own axis in 24 h and around the sun in a year. If the temperature variations over a day are plotted for most of the parts of the world, the variation observed would be found close to sinusoidal (Fig. 16.3). Similarly, the variation of daily average ambient temperature plotted over the whole year is near sinusoidal (Fig. 16.4). It will be seen that if the daily variation is superimposed on the yearly variation, it will appear as 365 sinusoidal ripples on the yearly sinusoidal variation curve (Fig. 16.5). The monthly variation which is only one-twelfth part of the yearly ambient variation curve, could be treated as a straight line variation for all practical purposes.

16.4.2 Availability of Ambient Temperature Data for a Location

The data on ambients for a given location can normally be obtained from the records maintained by the Meteorological Departments for the past several years located in Kolkata, Delhi, Madras, Nagpur and

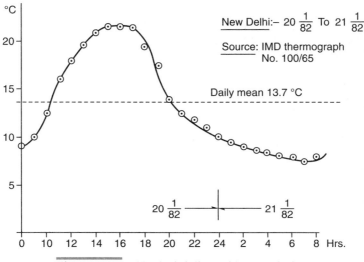

Figure 16.3 *Typical daily ambient variation.*

Loading and Life of Transformers

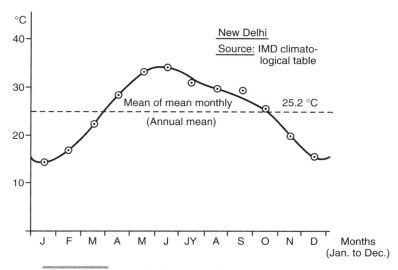

Figure 16.4 *Typical mean ambient variation over a year.*

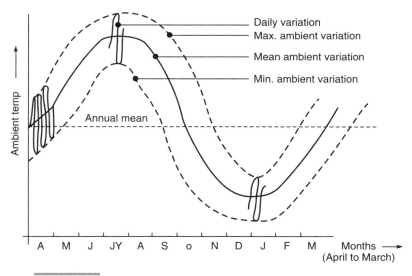

Figure 16.5 *Pattern of ambient variation over one whole year.*

Poona. Information on the *extreme* maximum and minimum ambient over a number of years, *mean* of daily maximum and minimum over a month, *highest* and *lowest* in a month is available for several locations in British Meteorological office publication *Tables of relevant humidity and precipitation for the world*, and also in the

Indian Meteorologically Department Climatological Tables of Observatories for 228 locations based on 30 years' records from 1931 to 1960.

In the following paragraphs, the use and interpretation of the available data is given towards developing ambient, load and life relationship.

16.5 Relationship Between Weighted Ambient, Winding Rise and Hot-Spot Temperatures

For evaluation of optimum loading with life of transformer as basis, it is necessary to understand the relationship between ambient, winding rise and winding hot-spot.

For ready reference, ambient temperatures as specified in IEC : 76-1 (1976) and IS : 2026 (Part I)–1977 are given below:

➢ **Table 16.2** Reference Ambients

	IS : 2026	IEC–76
1. Max. ambient air temperature	50°C	40°C
2. Max. daily average air temperature	40°C	30°C
3. Max. yearly weighted air temperature	32°C	20°C
4. Max. cooling water temperature	30°C	25°C
5. Average cooling water temperature in a day	25°C	—

Also, for ready reference, temperature rise limits for oil immersed transformers with class A insulation as specified in IEC 76-2 (1976) and IS : 2026 (Part II)–1977 are:

➢ **Table 16.3** Reference Temperature Rises

I. Air Cooled		IS : 2026	IEC 76
1. Winding rise	(a) ONAN ONAF		
	ONAF (Non-directed)	55°C	65°C
	(b) ODAF (Directed)	60°C	70°C
2. Top oil	(a) When sealed or with conservator	50°C	60°C
	(b) When not so sealed or without conservator	45°C	55°C

Contd.

➤ **Table 16.3** Contd.

		IS: 2026	IEC 76
II. Water Cooled			
1. Winding rise	(a) OFWF (Non-directed)	60°C	65°C
	(b) ODWF (Directed)	65°C	70°C
2. Top oil rise	(a) When sealed or with conservator	55°C	60°C
	(b) When not so sealed or without conservator	50°C	55°C

Basically, the cooling medium temperature as laid down by IS : 2026 (Part I)–1977 is 32°C and by IEC 76–1 (1976) is 20°C, over which deviations are provided for, in such a way that the increased use of life when operating with cooling medium temperature above these values (as in summer) is balanced by the reduced use of life when operating below these values (as in winter).

Experience indicates that normal life is some tens of years. It cannot be stated precisely, because it may vary even between identical units, owing in particular to operating factors which may differ from one unit to another.

16.5.1 Ambient–Winding Hot-Spot Relationship

A simple thermal diagram of relationship between ambient, winding and oil rises and winding hot-spot is shown in Fig. 16.6. The actual distribution of temperature is complex, but the difference is not considered sufficiently significant to invalidate the method. The following assumptions are made:

(a) The oil temperature increases linearly up the winding.
(b) The average oil temperature rise is the same for all the windings of the same leg.
(c) The difference of temperature between the oil at the top of the winding (assumed equal to that of the top oil) and the oil at the bottom of the winding (assumed equal to that at the bottom of the cooler) is the same for all windings.
(d) The average temperature rise of the copper at any position up the winding, increases linearly parallel to the oil temperature rise, with a constant difference $\Delta\theta_{wo}$ between the two straight lines: see Fig. 16.7 ($\Delta\theta_{wo}$ being the difference between the average temperature rise by resistance and the average oil temperature rise).

(e) The average temperature rise of the top portion of the winding will be the sum of top oil temperature and $\Delta\theta_{wo}$.

(f) The hot-spot temperature rise is higher than the average temperature rise of the top portion of the winding. To take account of the difference between these two temperature rises, a value of 0.1 $\Delta\theta_{wo}$ is assumed for natural oil circulation. [Forced oil circulation has been dealt with in IEC Publication 354 (1972) Appendix A, Sub-clause 1.2.2.] Thus the temperature rise of the hot-spot is equal to the top oil temperature rise plus 1.1 $\Delta\theta_{wo}$.

The ambient-winding hot-spots in IS and IEC specifications on transformers are related to the reference temperature rises as follows:

I. For Natural Oil Circulation ONAN/ONAF

➢ Table 16.4

	Description	IS : 2026	IEC 76
(a)	Max. yearly weighted temp.	32°C	20°C
(b)	Top oil rise	*45/50°C	*55°C/60°C
(c)	Average winding rise by resistance (non-directed)	55°C	65°C
(d)	Mean oil rise ONAN/ONAF assumed 0.8 × top oil rise	36/40°C	44/48°C
(e)	Winding gradient $\Delta\theta_{wo}$ (c − d)	19/15°C	21/17°C
(f)	Hot-spot gradient 10% of $\Delta\theta_{wo}$	1.9/1.5°C	2.1/1.7°C
(g)	Hot-spot temperature at rated load (a + b + e + f)	97.9/98.5°C	98.1/98.7°C
(h)	Hot-spot temperature for unity life	98°C	98°C

* For non-conservator or unsealed transformers.

II. For Forced Oil Circulation OFAF, OFWF

Generally, in forced oil cooled transformers the difference between the inlet the outlet oil temperature rise will be less than with natural circulation. Therefore, with winding temperature rise of 65°C (IEC) and 55°C (IS) the hot-spot temperature will be lower than 98°C. However, for all practical purposes, 98°C can be considered as applicable for forced cooled (non-directed) transformers also.

However, as it may be more economical to use higher current density and lower oil rise, for the purpose of framing IEC loading guide, top oil temperature rise of 40°C and a hot-spot temperature rise of 78°C at rated load has been assumed as most onerous. Yet, another difference in thermal time constant of such transformers which is assumed as 2 hr and not 3 hr as for natural cooled transformers, would make the loading guide tables different. The IS : 6600–1972 Guide for loading oil immersed transformers does not cover the forced oil cooled transformers at present.

The patterns of variation of temperatures in a transformer with constant load and varying temperatures and also with varying loads and constant ambient have been shown in Fig. 16.6(a) and 16.6(b). In fact, in practice the variation will be a combination of both varying load and varying ambient (Fig. 16.6(b)). A simplified load diagram for cyclic duty is given in Fig. 16.7(b).

16.5.2 Rated Load and Weighted Ambient

The transformer name plate rating is based on the load that can be carried continuously without affecting the normal life expectancy of a transformer at *annual weighted ambient air temperature*, provided the temperature of the ambient air and the oil and winding rises are within the specified limits of IS : 2026/IEC 76 as applicable to

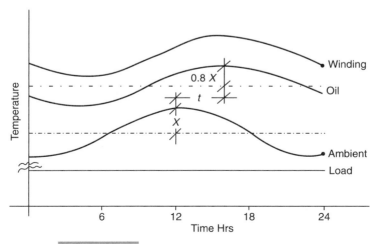

Figure 16.6(a) *Steady load v. varying ambient.*

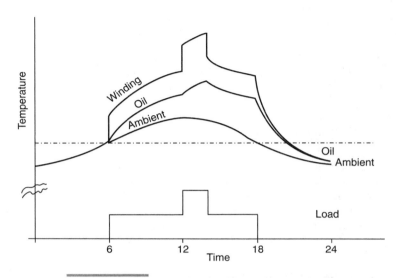

Figure 16.6(b) *Varying load v. varying ambient.*

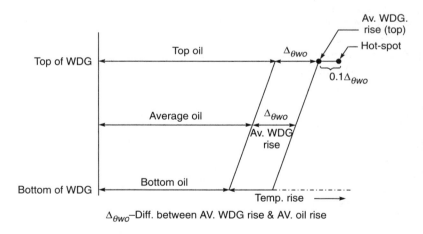

Figure 16.7(a) *Constant ambient and varying load.*

a particular case. The winding hot-spot temperature at the reference ambient temperatures will be 98°C, which will rise and fall with the variation of ambient air temperature, but to a lesser extent, depending upon the thermal time constant of the transformer and with a certain time lag. Even though the law of insulation ageing is not linear, for any variation of hot-spot temperature above and below

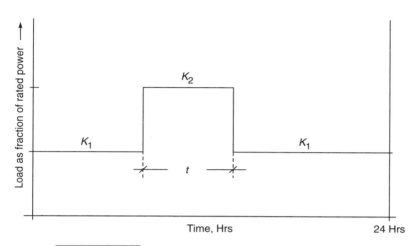

Figure 1.6.7(b) *Simplified load diagram for cyclic duty.*

98°C hot-spot, the expected life of transformer will remain unaffected over the normal as the hot-spot is based on the weighted ambient temperature, which by definition takes into account the different rates of ageing at the elevated and subnormal temperatures. In other words, if the transformer loading, whether daily, monthly or in any other convenient combination of time periods, is adopted on the basis of corresponding weighted ambient temperatures applicable over the period, the actual ambient variation would be of no consequence and can, therefore, be ignored. At the same time, it is necessary that whatever loading scheme, whether daily, monthly or in any other combination is followed, it should be carried throughout the entire one year, as the reference annual weighted ambient temperature is based on one full cycle of ambient temperature variation, resulting from one complete rotation of the earth round the sun. It is also important not to confuse the weighted ambient with the arithmetic mean ambient temperature. The difference between the two has been dealt with in the following paragraphs.

16.5.3 Difference between Arithmetic Mean and Weighted Mean Temperature

The essential difference between the two is that, whereas the *weighted ambient* is tied to the laws of insulation ageing, the *arithmetic mean ambient* is not. A close look at the curve in Fig. 16.2 will show that for an equal variation of hot-spot temperature over a given period on either side of 98°C hot-spot, will result in a net loss of insulation life over the normal, as the rate of ageing during excess temperature is more rapid than with the temperature below 98°C. In other words, the resultant rate of loss of life over the entire period will be more than the normal rate of loss of life at 98°C. To illustrate this point, consider temperature plot in Fig. 16.8. The ambient variation in a day is such that it results in a transformer hot-spot of 104°C for 12 hr and 92°C for the rest of 12 hr. Based on the relationship between the hot-spot temperature and the rate of loss of life as shown in Fig. 16.2, the average rate of loss of life in this case over a period of 24 hr will be

$$\frac{12 \times 2 + 12 \times 0.5}{24} = 1.25$$

units. The corresponding value of hot-spot temperature for 1.25 rate of loss of life from Fig. 16.2 will be 100°C and not the arithmetic mean temperature of 98°C. In other words, for the same loss of life, the value of *weighted ambient temperature*, will be 100°C–98°C = 2°C higher than the *arithmetic mean value* in this case.

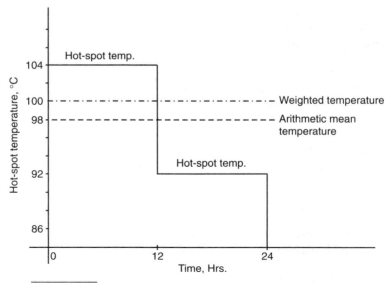

Figure 16.8 *Weighted ambient and arithmetic mean ambient.*

16.5.4 Correction Factors and Range of Temperature Variation

(a) Correction Factor

This difference between the *weighted ambient* and the *arithmetic mean ambient* has been termed as *correction factor*. Because of the nonlinearity of the law of ageing, the correction factors will be different for different shapes of ambient variation curves.

(b) Range of Temperature Variation

It will be observed from the plots of the daily and yearly ambient temperatures, that the pattern of variation over a day and a year is near *sinusoidal* (Figs 16.3 and 16.4) and over the month it can be approximated to a *straight line* (Fig. 16.9). The difference between the extremes of the ambient variation has been termed as the *range of ambient variation*.

(c) Correction Factors for Sinusoidal and Straight Line Variation of Ambients

The correction factors for *sinusoidal* and *straight line* variation of ambients which are necessary to calculate the weighted ambient have

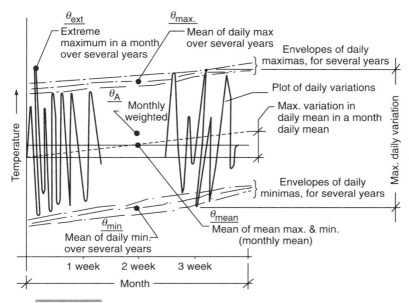

Figure 16.9 *Ambient variation over a month for several years.*

been worked out mathematically in Appendix I. The correction factors for various *ranges of ambient variations; sinusoidal* or *straight line* pattern, have been given in Fig. 16.10 (Curve A for *straight line variation* and curve B for *sinusoidal variation*).

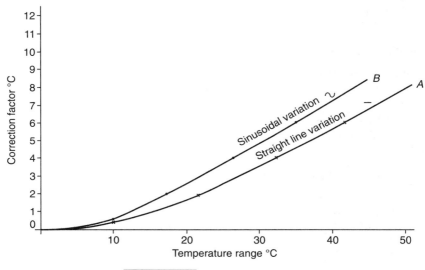

Figure 16.10 *Correction factor.*

(d) *Correction Factors for Stepped Variation of Weighted Ambients*

To arrive at the value of the weighted ambient applicable to the entire period covered by the two groups of months with two different weighted ambients (Appendix I), an appropriate correction from Fig. 16.11 is to be subtracted from the higher of the two weighted ambients. Following this approach further will give the value of annual weighted ambient for the place.

16.6 Determination of Weighted Ambient Temperature

16.6.1 Annual Weighted Ambient Air Temperature in India

No ready reference or documented information on this subject is available except one paper by Gupta and Awasthi. The practical difficulties arise in working out annual weighted ambient, because

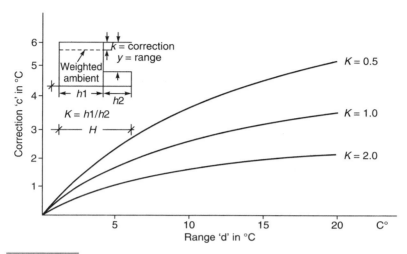

Figure 16.11 *Correction factor for stepped variation of weighted ambient.*

it is extremely difficult to predict the weather conditions from one year to the next with any degree of accuracy. The safest procedure, therefore, is to assume that the *weighted ambient temperature* adopted will be the highest, which has ever occurred, and if higher temperatures do occur, they will be only slightly higher. The weighted ambient temperatures can be calculated from the data on weather over several years available in the meteorological offices. For India, the annual mean temperature has been worked out as 26.7°C (say 27°C), by Shri H.R. Kulkarni in his paper on *Standardization of ambient temperatures*. The published records of temperatures in India also indicate that the *maximum variation in the daily mean ambient temperature* over a year is not likely to exceed 23°C and the *maximum variation of ambient air temperature in a day* by more than 23°C. Therefore, applying correction factor from curves in Fig. 16.10, the conservative correction on account of these two variations, will be about 5°C, assuming the two variations as sinusoidal. Actually, the daily variation in ambient temperature affects transformer temperature by about 80% due to the thermal time constant of the transformer.[1]

Therefore, the correction for daily variation can be applied from the curve *A* for the straight line variation (Fig. 16.10) which is 80% of the curve *B* values for the sinusoidal variation. The maximum annual weighted ambient for India, therefore, can thus be safely fixed at 27°C + 5°C = 32°C. [IS : 2026 (Part I)–1977 and IS : 6600–1972].

16.6.2 Monthly Weighted Air Ambient Temperature for a Given Location

It is essential to know the weighted air ambient temperature for a given location for the entire one year or parts thereof, normally months, over which optimum loading of a transformer could be determined. Considering that most of the ambient records available with the Meteorological Department, are either monthly or yearly, in this para. a method has been suggested for determining the weighted air ambient temperature employing the available data.

The weighted ambient air temperature for a month is based on the following three important ambient values (Ref. Fig. 16.9):

(a) Monthly mean value of the ambient
(b) Maximum variation in the daily mean ambient over a month
(c) Maximum ambient variation in any day of the month

It will be reasonably safe to consider the worst of the three values over a number of years, which can be obtained from the IMD records. The most conservative value of the *weighted ambient for the month* will be the sum of the *monthly mean ambient* and the appropriate correction factors for the two variations in ambient referred to above. Whereas correction for variation in the daily mean ambient temperature over a month which could be approximated to a straight line variation will be applied from curve A of Fig. 16.10, the correction for sinusoidal variation of ambient temperature over a day will also be applied from curve A and not curve B of Fig. 16.10. This is based on investigations, which have shown that the actual variations in the transformer temperatures are about 80% of the total variation in the ambient air temperature in a day, due to the thermal time constant of the transformer. As the correction factors in curve A are about 80% of those from curve B, correction for daily variation could be safely applied from curve A. It will be observed from Fig. 16.9 that for a month, daily variation touches maximum and minimum every day. The mean of daily maxima and minima for the month occurring over a number of years are given in the IMD tables, from which monthly mean ambient temperature can be worked out. But utilizing the data on mean of highest temperatures in the given month, it is possible to get an approximate idea of (a) the maximum variation in the daily mean over one month and (b) the maximum daily variation over one month. Correction factors for the variations (Fig. 16.10) when added to the *mean monthly ambient* will give a reasonably safe monthly weighted ambient temperatures. (Ref. Appendix II for worked out examples.)

16.6.3 Weighted Ambient for a Group of Months in a Year

To obtain the value of *weighted ambient for a group of months having two different weighted ambients*, a further correction is necessary. The mathematical derivation is given in Appendix I. Curves in Fig. 16.11 give correction factors. The correction, read from the appropriate curve is to be deducted from the higher of the two weighted ambients to arrive at the resultant weighted ambient, applicable to the total period of time of the two load groups.

However, it is important that whatever scheme of loading for a transformer is adopted, it is followed for the entire period for which the weighted ambient is applicable and not a part thereof.

It will be seen from Appendix III that the weighted ambient temperatures are considerably lower in winter months than in the summer. It will, therefore, be advantageous to load a transformer on the basis of *monthly weighted ambient temperatures* where the heavy peak loads are met in winter, otherwise to cater for summer peak loads it would be more advantageous to load the transformers on annual weighted ambient basis. This point has been explained in the worked out tables for New Delhi at Appendix II (Sheet 3) and also Fig. 16.12.

16.6.4 Ambient Temperature to be Considered in Tables of Loading Guides. Determination of Effective or Weighted Value of Cooling Medium Ambient Temperature

If θ_a varies appreciably during the high loading time t, then a weighted value of θ_a should be used, because the weighted ambient temperature will be higher than the arithmetic average.

Consider operation at constant load with a varying ambient temperature θ_a for a given period t. The weighted ambient temperature θ'_a during that period is given by the formula

$$2^{\theta'_a/6} = \frac{1}{t} \int_0^1 2^{\theta_a/6} dt$$

If the time t is divided into N equal intervals, the preceding formula becomes

$$2^{\theta'_a/6} = \frac{1}{N} \sum_1^N 2^{\theta_a/6}$$

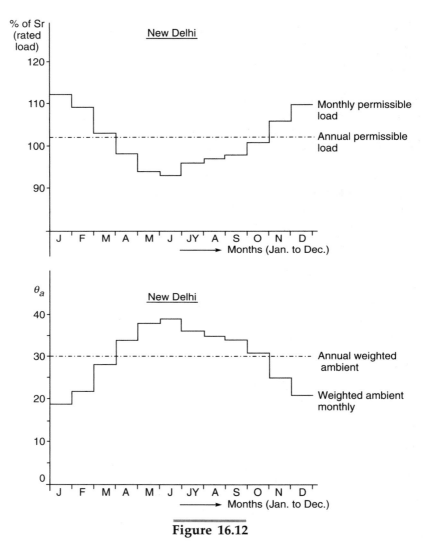

Figure 16.12

From this, the weighted ambient temperature

$$\theta'_a = 6 \log 2 \left(\frac{1}{N} \sum_1^N 2^{\theta a/6} \right)$$

which is practically equal to

$$= 20 \log_{10} \left[\frac{1}{N} \sum_1^N 10^{\theta a/20} \right] \quad (5)$$

For a transformer operating on constant load throughout the year, a weighted ambient temperature can be fixed once and for all in a given climate. This must be such that the faster deterioration during summer is exactly balanced by the slower deterioration in winter. In a temperature climate it is found that the annual weighted mean is generally 5°C higher than the arithmetic mean.

Example 16.1 Using monthly average values (more accurately using monthly weighted values) for θ_a:

$$\text{Annual weighted ambient temperature} = 20 \log_{10} \left[\frac{1}{12} \sum_{1}^{12} 10^{\theta a/20} \right]$$

θ_a = 30°C for 2 months
    20°C for 4 months
    10°C for 4 months
    0°C for 2 months

Average 15°C Weighted average 19.8°C

16.7 Relationship between Weighted Ambient and Load

Once the weighted ambient has been determined for a given location, the corresponding loading capabilities of a transformer can be worked out from the following relationship between weighted ambient, oil and winding rises and load.

$$\theta_c = \theta_a + \Delta\theta_{b0} + (\Delta\theta_b - \Delta\theta_{b0})(1 - e^{-t/\tau}) + (\Delta\theta_{cr} - \Delta\theta_{br})k^{2y}$$

$$\Delta\theta_b = \Delta\theta_{br} \left(\frac{1 + dk^2}{1 + d} \right)^x$$

where
 θ_a = Weighted ambient temperature (cooling medium)
 θ_c = Winding hot-spot temperature at K load = $\theta_a + \Delta\theta_c$
 θ_{cr} = Winding hot-spot temperature at rated load
 $\Delta\theta_c$ = Winding hot-spot temperature rise at K load
 $\Delta\theta_{cr}$ = Winding hot-spot temperature rise at rated load
 θ_{bo} = Original oil temperature
 θ_b = Top oil temperature at K load = $\theta_a + \Delta\theta_b$
 θ_{br} = Top oil temperature at rated load
 $\Delta\theta_b$ = Top oil temperature rise at load K
 $\Delta\theta_{br}$ = Top oil temperature rise at rated power
 t = Time duration in hours of any load power
 τ = Oil-air thermal time constant of transformer at rated load in hours

K = Ratio of any load/rated load = $\dfrac{S}{S_r}$

K_1 = Ratio of initial load/rated load = $\dfrac{S_1}{S_r}$

K_2 = Ratio of permissible load/rated load = $\dfrac{S_2}{S_r}$

x = Exponent for oil rise
y = Exponent for winding rise
d = Ratio of load loss at rated load/no-load loss (assumed 5 for the guides)
a = Subscript representing ambient (cooling medium)
b = Subscript representing top oil
c = Subscript representing hot-spot of winding
r = Subscript representing rated value

From the equation, the permissible load can be worked out by substituting the values for the various parameters. For convenience, a load versus weighted ambient curve has been plotted (Fig. 16.13) which applies to ONAN/ONAF/OFAF cooling with nondirected flow in transformers with or without conservator or sealing. Following similar lines of approach, other weighted ambient versus load curves can be drawn for different types of transformer coolings as above relationship is universal and applicable to any type of cooling.

16.8 Alternative Approach for the Calculation of Weighted Ambient

16.8.1 Monthly and Annual Weighted Ambients

The approach given in Sec. 16.7 above requires considerable data and lengthy calculations. A simpler, sufficiently conservative and practical approach could be adopted by adding *fixed correction* of 5°C (which allows for maximum variation in the mean and maximum variations in daily temperature over a month or year 23°C each) to the maximum monthly mean ambient temperature to obtain the *monthly weighted ambient temperature*. The value of maximum mean ambient temperature for a month, can be obtained from the IMD records by taking a mean of the mean maximum and minimum daily temperature for each month. A 5°C correction is safe enough as brought out by the worked out examples in this chapter.

Loading and Life of Transformers 401

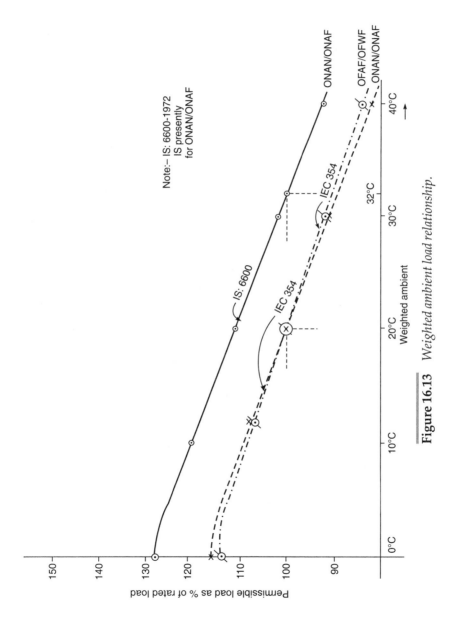

Figure 16.13 *Weighted ambient load relationship.*

In fact correction of 5°C will be more conservative for places near the sea, where temperature variations are less than the plains. Appendix III is based on the above approach and could be used for optimizing the use of transformers in the 48 locations. Similar loading tables can be worked out for another 180 locations in India from the available temperature records in the IMD climatological tables, of observatories in India (1931–60).

16.8.2 Tables for Permissible Loads on a Transformer

A few worked out examples have been given in Appendix II, for determining the monthly and annual weighted ambients and loads following the more detailed approach dealt with in Sec. 16.7.

Permissible loads on a transformer to IS : 2026 with ONAN/ONAF cooling, have been worked out in detail for 48 locations in India by following alternative approach for calculating the weighted ambients given in Sec. 16.8. The 48 locations have been chosen from the IMD climatological tables, which provide ambient data for about 230 locations, so as to cover most of the states in India. In the choice of locations, coastal areas, extremely hot desert areas and some hill stations have been included to bring out their characteristic pattern. The loading tables which are sufficiently safe and conservative, provide values of monthly and annual weighted temperature and corresponding loads for the 48 locations on the assumption, that the transformers with ONAN/ONAF cooling, strictly comply with IS : 2026.

Some of the interesting observations which emerge from the Appendix III permissible loads are as follows:

(a) Out of 48 locations the permissible annual load for 43 locations is within ±2% of the rated capacity of transformers.

(b) Where peak-loads are heavier in winter, than in summer months, it will be found advantageous to follow *permissible monthly loads*; but where *summer loads* are heavier than winter months, use of *permissible annual load may be more advantageous*.

(c) For determining permissible daily loads on a transformer, IS : 6600 *Guide for loading for oil immersed transformers* can be followed using table for appropriate monthly weighted ambient.

(d) Coastal locations do not show any appreciable difference in monthly permissible loads over the year compared to the locations in the interior.

(e) The maximum annual weighted ambient is 34°C for Madurai and minimum is 18.4°C for Srinagar.

(f) The maximum monthly weighted ambient is 41.2°C in May for Ramagundam and Kota, followed closely by Jhansi, Nagpur and Bikaner. The value of minimum monthly weighted ambient temperature is 6°C in January for Srinagar.

(g) For Kolkata, Cuttack, Khammam, Madras, Madurai, Ramagundam, Surat and Tiruchy, the weighted monthly ambient is more than 32°C for eight months, the longest period in a year.

(h) The highest extreme ambient temperature of 49.4°C occurred in May in Bikaner, followed by 48.3°C in May/June in Agra, Hissar, Lucknow and Ludhiana.

(i) The permissible loads given in the Annexure III sheet 1 and 2 are safe loads for transformers, based on weighted ambients and indicated as a percentage of the rate value. As the correction factor of 5°C to arrive at the monthly weighted ambient is over-applied by 2°C, the permissible *monthly loading* can be increased by another 2% over the figures given in the table, if such need arises in actual use.

16.8.3 Temperature Indicator Setting under Overload Conditions

Most of the transformers are provided with oil temperature indicators (OTI) for measuring hot top oil temperature of a transformer. Winding temperature indicators (WTI) are also used to give an idea of the winding hot-spot temperature by thermal image method.

Under steady rated load conditions and ambients assumed at reference level (IS/IEC), the top oil temperature will be between 77°C and 82°C depending upon whether transformer is without or with conservator, and winding hot-spot temperature of 98°C. In practice, however, neither the loads are steady nor the ambients. According to loading guides which are based on normal cyclic duty, load current should not exceed 1.5 times the rated value. Further, overloads shall not result in hot-spot temperature exceeding 140°C and top oil temperature 115°C. (The limit of 115°C for the temperature has been set bearing in mind that the oil may overflow at oil temperatures above normal.)

As already discussed in the earlier parts of this chapter, any variation in the daily air ambient temperature results in about 80% variation in the transformer temperature. This will be considered while working out ultimate top oil and hot-spot temperatures likely to occur during the period under consideration. For loading conditions as laid down in IS and IEC loading guides, it may be safer to set the OTI and WTI alarm, 100°C OTI and 115°C WTI and trip OTI 105°C and WTI 125°C. This will allow a sufficiently safe margin for any extreme cooling medium temperature condition, which may occur once in a while over the limits, on the basis of which the weighted ambient was calculated. For setting of OTI and WTIs, manufacturers' recommendations could be followed. And if need arises, they could be altered by calculating the limits under any other load and ambient conditions.

16.8.4 Effect of Altitude

The transformer standards lay down the limits for the operation of transformers at different altitudes because of the variation of these limits with the density of the air. It is to be noted that in the preparation of tables at Appendix III, no allowance has been made for altitude for locations like Shillong, Simla, Srinagar, Kathmandu and other hill stations. Therefore, before using these tables a suitable allowance/correction will have to be made for altitude.

16.8.5 Limitations Imposed by Other Considerations

The permissible loading of a transformer as discussed in this chapter and in the guides, is restricted to thermal considerations. Other economic factors depending upon capitalization of losses, etc., will have to be considered separately. The transformer that has been operating at loads higher than the rated may not comply with the thermal requirements on specified short-circuit limits in IEC or IS. Before allowing over-loads it will also be necessary to check the ratings of the terminal outlets, tapchange devices and similar other attachments.

The loading possibility of various fittings should be determined by referring to the manufacturer and if there are constraints, lower limit of loading and duration will have to be adopted. It would also be necessary to ascertain the thermal capability of the associated equipment like cables, bus-bars, circuit breakers, current transformers, etc. It has also to be ensured that the winding hot-spot temperature does not exceed 140°C owing to accelerated deterioration effects, either because the formation of deterioration products is too fast for them to be taken away by the oil or because a gaseous phase is started, sufficiently rapid to lead to over-saturation and the formation of bubbles which may endanger the electric strength. Limitation of conservator size or air cells, may also restrict the overload. The present guides are applicable to transformers up to 100 MVA ratings. However, work is being done by an IEEE task force to develop a loading guide for transformers above 100 MVA ratings, for Working Group on Guides for Loading Insulation Life Sub-committee of IEEE Transformer Committee.

In recent times a strong interest has been shown in the functional life characteristics of large transformers loaded above name plate ratings. Cost would be prohibitive for performing life test on transformers above 100 MVA ratings, but some research projects have been started towards gaining an insight into factors affecting loadability. The principal factors influencing the approach adopted by IEEE Task Force are:

(a) Stray flux heating was identified to be a possible limitation to transformers rated above 33 MVA (single-phase) and 100 MVA (three-phase) when operated in excess of name plate rating. Stray flux heating, which is function of both the MVA rating and the leakage reactance of the transformer, is caused by leakage flux which finds a path outside the iron core. The heat generation typically occurs in the tank and structural parts of the transformer, but can also occur in the core and the winding leads.

(b) Research work performed in the areas of the combined effects of thermal ageing and short-circuit stresses on dielectric strength leads to a conclusion that life expectancy could be reduced in larger transformers with high short-circuit forces.

(c) The phenomenon of bubble evolution from cellulose insula-tion at overload temperatures is presently an area of major research. The results of research to date indicate that bubbling does significantly reduce both power frequency and impulse strength of transformer insulation and is recognized as an important limitation when operating above nameplate rating.

(d) Because the large transformers usually have higher voltage ratings, they tend to have greater average dielectric stresses and larger volumes of insulation under stress. Also, they are often specified with reduced steps of BIL. These factors, together inherently reduce the margins for dielectric deterioration.

(e) There is a wide variation in user's requirements for loading beyond nameplate rating for the large transformers. In addition, there appears to be a trend to use planned loading above nameplate rating as contrasted to the more conser-vative use of emergency loading.

(f) Many misunderstandings and problems have arisen from reference to the present loading guide in a transformer specification to identify loading capability requirements. The present guide does not convey the necessary information required by a transformer designer, to provide the above nameplate rating capabilities in the transformer that are desired by the user.

(g) The consequences of failure of these large transformers are becoming a matter of more concern with the rapidly increasing outage and repair cost. Hence, it is of great importance that the most complete and up-to-date infor-mation should be available for planning loads above nameplate ratings.

16.8.6 Effect of Loading beyond Nameplate Rating

When a transformer is designed for operation at a given rating level, stress factors are selected to achieve a satisfactory life expectancy at that rating level. If the rating level is exceeded, there is some en-croachment on margins and the question of risk of premature failure must be considered. While the quantification of the risk factor may be difficult, as a first step it is necessary to identify the potential effects from increased loading. Some of the important areas of concern during transformer loading beyond nameplate rating are:

(a) Evolution of gas from overheated insulation on winding and lead conductors which may jeopardize dielectric integrity. This factor is a relatively new concern which has been addressed in the recent literature.

(b) Evolution of gas from insulation adjacent to metallic parts heated by stray electromagnetic flux.

(c) Deterioration of the electrical and mechanical properties of conductor insulation from thermal ageing. This is the traditional *life* characteristic on which calculations are furnished in existing guides.
(d) Cumulative thermal degradation of mechanical properties of structural insulation parts, plus some temporary reduction of mechanical capability at elevated temperatures.
(e) Permanent deformations of conductors, insulation materials or structural parts produced by thermal expansion at overload temperatures.
(f) Gasket rupture in bushings produced by internal pressure buildup at current levels above rated capability.
(g) Degrading contact resistance in tapchangers resulting from buildup of oil decomposition products at a localized contact point hot-spot. This could only be a concern with prolonged operation beyond nameplate rating.
(h) Ancillary equipment internal to the transformer which may be subjected to effects similar to those mentioned above.
(i) Oil spills produced by oil expansion during operation at temperatures above the design value. This is an inconvenience rather than an operating risk.

16.9 Transformer Loading Guides

Loading guides for power transformers date back to 1942 with very little change in basic philosophy. In the currently published guides, the emphasis is still on a long-time end of life failure mode associated with the thermal deterioration of mechanical and dielectric properties of solid insulation in the windings. The useful *life* of a transformer has been arbitrarily equated to the loss of 50% of the initial tensile strength of winding conductor insulation. Consumption of *life* may be calculated according to equations or tables given in the guides based on Arrhenius type relationship between the degradation rate and temperature. Although there are other stresses like mechanical, dielectric and thermal produced by stray fluxes, which increase with MVA rating of transformers, consideration of these stress factors has generated an agreement in the industry, that a reasonable point of separation for the loading guide would be 100 MVA. The present guides, therefore, are limited. In the available transformer loading guides the basic approach is the same, except that the loading tables have been modified to suit the local conditions in a country. Some of the salient factors of relevance to this chapter based on the study of IEC 354 (1972) and IS : 6600-1972 have been discussed in the paragraphs below:

(a) The guides are applicable to transformers up to 100 MVA, which complying with the relevant IEC 76 or IS : 2026 for transformers are operated within the specified limits. Any deviation must be considered carefully and such changes made in loading as become necessary for the safety of the transformers and other associated equipment.

(b) For transformers above 100 MVA, advice of manufacturer should be followed. (IEEE task force is working on the guide to loading of oil immersed transformers above 100 MVA and also IEC Pub. 354 is being revised to include transformers above 100 MVA.)

(c) Under no circumstances, hot-spot temperature be allowed to exceed 140°C beyond which gaseous phase starts and may endanger the electric strength.

(d) Top oil temperature should be kept within 115°C and matched to suit the expansion space for oil.

(e) To use the guide, knowledge of weighted ambient temperatures: daily, monthly or annual for a given location, is necessary. Practically, these values can easily to worked out from the meteorological records for various months and the year as discussed in this chapter. As the loading table in the guide are based on annual weighted ambient temperature, the daily loads on the transformer can be met by referring to the appropriate table in the guide, which corresponds to the annual weighted ambient, where transformer is actually located.

(f) Similarly, daily loads can be met by referring to the tables in guide on the basis of monthly weighted ambient. Loading on the basis of monthly weighted ambient then will have to be followed for the entire one year and not changed to annual weighted in the middle, as integration of load-loss relation-ship is over one full year.

(g) The transformer rated ONAN and ONAF can be loaded as per appropriate set of tables in the guide, in terms of rated power for ONAN, if fans are not working and rated power for ONAF if fans are working. Similarly, transformers with both ONAN and OFAF cooling can be dealt with.

(h) Tables in the loading guides give permissible cyclic daily duty with a normal life consumption of 24 hr per day, equal to that consumed during 24 hr at 98°C hot-spot. If actual load diagram has two or more periods of high load separated by periods of low load, the high-load time can be taken as the sum of all high-load times. This condition of intermittent high loads is less onerous than single overload for the same total time.

(i) For a given load cycle, and given ambient conditions, tables can be used to determine the rating of the transformer needed for the load.

16.10 Loading by Hot-Spot Temperature Measured by WTI

With the knowledge of *relative rate of use of life* at any hot-spot temperature as discussed in Sec. 16.3 of this chapter, for practical purposes, consumption of life on any day could be worked out for a transformer for any loading and ambient condition by integrating the consumption of life based on the record of readings of the winding temperature indicator at regular intervals over the day. The winding temperature indicator indicates thermal image of the winding hot-spot temperature and can be considered sufficiently accurate when it is corrected to the actual measured values based on temperature rise test on the transformer. In adopting this approach, care should be taken to ensure:

(a) Oil temperature and winding temperature at any point of time do not exceed the specified limits of 115°C and 140°C respectively.

(b) Upper limit of load as dictated by the loading limits of other associated components shall not be exceeded. In any case, loads above 150% of the rated are special emergency conditions.

(c) Process of life integration when done manually should be over a day, in such a way that resultant consumption of life is related to one day's life at 98°C hot-spot (Normal day).

(d) It would be safer to adopt the highest value of hot-spot reached in any time interval for manual integration of life.
(e) The above method of life integration by using WTI readings takes care of any changes in ambient and load and effect of thermal mass.

Example 16.2 Calculation for working out the life of a transformer consumed on a day, on the basis of hourly WTI readings is given in the table below:

Hours	θ_c = Hot-spot (Hourly WTI reading °C)	Rate of using life	Life used in hours
1	80	0.125	0.125
2	78	0.100	0.10
3	79	0.11	0.11
4	80	0.125	0.125
5	81	0.13	0.13
6	86	0.25	0.25
7	93	0.55	0.55
8	98	1.00	1.00
9	104	2.00	2.00
10	104	2.00	2.00
11	101	1.40	1.40
12	93	0.55	0.55
13	98	1.00	1.00
14	103	1.70	1.70
15	104	2.00	2.00
16	104	2.00	2.00
17	98	1.00	1.00
18	93	0.55	0.55
19	98	1.00	1.00
20	101	1.40	1.40
21	101	1.40	1.40
22	95	1.00	1.00
23	93	0.55	0.55
24	83	0.20	0.20
		Total	22.69 Hr

From the above table it will be seen that against the normal allowed consumption of life of 24 hr actual consumption has been 22.69 hr, which leaves a balance of 1.31 hr of unused life.

16.11 Selection and Use of a Transformer

In the guides to the loading, the load diagrams have been simplified into low-load and high-load periods. Based on the value of weighted ambient θ_a, appropriate loading tables in the guide can be used. However, if θ_a lies between two tables, either select the nearest one above or interpolate between the closest two tables.

For transformers rated for both ONAN and ONAF cooling, for ONAN cooling use appropriate tables in terms of the rated power for ONAN cooling, and in terms of the rated power of ONAF cooling, if the fans are brought into operation. Similarly, transformers with ONAN and OFAF cooling will be dealt with or without oil pumps in operation.

It is recommended that pumps and fans are put into service before the overloading occurs, in order to have the winding hot-spot temperature low enough to slow down the ageing process. The power taken by these auxiliaries is at least partially compensated for by the decrease in load loss, resulting from a low temperature. Following worked out examples will help towards a better understanding of the concept of life and loading:

Example 16.3 Determine maximum permissible load which a 15 MVA, ONAN cooled transformer to IS : 2026 can take for 6 hr, if the initial load on it was 12 MVA. The weighted ambient temperature is 20°C.

Cooling type	:	ONAN
θ_a	:	20°C
K_1	:	0.8 (12 MVA of rated 15 MVA)
*t	:	6 hr

*(Designated h in IS : 6600)
K_2 (from Table 3 of IS : 6600) = 1.25
∴ Maximum permissible = 1.25 × 15 = 18.75 MVA load for 6 hr.

The overload will be reduced to 12 MVA for the remaining part of the day.

Example 16.4 Same transformer as in Example 16.3, but with initial load of 7.5 MVA. Determine the maximum permissible load for 2 hr.

Cooling type	:	ONAN
θ_a	:	20°C
K_1	:	0.5 (7.5 MVA of rated 15 MVA)
t	:	2 hr

$\therefore$ K_2 (from Table 3 of IS : 6600) = 1.55

But as the upper limit of load is restricted to 1.5, the maximum permissible load the transformer could be allowed to carry for

$$2 \text{ hr} = 1.5 \times 15$$
$$= 22.5 \text{ MVA}$$

before load is reduced to 7.5 MVA for the remaining part of the day.

Example 16.5 Find out the rating of ONAN cooled transformer to IS : 2026, which will cater for a lean load of 8 MVA for 18 hr and 14 MVA for balance 6 hr in a day in a location with weighted annual ambient temperature of 32°C.

$$\frac{K_2}{K_1} = \frac{14}{8} = 1.75$$

A life with $K_2/K_1 = 1.75$ can be plotted intersecting the curves in Table 5 of IS : 6600. Where the line intersects the 6 hr curve, the K_2 reads = 1.14.

$$S_r = \text{(Normal transformer rating)} = \frac{S_2}{K_2} = \frac{14}{1.14} = 12.28 \text{ MVA}.$$

Example 16.6 What will be the normal rating of ONAN/ONAF cooled transformer to IS : 2026, which will deliver continuously a load of 12.7 MVA at a location with annual weighted temperature of 30°C.

Cooling type	:	ONAN/ONAF
θ_a	:	30°C
t	:	24
K_2	:	1.02 (From Table 4 of IS : 6600)

$$K_2 = \frac{S_2}{S_r}$$

$$S_r = \frac{S_2}{K_r} = \frac{12.8}{1.02} = 12.5 \text{ MVA}$$

Loading and Life of Transformers **413**

Example 16.7 A 20 MVA ONAN/ONAF transformer to IS : 2026 is located in New Delhi, where peak loads are heavier in winter than in summer. Find out the safe permissible loads it can carry in each month.

Method 1
The permissible load curve vis-à-vis weighted ambient is given in Fig. 16.13 (also refer Appendix II, Sheet 2). This is based on addition of a flat correction of 5°C on the mean monthly ambient to obtain monthly weighted ambient and corresponding permissible load from curve in Fig. 16.13.

Method 2
The monthly weighted ambients can be determined by adding corrections worked out on the basis of IMD records (Ref. Appendix II, Sheet 3) and permissible load from curve in Fig. 16.13. This method permits slightly higher overloads than method 1.

REFERENCES

1. *Equivalent Ambient Temperature for Loading Transformers* by W.C. Sealey of Allis Chalmers (USA).
2. *Transformer Engineering, A Treatise On Theory, Operation and Application of Transformers* by GE (USA).
3. *Transformers for Electric Power Industry* by Westinghouse (USA).
4. IS : 2026 : *Specification for Power Transformers.*
5. IS : 6600 : *Guide for Loading of Power Transformers.*
6. Kulkarni, H.R., *Paper on Standardization of Ambient Temperatures.*
7. *IEC Loading Guide for Oil Immersed Transformers*, IEC publication 354.
8. Gupta, D.P., and L.C. Awasthy, *Weighted Ambient Temperature as the Basis for Loading Transformer*, read in CBIP Research Session at Jaipur.
9. *Progress Report on a Guide for Loading Power Transformers Rated in excess of 100 MVA*, IEEE Paper Vol. PAS 100, 8 Aug. 1981.

Appendix I
Correction Factors

Sheet No. 1

A. 1.1 Correction Factor for Sinusoidal Variation

If ambient temperature varies in a sine wave (Fig. A1.1), with duration of cycle as H, the weighted ambient (correction over base temperature) denoted by x can be worked out from the following relationship:

$$2^{x/6} = \frac{1}{H} \int_0^H 2^{\theta/6} \cdot dh \qquad (i)$$

where $\theta = \theta_m \sin h$
$H = 2\pi$

Therefore, Eq. (i) can be re-written as

$$2^{x/6} = \frac{1}{2\pi} \int_0^{2\pi} 2^{\theta_m \sin h/6} dh \qquad (ii)$$

The expansion of $2^{\theta_m \sin h}$ is :

$$2^{\theta_m \sin h/6} = 1 + \frac{\theta_m \sin h}{6} \log_e 2 + \frac{\theta_m \sin^2 h}{6^2 \angle 2} (\log_e 2)^2 + ...$$

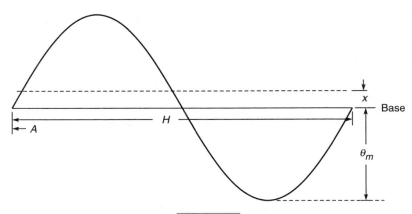

Figure A1.1

Neglecting higher powers of θ_m and solving Eq. (ii), value of x is given by
$$x = 19.93 \log_{10}(1 + 0.0033 \theta_m^2) \quad \text{(iii)}$$
For convenience, results have been plotted as curve B of Fig. 16.10.

Sheet No. 2

A.1.2 Correction Factor for Straight Line Variation

A similar approach as above can be followed for working out a correction for straight line variation also.

In the case of straight line variation as shown in Fig. A.1.2, at any instant the value of θ can be written as
$$\theta = \frac{2\theta_m}{H} h$$
(value of h at any instant must bear the correct sign). If x is the weighted ambient temperature (correction over base temperature), it can be represented by the following relationship:

$$2^{x/6} = \frac{1}{H} \int_{-H/2}^{+H/2} 2^{\theta/6} \cdot dh \quad \text{(i)}$$

or

$$2^{x/6} = \frac{1}{H} \int_{-H/2}^{+H/2} 2^{\frac{2\theta_m h}{6H}} \cdot dh \quad \text{(ii)}$$

Function $\dfrac{2\theta_m}{6H} h$ can be expanded and written as follows by ignoring higher powers of θ_m:

$$2^{2\theta_m/6H \times h} = 1 + \frac{2\theta_m}{6H}(\log_e 2) h + \frac{4\theta_m^2 (\log_e 2)^2}{36 H^2 \, 2} h^2 + \dots$$

Substituting value of $2^{2\theta_m h/6H}$ in Eq. (ii) and solving for x:
$$x = 19.93 \log_{10}(1 + 0.0022\, \theta_m^2) \quad \text{(iii)}$$
for convenience of working, results have been plotted as curve A of Fig. 16.10.

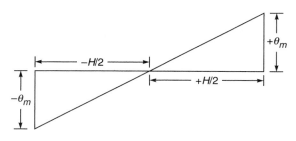

Figure A.1.2

A.1.3 Correction Factor for Stepped Variation

If two weighted ambient temperatures θ_m and $\theta_m - y$ are maintained for a period h_1 and h_2 respectively, and x denotes the factor which should be subtracted from θ_m to get the weighted ambient applicable to total time H, the relationship between various parameters is

Sheet No. 3

$$2^{(\theta_m - x)/6} = \frac{1}{H}\left\{\int_0^{h_1} 2^{\theta_m/6}\, dh + \int_{h_1}^{H} 2^{(\theta_m - y)/6}\, dh\right\} \tag{i}$$

Assuming $h_1/h_2 = K$ and rewriting Eq. (i)

$$2^{(\theta_m - x)/6} = \frac{1}{(1+k)h_2}\{2^{\theta_m/6}\, kh_2 + 2^{(\theta_m - y)/6} h_2\}$$

$$= \frac{2^{\theta_m/6}}{1+k}\{k + 2^{-y/6}\} \tag{ii}$$

Expanding $2^{y/6} = 1 + \dfrac{y}{6}\log 2 + \ldots = 1 + 0.115\, y$

Rewrite Eq. (ii) as

$$2^{-x/6} = \left\{1 - \frac{1}{(1+k)(1+8.7/y)}\right\}$$

$$x = -19.93 \log\left\{1 - \frac{1}{(1+k)(1+8.7/y)}\right\} \tag{iii}$$

For convenience, working results have been plotted as curves in Fig. 16.11.

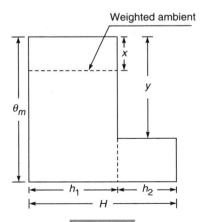

Figure A.1.3

Loading and Life of Transformers

APPENDIX II

Sheet No. 1

	Location - Agra Source - IMD Table Page 115													Appendix -II Sheet No-1
Sr. No.	Description	J	F	M	A	M	J	JY	A	S	O	N	D	Annual
A.	Mean of daily max.	22.7	25.7	31.9	37.7	41.8	40.5	34.8	32.8	33.2	33.3	29.2	24.1	32.3°C
B.	Mean of daily min.	7.4	10.3	15.7	21.6	27.2	29.5	27.0	25.8	24.6	19.1	12.0	8.2	19.0°C
C.	Mean monthly	15.1	18.0	23.8	29.6	34.5	35.0	30.9	29.3	28.9	26.2	20.6	16.2	25.7°C
D.	Max of variation in Mean in the month	2.9	5.8	5.8	4.8	0.5	4.1	1.6	0.4	2.7	5.6	4.4	1.4	
E.	Mean of highest in the month over several years	27.2	31.5	37.8	42.5	45.4	45.0	40.7	35.9	36.5	36.4	33.0	28.1	
F.	Max. daily variation $2 \times (E - C - D/2)$	21.4	21.2	22.2	21.0	21.4	16.0	18.0	12.8	12.4	14.8	20.4	22.4	
G.	Correction for 'D'	0.1	0.2	0.2	0.2	–	0.1	–	–	–	0.2	0.2	–	
H.	Correction for 'F' 80% of curve	1.9	1.9	2.1	1.8	1.9	1.1	1.4	0.6	0.6	0.9	1.7	2.1	
I.	Weighted ambient C + G + H = G + H assumed 5°C =	17.1 (20.1)	20.1 (23.0)	26.1 (28.8)	31.6 (34.6)	36.4 (39.5)	36.2 (40.0)	32.3 (35.9)	29.9 (34.3)	29.5 (33.9)	27.3 (31.2)	22.5 (25.6)	18.3 (21.2)	

J.	Correction for max. variation in mean (monthly) over one whole year. C max - C min	$35 - 15.1 = 19.9°C$	~ (Curve B) = 2.6°C
K.	Correction for max. daily variation in one whole year F max.	22.4°C	(Curve A) = 21°C Total correction = 4.7°C
L.	Annual weighted C + J + K =	25.7 + 4.7	= 30.4°C (30.7°C)

Values in ( ) by adding 5°C correction

Monthly Plot — 30 days

Yearly Plot — 12 months

Daily, F max, Mean monthly

Transformers

Sheet No. 2

	Location : Bombay Source : IMD table, Page 211												Appendix II Sheet No. 2
Ser. No.	Description	J	F	M	A	M	J	JY	A	S	O	N	D	Annual
A.	Mean of daily max.	29.1	29.5	31.0	32.3	33.3	31.9	29.8	29.5	30.1	31.9	32.3	30.9	31.0
B.	Mean of daily min.	19.0	20.3	22.7	25.1	26.9	26.3	25.1	24.8	24.7	24.6	22.8	20.8	23.6
C.	Mean of monthly	24.2	24.9	26.8	28.7	30.1	29.1	27.4	27.2	27.4	28.2	27.6	25.8	27.3
D.	Max. variation in Mean in the month	0.7	1.9	1.9	1.4	1.0	1.7	0.2	0.2	0.8	0.6	1.8	1.6	
E.	Mean of highest in the month over several years	33.0	33.9	35.0	34.8	34.6	34.4	31.8	31.1	31.7	35.1	34.8	33.7	36.7
F.	Max daily variation $2 \times (E - C - D/2)$	16.8	16.0	14.4	10.8	8.0	9.0	9.0	7.6	7.8	13.2	12.6	14.2	
G.	Correction from 'D'	—	—	—	—	—	—	—	—	—	—	—	—	
H.	Correction for 'F' 80% of curve	1.2	1.1	0.9	0.4	0.2	0.3	0.3	0.2	0.2	0.7	0.6	0.8	
I.	Weighted ambient $C + G + H =$	25.4 (29.2)	26.0 (29.9)	27.7 (31.8)	29.1 (33.7)	30.3 (35.1)	29.4 (34.1)	27.7 (32.4)	27.4 (32.2)	27.6 (32.4)	28.9 (33.2)	28.2 (32.6)	26.6 (30.8)	
J.	Correction for max. variation in mean (monthly) over one whole year													
	C max – C min.	$30.1 - 24.2 = 5.9°C$						$CF = 0.2°C$						
K.	Correction for max. daily variation over one whole year													
	F max.	16.8°C												
								$CF = 1.2°C$						
								Total corr. 1.4°C						
L.	Annual weighted $C + J + K =$													28.7°C (32.3°C)

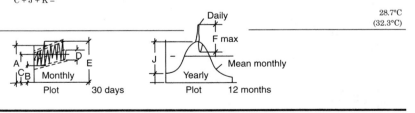

Loading and Life of Transformers

Sheet No. 3

	Location – New Delhi Source :- IMD table page 127												Appendix- II Sheet No-3
Sr. No.	Description	J	F	M	A	M	J	JY	A	S	O	N	D	Annual
A.	Mean of daily max.	21.3	23.6	30.2	36.2	40.5	39.9	35.3	33.7	34.1	33.1	28.7	23.4	31.7
B.	Mean of daily min	7.3	10.1	15.1	21.0	26.6	28.7	27.2	26.1	24.6	18.7	11.8	8.0	18.8
C.	Mean monthly	14.3	16.8	22.6	28.6	33.5	34.3	31.2	29.9	29.4	25.9	20.2	15.7	25.2
D.	Max. variation in mean in the month	2.5	5.8	6.0	4.9	0.8	3.1	1.3	0.5	3.5	5.7	4.5	1.4	
E.	Mean of highest in the month over several years	27.0	29.6	35.8	41.2	44.3	44.1	40.3	37.2	37.0	36.3	32.4	26.0	44.9
F.	Max. daily variation $2 \times (E - C - D/2)$	23.0	19.8	20.4	20.2	20.8	16.4	17.0	14.2	11.7	15.2	20.0	19.2	
G.	Correction for D	–	0.1	0.1	0.1	–	0.1	–	–	0.1	0.1	0.1	–	
H.	Correction for F 80% of curve	2.2	1.7	1.8	1.8	1.9	1.1	1.3	0.8	0.5	1.0	1.7	1.6	
I.	Weighted ambient C + G + H =	16.5	18.6	24.5	30.5	35.4	35.5	32.5	30.7	30.0	27.0	22.0	17.3	
J.	Correction for max. variation in mean (monthly) over one whole year. C max – C min:	34.3 – 14.3 = 20°C					CF = 2.6°C							
K.	Correction for max. daily variation over one whole year F. max	23.0°C					CF = 2.2°C							
							Total correction = 4.8							
L.	Annual weighted C + J + K = 30.2													30.0 30.2

Daily Monthly Plot 30 days

F max Mean monthly Yearly Plot 12 months

420   *Transformers*

Sheet No. 4

Location : Srinagar Source IMD table page 131													Appendix II Sheet No– 4
Sr. No. Description	J	F	M	A	M	J	JY	A	S	O	N	D	Annual
A. Mean of daily max.	4.4	7.9	13.4	19.3	24.6	29.6	30.8	29.9	28.3	22.6	15.5	8.8	19.5
B. Mean of daily min.	–2.3	–0.8	3.5	7.4	11.2	14.4	18.4	17.9	12.7	5.7	–0.1	–1.8	7.2
C. Mean monthly	1.0	3.6	8.4	13.4	17.9	21.7	24.6	23.9	20.5	14.2	7.7	3.5	13.4
D. Max. variation in mean in the month	2.6	4.8	5.0	4.5	4.5	3.8	2.9	0.7	3.4	6.3	6.5	4.2	
E. Mean of highest in the month over several years	9.4	13.7	20.1	26.3	31.5	34.4	35.5	34.0	32.6	28.5	20.4	14.4	36.0
F. Max. daily variation $2 \times (E - C - D/2)$	14.2	15.4	18.4	21.4	22.8	21.6	18.8	19.4	20.8	22.2	19.0	17.6	
G. Correction for D	–	0.1	0.1	0.1	0.1	0.1	–	–	0.1	0.2	0.2	0.1	
H. Correction for F 80% of curve	0.8	1.0	1.5	1.9	2.1	2.0	1.6	1.7	1.8	2.0	1.6	1.3	
I. Weighted Ambient $C + G + H =$	1.8	4.7	10.0	15.4	20.1	23.8	26.2	25.6	22.4	16.4	9.5	4.9	
J. Correction for max variation in mean (monthly) over one whole year C max – C min	24.6 – 1.0 = 23.6						$CF = 3.3°C$						
K. Correction for max daily variation over one whole year							$CF = 2.1°C$ 22.8°C						
F. Max							5.4						
L. Annual weighted $C + J + K =$													18.8 °C

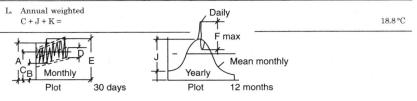

APPENDIX III

Loading Table Transformers ONAN/ONAF Cooled (To IS : 2026)

Sheet No. 1

Location	Jan	Feb	Mar	Apr	May	June	July	Aug	Sep	Oct	Nov	Dec	Annual
Agra	111	108	102	97	92	92	96	98	98	101	105	110	101
Ahmedabad	106	103	100	96	94	94	98	99	98	99	102	105	100
Ajmer	111	108	103	98	96	94	98	99	99	102	106	110	102
Allahabad	110	107	102	96	92	93	97	98	98	100	105	109	101
Ambala	112	110	104	99	94	93	96	97	97	102	107	111	102
Amritsar	114	111	107	102	97	94	96	97	98	102	108	113	102
Bangalore	105	102	101	100	100	102	103	103	103	103	104	105	102
Baroda	106	102	100	96	94	95	98	99	99	100	102	105	100
Belgaum	104	102	100	99	100	102	103	103	103	102	103	104	102
Bhopal	108	105	102	98	93	96	100	101	101	102	105	108	102
Bikaner	112	109	103	98	92	91	94	97	96	100	106	111	101
Bombay	102	102	100	98	97	98	100	100	100	99	99	101	100
Calcutta	106	103	99	96	96	97	98	98	98	99	102	105	100
Coimbatore	102	101	99	98	98	100	101	101	101	101	102	102	101
Cuttack	104	102	98	95	94	96	98	98	98	99	102	104	99
Cochin	100	100	98	98	99	100	101	101	101	100	100	100	100
Dehradun	113	111	107	102	98	98	100	101	102	104	109	112	104
Dhanbad	108	105	101	96	95	96	99	99	99	101	104	107	101
Gauhati	108	106	103	101	100	99	98	98	98	101	104	107	102
Hissar	112	109	104	99	94	92	95	96	97	101	106	111	102
Hyderabad	104	102	100	97	95	98	101	101	101	102	104	105	101
Jabalpur	108	106	102	98	93	95	100	100	100	102	106	108	102
Jaipur	111	108	103	98	94	94	97	99	99	101	105	109	102
Jammu	113	110	105	100	95	93	96	98	98	102	106	111	102

Notes: Safe loads for transformers based on Weighted Ambient (Figures in %).

Sheet No. 2

Location	Jan	Feb	Mar	Apr	May	June	July	Aug	Sep	Oct	Nov	Dec	Annual
Jhansi	109	106	102	96	91	92	97	99	99	101	105	109	101
Kathmandu	116	114	109	106	103	102	102	102	103	106	111	115	107
Khammam	102	100	97	95	92	95	98	99	98	99	102	103	98
Kota	108	105	100	95	91	92	97	98	98	99	103	107	100
Karnool	102	100	97	94	93	97	99	99	99	100	102	104	100
Lucknow	110	107	102	97	93	93	97	98	98	101	105	109	101
Ludhiana	113	110	105	100	94	93	96	97	97	101	107	111	102
Madras	102	101	99	97	94	95	96	97	97	99	101	102	98
Madurai	101	100	98	96	95	96	96	97	97	98	100	101	98
Marmugao	101	102	100	98	98	100	101	101	101	100	100	101	100
Mysore	104	102	100	99	100	102	103	102	102	102	103	104	102
Nagpur	105	102	99	95	91	94	99	100	100	99	104	106	100
New Delhi	112	109	103	98	94	93	96	97	98	101	106	110	102
Patna	109	106	101	97	95	95	97	98	98	100	103	108	101
Poona	105	103	101	98	97	100	102	102	102	101	103	105	102
Puri	104	102	100	98	97	97	98	98	98	99	102	104	100
Ramagundam	102	101	97	93	91	94	99	99	99	99	102	103	98
Shillong	116	114	110	107	106	105	105	105	106	109	113	116	109
Simla	120	118	115	111	107	106	108	108	109	112	115	117	112
Srinagar	124	122	117	113	108	104	102	102	105	112	117	122	113
Surat	103	102	99	96	96	97	99	99	99	98	100	102	99
Truchy	102	100	98	96	95	95	95	97	97	99	101	102	98
Trivandrum	100	100	99	98	99	100	101	101	100	100	100	100	100
Vizag	103	102	100	98	96	96	98	98	98	99	102	103	100

Notes: Safe loads for transformers based on Weighted Ambient (Figures in %).

Sheet No. 3

S. No.	Location		Jan	Feb	Mar	Apr	May	June	July	Aug	Sep	Oct	Nov	Dec	Annual
1.	Agra	θ_{ext}	31.1	35.6	42.8	45.0	47.2	48.3	45.6	42.2	40.6	41.1	36.1	30.0	48.3
		θ_{max}	22.7	25.7	31.9	37.7	41.8	40.5	34.8	32.8	33.2	33.3	29.2	24.1	32.3
		θ_{min}	7.4	10.3	15.7	21.6	27.2	29.5	27.0	25.8	24.6	19.1	12.0	8.2	19.0
		θ_{mean}	15.1	18.0	23.8	29.6	34.5	35.0	30.9	29.3	28.9	26.2	20.6	16.2	25.7
		θ_a	20.1	23.0	28.8	34.6	39.5	40.0	35.9	34.3	33.9	31.2	25.6	21.2	30.7
		% Rated Load	111	108	102	97	92	92	96	98	98	101	105	110	101
2.	Ahmedabad	θ_{ext}	36.1	40.6	43.9	46.2	47.8	47.2	42.2	38.9	41.7	42.8	38.9	35.6	47.8
		θ_{max}	28.7	31.0	35.7	39.7	40.7	38.0	33.2	31.8	33.1	35.6	33.0	29.6	34.2
		θ_{min}	11.9	14.5	18.6	23.0	26.3	27.4	25.7	24.6	24.2	21.2	16.1	12.6	20.5
		θ_{mean}	20.3	22.75	27.2	31.35	33.5	32.7	29.4	28.2	28.6	28.4	24.6	21.1	27.4
		θ_a	25.3	27.8	32.2	36.4	38.5	37.7	34.4	33.2	33.6	33.4	29.6	26.1	32.4
		% Rated Load	106	103	100	96	94	94	98	99	98	99	102	105	100
3.	Ajmer	θ_{ext}	31.7	35.6	41.7	44.6	45.6	45.6	44.4	40.0	40.6	38.9	35.0	31.1	45.6
		θ_{max}	22.2	25.3	30.7	35.9	39.5	38.1	33.2	30.9	32.1	32.9	28.9	24.4	31.2
		θ_{min}	7.3	9.9	15.7	21.9	27.3	27.7	25.6	24.3	23.7	17.8	10.9	7.7	18.3
		θ_{mean}	14.8	17.6	23.2	28.9	30.7	32.9	29.4	27.6	27.9	25.4	19.9	16.0	24.8
		θ_a	19.8	22.6	28.2	33.9	35.7	37.9	34.4	32.6	32.9	30.4	24.9	21.0	29.8
		% Rated Load	111	108	103	98	96	94	98	99	99	102	106	110	102

Sheet No. 4

S. No.	Location		Jan	Feb	Mar	Apr	May	June	July	Aug	Sep	Oct	Nov	Dec	Annual
4.	Allahabad	θ_{ext}	31.1	36.1	41.7	45.0	47.2	47.8	45.6	40.0	39.4	40.6	35.6	31.3	47.8
		θ_{max}	23.7	26.7	33.3	38.8	42.1	39.8	33.6	32.6	32.8	32.6	29.0	24.8	32.4
		θ_{min}	9.1	11.6	17.0	22.5	27.4	28.9	26.6	26.0	25.2	20.4	13.1	9.3	19.8
		θ_{mean}	16.4	19.2	25.2	30.6	34.8	34.4	30.1	29.3	29.0	26.5	21.0	17.0	26.1
		θ_a	21.4	24.2	30.2	35.6	39.8	39.4	35.1	34.3	34.0	31.5	26.0	22.0	31.1
		% Rated Load	110	107	102	96	92	93	97	98	98	100	105	109	101
5.	Ambala	θ_{ext}	28.9	33.9	41.7	45.0	47.8	47.8	46.7	43.9	40.6	39.4	35.6	29.4	47.8
		θ_{max}	20.8	23.8	29.6	36.2	40.8	40.5	35.2	33.8	35.4	33.2	28.6	23.2	31.8
		θ_{min}	6.8	8.5	14.1	19.7	24.9	27.3	26.0	25.4	23.9	16.4	10.2	7.1	17.5
		θ_{mean}	13.8	16.2	21.8	28.0	32.8	33.9	30.6	29.6	29.6	24.8	19.4	15.2	24.6
		θ_a	18.8	21.2	26.8	33.0	37.8	38.9	35.6	34.6	34.6	29.8	24.4	20.2	29.6
		% Rated Load	112	110	104	99	94	93	96	97	97	102	107	111	102
6.	Amritsar	θ_{ext}	25.0	32.2	35.6	43.3	46.1	46.7	45.6	40.0	40.6	38.3	32.2	27.7	46.7
		θ_{max}	18.6	22.6	27.5	34.2	38.9	40.4	35.6	34.2	34.4	31.9	26.5	24.4	30.5
		θ_{min}	4.5	6.5	11.5	16.2	21.4	25.2	25.9	25.3	23.8	16.6	8.8	5.0	15.9
		θ_{mean}	11.6	14.6	19.5	25.2	30.2	32.8	30.8	29.8	28.8	24.2	17.6	13.2	23.2
		θ_a	16.6	19.6	24.5	30.2	35.2	37.8	35.8	34.8	33.8	29.2	22.6	18.2	28.2
		% Rated Load	114	111	107	102	97	94	96	97	98	102	108	113	102

Loading and Life of Transformers

Sheet No. 5

S. No.	Location		Jan	Feb	Mar	Apr	May	June	July	Aug	Sep	Oct	Nov	Dec	Annual
7.	Bangalore	θ_{ext}	32.2	34.4	37.2	38.3	38.9	37.8	33.3	33.3	33.3	32.2	31.1	31.1	38.9
		θ_{max}	26.9	29.7	32.3	33.4	32.7	28.9	27.2	27.3	27.6	27.5	26.3	25.7	28.8
		θ_{min}	15.0	16.5	19.0	21.2	21.1	19.7	19.2	19.2	18.9	18.9	17.2	15.3	18.4
		θ_{mean}	21.0	23.1	25.6	27.3	26.9	24.3	23.2	23.2	23.2	23.2	21.8	20.5	23.6
		θ_a	26.0	28.1	30.6	32.3	31.9	29.3	28.2	28.2	28.2	28.2	26.8	25.5	28.6
		% Rated Load	105	102	101	100	100	102	103	103	103	103	104	105	102
8.	Baroda	θ_{ext}	35.6	41.7	43.3	45.9	46.7	45.6	40.0	37.2	41.1	41.7	38.3	36.1	46.7
		θ_{max}	30.1	32.4	36.6	39.9	40.7	37.2	32.4	31.5	32.6	35.0	33.4	31.0	34.4
		θ_{min}	10.8	12.7	16.6	21.7	26.1	27.1	25.4	24.8	24.1	19.9	14.3	11.4	19.6
		θ_{mean}	20.4	22.6	26.6	30.8	33.4	32.2	28.9	28.2	28.4	27.4	23.8	21.2	27.0
		θ_a	25.4	27.6	31.6	35.8	38.4	37.2	33.9	33.2	33.4	32.4	28.8	26.2	32.0
		% Rated Load	106	102	100	96	94	95	98	99	99	100	102	105	100
9.	Belgaum	θ_{ext}	33.3	37.2	39.4	39.5	40.6	37.2	32.0	31.7	33.9	33.6	32.8	34.6	40.6
		θ_{max}	30.1	32.2	35.0	35.7	34.0	27.5	25.2	25.6	27.0	30.1	29.3	29.3	30.1
		θ_{min}	14.0	15.1	18.0	19.5	20.6	20.6	19.8	19.4	19.0	18.6	17.1	13.9	18.0
		θ_{mean}	22.0	23.6	26.5	27.6	27.3	24.0	22.5	22.5	23.0	24.4	23.2	21.6	24.0
		θ_a	27.0	28.6	31.5	32.6	32.3	29.0	27.5	27.5	28.0	29.4	28.2	26.6	29.0
		% Rated Load	104	102	100	99	100	102	103	103	103	102	103	104	102

Transformers

Sheet No. 6

S. No.	Location		Jan	Feb	Mar	Apr	May	June	July	Aug	Sep	Oct	Nov	Dec	Annual
10.	Bhopal	θ_{ext}	32.2	36.1	40.0	44.2	45.6	43.9	40.6	35.0	36.1	37.8	33.3	32.8	45.6
		θ_{max}	25.7	28.5	33.6	37.8	40.7	36.9	29.9	28.6	30.1	31.3	28.5	26.1	31.5
		θ_{min}	10.4	12.5	17.1	21.2	26.4	25.4	23.2	22.5	21.9	18.0	13.3	10.6	18.5
		θ_{mean}	18.0	21.5	25.4	29.5	33.6	31.2	26.6	25.6	26.0	24.6	20.9	18.4	25.0
		θ_a	23.0	26.5	30.4	34.5	38.6	36.2	31.6	30.6	31.0	29.6	25.9	23.4	30.0
		% Rated Load	108	105	102	98	93	96	100	101	101	102	105	108	102
11.	Bikaner	θ_{ext}	31.1	37.2	42.8	47.2	49.4	48.9	47.2	43.3	43.9	42.2	37.2	32.2	49.4
		θ_{max}	22.3	26.1	31.8	37.6	42.0	41.7	38.9	36.1	36.6	35.3	30.3	24.7	33.6
		θ_{min}	5.0	8.2	14.4	20.8	27.2	29.3	28.0	26.6	25.2	18.6	10.0	5.6	18.2
		θ_{mean}	13.6	17.2	23.1	29.2	34.6	35.5	33.5	31.4	30.9	27.0	20.2	15.2	25.9
		θ_a	18.6	22.2	28.1	34.2	39.6	40.5	38.5	35.4	35.9	32.0	25.2	20.2	30.9
		% Rated Load	112	109	103	98	92	91	94	97	96	100	106	111	210
12.	Bombay	θ_{ext}	35.0	38.3	39.7	40.6	36.2	37.2	35.6	32.3	35.0	36.6	36.2	35.1	40.6
		θ_{max}	29.1	29.5	31.0	32.3	33.3	31.9	29.8	29.5	30.1	31.9	32.3	30.9	31.0
		θ_{min}	19.4	20.3	22.7	25.1	26.9	26.3	25.1	24.8	24.7	24.6	22.8	20.8	23.6
		θ_{mean}	24.2	24.9	26.8	28.7	30.1	29.1	27.4	27.2	27.4	28.2	27.6	25.8	27.3
		θ_a	29.2	29.9	31.8	33.7	35.1	34.1	32.4	32.2	32.4	33.2	32.6	30.8	32.3
		% Rated Load	102	102	100	98	97	98	100	100	100	99	99	101	100

Sheet No. 7

S. No.	Location		Jan	Feb	Mar	Apr	May	June	July	Aug	Sep	Oct	Nov	Dec	Annual
13.	Kolkata	θ_{ext}	31.9	36.7	41.1	43.3	43.7	43.9	36.7	36.1	36.1	35.6	33.9	30.6	43.9
		θ_{max}	26.8	29.5	34.3	36.3	35.8	34.1	32.0	32.0	32.3	31.8	29.5	27.0	31.8
		θ_{min}	13.6	16.5	21.5	25.0	26.5	26.7	26.3	26.3	26.1	23.9	18.4	14.2	22.1
		θ_{mean}	20.2	23.0	27.9	30.6	31.2	30.4	29.2	29.2	29.2	27.8	24.0	20.6	27.0
		θ_a	25.2	28.0	32.9	35.6	36.2	35.4	34.2	34.2	34.2	32.8	29.0	25.6	32.0
		% Rated Load	106	103	99	98	96	97	98	98	98	99	102	105	100
14.	Coimbatore	θ_{ext}	35.0	36.7	39.3	40.0	39.4	38.3	35.6	35.6	35.8	36.1	34.4	35.0	40.0
		θ_{max}	29.7	32.2	34.7	34.6	33.5	30.5	29.0	29.9	30.7	30.4	29.3	28.9	31.1
		θ_{min}	19.2	20.2	22.1	23.4	23.6	22.5	22.0	22.1	22.0	22.0	21.1	19.6	21.7
		θ_{mean}	24.4	26.2	28.4	29.0	28.6	26.5	25.5	26.0	26.4	26.2	25.2	24.2	26.4
		θ_a	29.4	31.2	33.4	34.0	33.6	31.5	30.5	31.0	31.4	31.2	30.2	29.2	31.4
		% Rated Load	102	101	99	98	98	100	101	101	101	101	102	102	101
15.	Cuttack	θ_{ext}	35.6	38.9	42.8	45.0	47.7	47.2	40.0	37.2	36.7	36.7	35.0	33.3	47.7
		θ_{max}	28.9	31.5	35.9	38.3	38.8	35.8	31.6	31.6	32.2	32.0	30.1	28.4	32.9
		θ_{min}	15.7	18.2	22.1	25.3	26.9	26.5	25.6	25.6	25.5	23.7	18.8	15.5	22.5
		θ_{mean}	22.3	24.8	29.0	31.8	32.8	31.2	31.2	28.6	28.8	27.8	24.4	22.0	27.7
		θ_a	27.3	29.8	34.0	36.8	37.8	36.2	33.6	33.6	33.8	32.8	29.4	27.0	32.7
		% Rated Load	104	102	98	95	94	96	98	98	98	99	102	104	99

428 *Transformers*

Sheet No. 8

S. No.	Location		Jan	Feb	Mar	Apr	May	June	July	Aug	Sep	Oct	Nov	Dec	Annual
16.	Cochin	θ_{ext}	33.3	34.0	34.4	34.1	34.6	32.7	31.7	32.2	31.1	32.2	32.8	32.8	34.6
		θ_{max}	30.6	30.7	31.3	31.4	30.9	29.0	28.1	28.1	28.3	29.2	29.8	30.3	29.8
		θ_{min}	23.2	24.3	25.8	26.0	25.7	24.1	23.7	24.0	24.2	24.2	24.1	23.5	24.4
		θ_{mean}	26.9	27.5	28.6	28.7	28.3	26.6	25.9	26.0	26.2	26.7	27.0	26.9	27.1
		θ_a	31.9	32.5	33.6	33.7	33.6	31.6	30.9	31.0	31.2	31.7	32.0	31.9	32.1
		% Rated Load	100	100	98	98	99	100	101	101	101	100	100	100	100
17.	Dehradun	θ_{ext}	26.1	29.2	37.2	40.6	42.8	43.9	40.6	37.2	34.4	36.1	30.1	27.2	43.9
		θ_{max}	19.1	21.4	26.4	32.1	36.2	35.3	30.4	29.5	29.6	28.2	24.7	20.9	27.8
		θ_{min}	6.1	8.2	12.4	17.0	21.5	23.6	23.1	22.7	21.3	16.1	10.3	7.00	15.8
		θ_{mean}	12.6	14.8	19.4	24.6	28.8	29.4	26.8	26.1	25.4	22.2	17.5	14.0	21.8
		θ_a	17.6	19.8	24.4	29.6	33.8	34.4	31.8	31.1	30.4	27.2	22.5	19.0	26.8
		% Rated Load	113	111	107	102	98	98	100	101	102	104	109	112	104
18.	Dhanbad	θ_{ext}	32.2	38.3	40.6	43.9	46.1	46.1	36.7	36.1	35.6	35.6	32.8	30.6	46.1
		θ_{max}	24.7	27.4	33.2	37.6	38.8	35.9	31.2	30.8	31.2	30.6	27.8	25.2	31.1
		θ_{min}	12.1	14.5	19.2	23.5	25.9	25.9	24.8	24.8	24.3	21.7	15.6	12.4	20.4
		θ_{mean}	18.4	21.0	26.2	30.6	32.4	30.9	28.0	27.8	27.8	26.2	21.7	18.8	25.8
		θ_a	23.4	26.0	31.2	35.6	37.4	35.9	33.0	32.8	32.8	31.2	26.7	23.8	30.8
		% Rated Load	108	105	101	96	95	96	99	99	99	101	104	107	101

Sheet No. 9

S. No.	Location		Jan	Feb	Mar	Apr	May	June	July	Aug	Sep	Oct	Nov	Dec	Annual
19.	Guahati	θ_{ext}	28.9	35.0	38.3	40.6	41.1	36.7	37.2	37.8	37.2	35.6	32.2	29.4	41.1
		θ_{max}	24.0	26.3	30.2	31.6	31.0	31.5	32.1	32.2	32.1	30.5	27.7	24.9	29.5
		θ_{min}	11.0	12.8	16.5	20.3	22.7	24.7	25.8	25.8	25.2	22.0	16.9	12.5	19.7
		θ_{mean}	17.5	19.6	23.4	26.0	26.9	28.1	29.0	29.0	28.6	26.2	22.3	18.7	24.6
		θ_a	22.5	24.6	28.4	31.0	31.9	33.1	34.0	34.0	33.6	31.2	27.3	23.7	29.6
		% Rated Load	108	106	103	101	100	99	98	98	98	101	104	107	102
20.	Hissar	θ_{ext}	30.6	34.4	45.6	47.9	48.3	47.8	47.2	43.3	42.2	41.7	36.7	33.6	48.3
		θ_{max}	21.7	25.0	30.7	37.0	41.6	41.3	37.3	35.5	35.7	34.6	29.6	24.1	32.8
		θ_{min}	5.5	8.1	13.3	19.0	24.6	27.7	27.3	26.1	23.9	17.4	9.8	6.0	17.4
		θ_{mean}	13.6	16.6	22.0	28.0	33.1	34.5	32.3	30.8	29.8	26.0	19.7	15.0	25.1
		θ_a	18.6	21.6	27.0	33.0	38.1	39.5	37.3	35.8	34.8	31.0	24.7	20.0	30.1
		% Rated Load	112	109	104	99	94	92	95	96	97	101	106	111	102
21.	Hyderabad	θ_{ext}	35.0	37.2	42.2	43.3	44.4	43.9	37.2	36.1	36.1	36.7	33.9	33.3	44.4
		θ_{max}	28.6	31.2	34.8	36.9	38.7	34.1	29.8	29.5	29.7	30.3	28.7	27.8	31.7
		θ_{min}	14.6	16.7	20.0	23.7	26.2	24.1	22.3	22.1	21.6	19.8	16.0	13.4	20.0
		θ_{mean}	21.6	24.0	27.4	30.3	32.4	29.1	26.0	25.8	25.6	25.0	22.4	20.6	25.8
		θ_a	26.6	29.0	32.4	35.3	37.4	34.1	31.0	30.8	30.6	30.0	27.4	25.6	30.8
		% Rated Load	104	102	100	97	95	98	101	101	101	102	104	105	101

430   *Transformers*

Sheet No. 10

S. No.	Location		Jan	Feb	Mar	Apr	May	June	July	Aug	Sep	Oct	Nov	Dec	Annual
22.	Jabalpur	θ_{ext}	32.8	37.2	41.1	45.6	46.7	46.1	41.7	35.0	35.6	36.7	33.9	32.8	46.7
		θ_{max}	26.1	28.9	34.0	38.5	41.9	37.6	30.3	29.5	30.8	31.4	28.9	26.9	32.1
		θ_{min}	9.8	11.4	15.5	20.5	25.9	26.4	23.9	23.6	23.1	18.4	11.7	9.0	18.3
		θ_{mean}	18.0	20.2	24.8	29.5	33.9	32.0	27.1	26.6	27.0	24.9	20.3	18.0	25.2
		θ_a	23.0	25.2	29.8	34.5	38.9	37.0	32.1	31.6	32.0	29.9	25.3	23.0	30.2
		% Rated Load	108	106	102	98	93	95	100	100	100	102	106	108	102
23.	Jaipur	θ_{ext}	31.7	36.7	42.8	44.9	47.8	47.2	46.7	41.7	41.7	40.0	36.1	31.1	47.8
		θ_{max}	22.0	25.4	30.9	36.5	40.6	39.2	34.1	31.9	33.2	33.2	29.0	24.4	31.7
		θ_{min}	8.3	10.7	15.5	21.0	25.8	27.3	25.6	24.3	23.0	18.3	12.0	9.1	18.4
		θ_{mean}	15.2	18.0	23.2	28.8	33.2	33.2	29.8	28.1	28.1	25.8	20.5	16.8	25.0
		θ_a	20.2	23.0	28.2	33.8	38.2	38.2	34.8	33.1	33.1	30.8	25.5	21.8	30.0
		% Rated Load	111	108	103	98	94	94	97	99	99	101	105	109	102
24.	Jammu	θ_{ext}	26.7	31.7	37.2	43.9	46.1	47.2	45.0	41.7	38.3	37.2	32.5	27.2	47.2
		θ_{max}	18.4	21.1	26.4	33.1	39.0	40.4	35.4	33.2	33.3	31.4	26.2	21.1	30.0
		θ_{min}	8.3	10.6	14.8	20.5	25.8	27.7	26.0	25.1	23.9	19.4	13.4	9.3	18.7
		θ_{mean}	13.4	15.8	20.6	26.8	32.4	34.0	30.7	29.2	28.6	25.4	19.8	15.2	24.4
		θ_a	18.4	20.8	25.6	31.8	37.4	39.0	35.7	34.2	33.6	30.4	24.8	20.2	29.4
		% Rated Load	113	110	105	100	95	93	96	98	98	102	106	111	102

Sheet No. 11

S. No.	Location		Jan	Feb	Mar	Apr	May	June	July	Aug	Sep	Oct	Nov	Dec	Annual
25.	Jhansi	θ_{ext}	33.3	37.8	43.3	45.6	47.2	47.8	45.6	42.2	40.6	40.6	36.1	32.8	47.8
		θ_{max}	24.1	27.5	33.5	38.9	42.6	40.4	33.5	31.7	32.5	33.3	29.7	25.5	32.8
		θ_{min}	9.2	11.7	17.4	23.3	28.8	29.3	25.9	24.9	24.1	19.5	13.1	9.3	19.7
		θ_{mean}	16.6	19.6	25.4	31.1	35.7	34.8	29.7	27.9	28.3	26.4	21.4	17.4	26.2
		θ_a	21.6	24.6	30.4	36.1	40.7	39.8	34.7	32.9	33.3	31.4	26.4	22.4	31.2
		% Rated Load	109	106	102	96	91	92	97	99	99	101	105	109	101
26.	Kathmandu (Nepal)	θ_{ext}	25.0	28.3	33.3	37.2	37.5	37.8	32.8	33.3	33.3	33.3	29.4	28.3	37.8
		θ_{max}	17.8	20.7	25.2	28.9	30.4	29.5	28.5	28.5	28.1	26.3	22.7	19.1	25.5
		θ_{min}	1.9	4.0	8.0	11.5	15.9	19.1	20.1	19.9	18.6	13.5	7.4	3.0	11.9
		θ_{mean}	9.8	12.4	16.6	20.2	23.2	24.3	24.3	24.2	23.4	19.9	15.0	11.0	18.7
		θ_a	14.8	17.4	21.6	25.2	28.2	29.3	29.3	29.2	28.4	24.9	20.0	16.0	23.7
		% Rated Load	116	114	109	106	103	102	102	102	103	106	111	115	107
27.	Khammam	θ_{ext}	35.0	38.9	43.3	45.0	47.2	46.7	39.4	37.8	37.2	37.4	33.9	33.4	47.2
		θ_{max}	31.0	33.6	36.8	39.0	41.3	37.6	32.6	32.2	32.6	32.5	30.6	30.1	34.2
		θ_{min}	17.6	20.0	23.1	25.9	28.1	27.2	24.9	24.7	24.4	22.9	19.1	16.7	22.9
		θ_{mean}	24.3	26.8	30.0	32.4	34.7	32.4	28.8	28.4	28.5	27.7	24.8	23.4	28.6
		θ_a	29.3	31.8	35.0	37.4	39.7	37.4	33.8	33.4	33.5	32.7	29.8	28.4	33.6
		% Rated Load	102	100	97	95	92	95	98	99	98	99	102	103	98

Sheet No. 12

S. No.	Location		Jan	Feb	Mar	Apr	May	June	July	Aug	Sep	Oct	Nov	Dec	Annual
28.	Kota	θ_{ext}	33.9	38.3	42.8	45.6	47.8	47.8	47.2	41.1	40.6	41.1	37.2	33.9	47.8
		θ_{max}	24.5	28.5	34.1	39.0	42.6	40.3	33.3	31.7	33.1	34.5	30.8	26.7	33.3
		θ_{min}	10.6	13.1	18.5	24.4	29.7	29.5	26.4	25.4	24.7	21.0	14.8	11.3	20.8
		θ_{mean}	17.6	20.8	27.3	31.7	36.2	34.9	29.8	28.6	28.9	27.8	22.8	19.0	27.0
		θ_a	22.6	25.8	32.3	36.7	41.2	39.9	34.8	33.6	33.9	32.8	27.8	24.0	32.0
		% Rated Load	108	105	100	95	91	92	97	98	98	99	103	107	100
29.	Karnool	θ_{ext}	36.1	38.9	41.7	44.4	45.6	44.4	38.3	37.8	37.8	38.3	36.1	34.4	45.6
		θ_{max}	31.3	34.3	37.5	39.3	40.0	35.6	32.5	32.1	31.9	32.4	31.0	30.3	34.0
		θ_{min}	17.0	19.3	22.5	26.0	27.2	25.0	23.8	23.5	23.3	22.4	19.2	16.6	22.1
		θ_{mean}	24.2	26.8	30.0	32.6	33.6	30.3	28.2	28.5	28.3	27.4	24.2	21.6	27.1
		θ_a	29.2	31.8	35.0	37.6	38.6	35.3	33.2	33.5	33.3	32.4	29.2	26.6	32.1
		% Rated Load	102	100	97	94	93	97	99	99	99	100	102	104	100
30.	Lucknow	θ_{ext}	30.6	35.0	41.7	45.6	47.2	48.3	45.6	38.9	39.4	40.0	35.0	33.3	48.3
		θ_{max}	23.3	26.4	32.9	38.3	41.2	39.3	33.6	32.5	33.0	32.8	29.3	24.8	32.3
		θ_{min}	8.9	11.5	16.3	21.8	26.5	28.0	26.6	26.0	25.1	19.8	12.7	9.1	19.4
		θ_{mean}	16.1	19.0	24.6	30.1	33.8	33.6	30.1	29.2	29.0	26.3	21.0	17.0	25.8
		θ_a	21.1	24.0	29.6	35.1	38.8	38.6	35.1	34.2	34.0	31.3	26.0	22.0	30.8
		% Rated Load	110	107	102	97	93	93	97	98	98	101	105	109	101

Sheet No. 13

Loading and Life of Transformers

S. No.	Location		Jan	Feb	Mar	Apr	May	June	July	Aug	Sep	Oct	Nov	Dec	Annual
31.	Ludhiana	θ_{ext}	28.9	33.3	41.1	46.1	48.3	47.9	47.8	44.4	41.7	40.0	35.0	29.4	48.3
		θ_{max}	20.2	23.3	29.0	36.0	41.2	41.1	36.0	34.7	35.3	33.9	28.8	22.9	31.9
		θ_{min}	5.8	8.4	12.9	18.5	24.2	27.1	26.7	26.1	23.9	17.5	10.1	6.2	17.3
		θ_{mean}	13.0	15.8	21.0	27.2	32.7	34.1	31.4	30.4	29.6	25.7	19.4	14.6	24.6
		θ_a	18.0	20.8	26.0	32.2	37.7	39.1	36.4	35.4	34.6	30.7	24.4	19.6	29.6
		% Rated Load	113	110	105	100	94	93	96	97	97	101	107	111	102
32.	Madras	θ_{ext}	32.8	36.7	40.6	42.8	45.0	43.3	41.1	40.0	38.9	39.4	34.4	32.8	45.0
		θ_{max}	28.8	30.6	32.7	34.9	37.6	37.3	35.2	34.5	33.9	31.8	29.2	28.2	32.9
		θ_{min}	20.3	21.1	23.1	26.0	27.8	27.6	26.3	25.8	25.4	24.4	22.5	21.0	24.3
		θ_{mean}	24.6	25.8	27.9	30.4	32.7	32.3	30.8	30.2	29.6	28.1	25.8	24.6	28.6
		θ_a	29.6	30.8	32.9	35.4	37.7	37.3	35.8	35.2	34.6	33.1	30.8	29.6	33.6
		% Rated Load	102	101	99	97	94	95	96	97	97	99	101	102	98
33.	Madurai	θ_{ext}	34.4	38.3	41.7	41.7	41.7	42.2	40.6	40.0	39.4	38.3	36.1	35.0	42.2
		θ_{max}	30.2	32.4	35.0	36.3	37.5	36.7	35.7	35.3	35.0	33.0	30.6	29.7	33.9
		θ_{min}	20.9	21.6	23.4	25.4	26.3	26.3	25.7	25.2	24.8	24.0	23.0	21.6	24.0
		θ_{mean}	25.6	27.0	29.2	30.8	31.9	31.5	30.7	30.2	29.9	28.5	26.8	25.6	29.0
		θ_a	30.6	32.0	34.2	35.8	36.9	36.5	35.7	35.2	34.9	33.5	31.8	30.6	34.0
		% Rated Load	101	100	98	96	95	96	96	97	97	98	100	101	98

434   Transformers

Sheet No. 14

S. No.	Location		Jan	Feb	Mar	Apr	May	June	July	Aug	Sep	Oct	Nov	Dec	Annual
34.	Marmugao	θ_{ext}	34.4	37.2	35.0	33.9	33.3	33.3	31.7	30.6	31.7	36.1	35.6	33.9	37.2
		θ_{max}	29.7	29.0	30.0	30.9	31.3	29.4	28.0	27.8	28.1	29.8	31.0	30.5	29.5
		θ_{min}	21.4	21.9	23.9	26.1	29.9	24.7	24.0	23.9	23.8	23.9	22.8	21.5	23.7
		θ_{mean}	25.6	25.4	27.0	28.5	29.1	27.0	26.0	25.8	26.0	26.8	26.9	26.0	26.6
		θ_a	30.6	30.4	32.0	33.5	34.1	32.0	31.0	30.8	31.0	31.8	31.9	31.0	31.6
		% Rated Load	101	102	100	98	98	100	101	101	101	100	100	101	100
35.	Mysore	θ_{ext}	32.8	36.1	37.8	39.4	37.8	37.2	33.3	33.9	33.3	32.8	32.2	31.7	39.4
		θ_{max}	28.3	31.2	33.5	34.0	32.6	28.9	27.3	27.9	28.7	28.4	27.4	27.0	29.6
		θ_{min}	16.4	18.2	20.2	21.4	21.2	20.2	19.7	19.6	19.3	19.6	18.3	16.5	19.2
		θ_{mean}	22.4	24.7	26.8	27.7	26.9	24.6	23.5	23.8	24.0	24.0	22.8	21.8	24.4
		θ_a	27.4	29.7	31.8	32.7	31.9	29.6	28.5	28.8	29.0	29.0	27.8	26.8	29.4
		% Rated Load	104	102	100	99	100	102	103	102	102	102	103	104	102
36.	Nagpur	θ_{ext}	35.0	38.9	45.0	46.1	47.8	47.2	40.6	37.8	38.9	38.3	35.6	33.9	47.8
		θ_{max}	28.6	32.5	36.4	39.7	42.8	38.4	31.2	30.4	31.5	31.9	29.9	28.7	33.5
		θ_{min}	12.7	15.1	19.1	23.9	28.4	26.9	24.0	23.7	23.1	20.0	14.1	12.1	20.3
		θ_{mean}	20.6	23.8	27.8	31.8	35.6	32.6	27.6	27.0	27.3	26.0	22.0	20.4	26.9
		θ_a	25.6	28.8	32.8	36.8	40.6	37.6	32.6	32.0	32.3	33.0	27.0	25.4	31.9
		% Rated Load	105	102	99	95	91	94	99	100	100	99	104	106	100

Loading and Life of Transformers

Sheet No. 15

S. No.	Location		Jan	Feb	Mar	Apr	May	June	July	Aug	Sep	Oct	Nov	Dec	Annual
37.	New Delhi	θ_{ext}	29.4	33.3	40.6	45.6	47.2	46.7	45.0	40.0	40.6	39.4	35.0	28.9	47.2
		θ_{max}	21.3	23.6	30.2	36.2	40.5	39.9	35.3	33.7	34.1	33.1	28.7	23.4	31.7
		θ_{min}	7.3	10.1	15.1	21.0	26.6	28.7	27.2	26.1	24.6	18.7	11.8	8.0	18.8
		θ_{mean}	14.3	16.8	22.6	28.6	33.5	34.3	31.2	29.9	29.4	25.9	20.2	15.7	25.2
		θ_a	19.3	21.8	27.6	33.6	38.5	39.3	36.2	34.9	34.4	30.9	25.2	20.7	30.2
		% Rated Load	112	109	103	98	94	93	96	97	98	101	106	110	102
38.	Patna	θ_{ext}	28.9	34.4	40.6	43.3	45.6	46.1	41.7	38.3	37.8	36.1	33.9	30.6	46.1
		θ_{max}	23.6	26.3	32.9	37.6	38.9	36.7	32.9	32.1	32.3	31.9	28.9	24.9	31.6
		θ_{min}	11.0	13.4	18.6	23.3	26.0	27.1	26.7	26.6	26.3	23.0	16.1	11.7	20.8
		θ_{mean}	17.3	19.8	25.8	30.4	32.4	31.9	29.8	29.4	29.3	27.4	22.5	18.3	26.2
		θ_a	22.3	24.8	30.8	35.4	37.4	36.9	34.8	34.4	34.3	32.4	27.5	23.3	31.2
		% Rated Load	109	106	101	97	95	95	97	98	98	100	103	108	101
39.	Poona	θ_{ext}	35.0	38.9	42.8	43.3	43.3	41.7	35.6	33.0	36.1	37.8	36.1	35.0	43.3
		θ_{max}	30.7	32.9	36.1	37.9	37.2	31.9	27.8	27.7	29.2	31.8	30.8	30.1	32.0
		θ_{min}	12.0	13.3	16.8	20.6	22.6	23.0	22.0	21.5	20.8	19.3	15.0	12.0	18.2
		θ_{mean}	21.4	23.1	26.4	29.2	29.9	27.4	24.9	24.6	25.0	25.6	22.9	21.0	25.1
		θ_a	26.4	28.1	31.4	34.2	34.9	32.4	29.9	29.6	30.0	30.6	27.9	26.0	30.1
		% Rated Load	105	103	101	98	97	100	102	102	102	101	103	105	102

436   *Transformers*

Sheet No. 16

S. No.	Location		Jan	Feb	Mar	Apr	May	June	July	Aug	Sep	Oct	Nov	Dec	Annual
40.	Puri	θ_{ext}	32.8	33.0	40.0	41.1	42.2	39.4	36.7	36.7	36.1	36.1	33.9	32.8	42.2
		θ_{max}	26.9	28.3	30.0	30.7	31.6	31.7	30.6	31.0	31.4	31.2	29.3	27.2	30.0
		θ_{min}	17.9	20.8	24.6	26.6	27.7	27.4	26.7	26.8	26.6	25.0	20.8	17.7	24.1
		θ_{mean}	22.4	24.6	27.3	28.6	29.6	29.6	28.6	28.9	29.0	28.1	25.0	22.4	27.0
		θ_a	27.4	29.6	32.3	33.6	34.6	34.6	33.6	33.6	34.0	33.1	30.0	27.4	32.0
		% Rated Load	104	102	100	98	97	97	98	98	98	99	102	104	100
41.	Ramgundam	θ_{ext}	33.9	38.9	42.8	44.6	47.2	47.2	40.0	37.4	37.3	36.4	35.2	34.1	47.2
		θ_{max}	31.1	34.1	37.7	40.3	42.8	38.6	32.1	31.3	32.0	32.5	30.7	30.2	34.5
		θ_{min}	16.1	18.8	22.7	26.9	29.7	28.2	24.7	24.4	24.4	22.8	17.5	15.0	22.6
		θ_{mean}	23.6	26.4	30.2	33.6	36.2	33.4	28.4	27.8	28.2	27.6	24.1	22.6	28.6
		θ_a	28.6	31.4	35.2	38.6	41.2	38.4	33.4	32.8	33.2	32.6	29.1	27.6	33.6
		% Rated Load	102	101	97	93	91	94	99	99	99	99	102	103	98
42.	Shillong	θ_{ext}	21.1	24.4	28.9	30.0	30.7	28.3	28.3	29.4	27.8	27.2	25.6	22.8	30.7
		θ_{max}	15.5	17.1	21.5	23.8	23.7	23.7	24.1	24.1	23.6	21.8	18.9	16.4	21.2
		θ_{min}	3.6	6.4	10.5	14.1	15.5	17.4	18.1	17.8	16.6	12.9	7.7	4.5	12.1
		θ_{mean}	9.6	11.8	16.0	19.0	19.6	20.6	21.1	21.0	20.1	17.4	13.3	10.4	16.6
		θ_a	14.6	16.8	21.0	24.0	24.6	25.6	26.1	26.0	25.1	22.4	18.3	15.4	21.6
		% Rated Load	116	114	110	107	106	105	105	105	106	109	113	116	109

Sheet No. 17

S. No.	Location		Jan	Feb	Mar	Apr	May	June	July	Aug	Sep	Oct	Nov	Dec	Annual
43.	Shimla	θ_{ext}	18.9	20.6	23.9	28.3	30.0	30.6	28.9	27.8	25.0	23.9	21.1	20.4	30.6
		θ_{max}	8.5	10.3	14.4	19.2	23.4	24.3	21.0	20.1	20.0	17.9	15.0	11.3	17.1
		θ_{min}	1.9	3.1	6.8	11.2	15.0	16.2	15.6	15.2	13.8	10.8	7.3	4.2	10.1
		θ_{mean}	5.2	6.7	10.6	15.2	19.2	20.2	18.3	17.6	16.9	14.3	11.2	7.8	13.6
		θ_a	10.2	11.7	15.6	20.2	24.2	25.2	23.3	22.6	21.9	19.3	16.2	12.8	18.6
		% Rated Load	120	118	115	111	107	106	108	108	109	112	115	117	112
44.	Srinagar	θ_{ext}	17.2	20.6	25.6	31.1	35.6	37.8	38.3	36.4	33.0	33.9	23.9	18.3	38.3
		θ_{max}	4.4	7.9	13.4	19.3	24.6	29.0	30.8	29.9	28.3	22.6	15.5	8.8	19.5
		θ_{min}	−2.3	−0.8	3.5	7.4	11.2	14.4	18.4	17.9	12.7	5.7	−0.1	−1.8	7.2
		θ_{mean}	1.0	3.6	8.4	13.4	17.9	21.7	24.6	23.9	20.5	14.2	7.7	3.5	13.4
		θ_a	6.0	8.6	13.4	18.4	22.9	26.7	29.6	28.9	25.5	19.2	12.7	8.5	18.4
		% Rated Load	124	122	177	133	108	104	102	102	105	112	117	122	113
45.	Surat	θ_{ext}	38.3	41.7	43.9	45.6	45.6	45.6	38.9	37.2	41.1	41.1	39.4	38.9	45.6
		θ_{max}	31.4	33.1	36.1	37.3	36.2	33.7	30.5	30.3	31.6	35.5	34.9	32.8	33.6
		θ_{min}	14.8	16.4	20.1	23.7	26.6	27.1	25.7	25.4	24.1	23.1	19.2	16.0	21.9
		θ_{mean}	23.1	24.8	28.1	30.5	31.4	30.4	28.1	27.8	27.8	29.3	27.0	24.4	27.8
		θ_a	28.1	29.8	33.1	35.5	36.4	35.4	33.1	32.8	32.8	34.3	32.0	29.4	32.8
		% Rated Load	103	102	99	96	96	97	99	99	99	98	100	102	99

Sheet No. 18

S. No.	Location		Jan	Feb	Mar	Apr	May	June	July	Aug	Sep	Oct	Nov	Dec	Annual
46.	Tiruchy	θ_{ext}	35.6	40.0	42.2	42.8	43.3	43.9	41.1	40.6	40.6	36.9	36.7	35.6	43.9
		θ_{max}	30.1	32.7	35.1	36.7	37.1	36.4	35.5	35.1	34.2	32.3	29.9	29.3	33.7
		θ_{min}	20.6	21.3	22.9	25.8	26.4	26.5	25.9	25.4	24.9	23.9	22.7	21.3	24.0
		θ_{mean}	25.4	27.0	29.0	31.2	31.8	31.4	30.7	30.2	29.6	28.1	26.3	25.3	28.8
		θ_a	30.4	32.0	34.0	36.2	36.8	36.4	35.7	35.2	34.6	33.1	31.3	30.3	33.8
		% Rated Load	102	100	98	96	95	95	95	97	97	99	101	102	98
47.	Trivandrum	θ_{ext}	34.4	35.0	36.2	35.0	35.2	34.4	31.7	32.8	33.8	32.8	33.9	34.4	36.2
		θ_{max}	31.3	31.7	32.5	32.4	31.6	29.4	29.1	29.4	29.9	29.9	30.1	30.9	30.7
		θ_{min}	22.3	22.9	24.2	25.1	25.0	23.6	23.2	23.3	23.3	23.4	23.1	22.5	23.5
		θ_{mean}	26.8	27.3	28.4	28.8	28.3	26.5	26.2	26.4	26.6	26.6	26.6	26.7	27.2
		θ_a	31.8	32.3	33.4	33.8	33.3	31.5	31.2	31.4	31.6	31.6	31.6	31.7	32.2
		% Rated Load	100	100	99	98	99	100	101	101	100	100	100	100	100
48.	Vizag	θ_{ext}	33.1	36.7	38.3	40.5	43.3	44.4	38.3	38.2	37.8	36.8	33.9	32.8	44.4
		θ_{max}	27.7	29.2	31.2	32.8	34.0	33.7	31.7	32.0	31.6	30.9	29.3	27.7	31.0
		θ_{min}	17.5	19.3	22.6	25.9	27.8	27.4	26.0	26.0	25.6	24.5	21.2	18.3	23.5
		θ_{mean}	22.6	24.2	26.9	29.4	30.9	30.6	28.8	29.0	28.6	27.7	25.2	23.0	27.2
		θ_a	27.6	29.2	31.9	34.4	35.9	35.6	33.8	34.0	33.6	32.7	30.2	28.0	32.2
		% Rated Load	103	102	100	98	96	96	98	98	98	99	102	103	100

CHAPTER 17

Erection and Commissioning

C.M. Shrivastava

The chapter describes some of the main precautions which must be taken during erection and commissioning of a transformer. Maintenance schedule has also been discussed so that continuous trouble-free service could be ensured.

17.1 Dispatch

After completing all contractual obligatory tests at works, transformer/reactor is made ready for dispatch.

When transport conditions permit, transformers are despatched

— fully assembled including fittings
— assembled without fittings, with inert gas (preferably dry nitrogen)
— assembled without fittings, with oil filled up to top yoke level
— partially assembled job for site assembly.

In a great majority of the cases accessories like radiators, bushings, explosion vent/pressure relief valve, dehydrating breather, rollers, Buchholz relay, conservator, pipe work, marshalling box are despatched separately. In some of the cases, tapchanger is also despatched separately. Transformer oil (if it is included in the order) is sent in separate sealed drums/tankers. When transformers are despatched with inert gas, positive pressure must be maintained throughout the period till gas is replaced by oil.

17.2 Inspection upon Arrival at Site

Immediately after transformer is received at site, it should be thoroughly examined externally for possible damages which may have occurred during transit. Nitrogen gas pressure (when filled during dispatch) should be checked. Positive pressure if not found indicates that there is a leakage, and there is a possibility of the moisture entering the tank during transit. This can be ascertained by measuring the dew point. The dew point measurement indicates the amount of surface moisture content in transformer insulation. As the insulation temperature and transformer gas pressure vary, the acceptable dew point will vary.

The various packages must also be checked. Internal inspection should be carried out to the extent possible through inspection covers. Particular attention should be paid to the connections, bolts, links, coil clamping bolts, tapchangers, current transformers and the general insulation.

Breakdown strength of oil of transformer tank (when the transformer is despatched filled with oil) and drums containing transformer oil (which have been despatched separately) should be examined carefully.

17.3 Handling

The following means are normally used for lifting operations

— overhead travelling crane or gantry crane
— jib crane
— derricks
— jacks and winches

The overhead travelling crane and jib cranes are obviously the most convenient and safe means. Precautions mentioned below must be adhered to:

— Transformer (main package) should be lifted only through lifting points provided for attaching the slings
— Cover must always be bolted in position

Transformer should be jacked up using the jacking pads specially provided for that purpose. Jacks should never be placed under any valves.

17.4 Installation

Following precautions should be taken before taking up erection

— Person going inside a transformer must wear clean clothes and clean synthetic-rubber-soled sandals or boots.
— Never stand directly on any part of the insulation.
— No one should be allowed on top of the transformer, unless he has emptied his pockets.
— All the tools and spanners used for erection should be securely tied with taps so that these could be recovered if dropped in, by accident.
— All components should be carefully cleaned outside separately, before erecting.

Fibrous material should not be used for cleaning. The presence of suspended fibrous material will reduce the electrical properties of transformer oil.

— Interior of the transformer should not be exposed to damp atmosphere as far as possible, to avoid condensation. In the event of a sudden change in the weather bringing rain or snow, provision must be made for closing the tank quickly and pressurizing it with nitrogen so as to preserve the insulation.
— Naked lights and flame should never be used near oil filled transformer. Smoking must not be allowed on the trans-former cover when the cover plates are open, nor in the vicinity of oil processing plant.
— Never allow any one to enter the transformer if adequate supply of air in the tank is not available.

17.4.1 Location and Site Preparation

No special foundation is necessary for the installation of a transformer except a level floor strong enough to support the weight and prevent accumulation of water. Foundation incorporating special oil drainage facilities during fire and emergency is recommended for large transformers. Transformers should be placed on the foundation so that easy access is available all around and diagram plates, thermometers, valves, oil gauges, etc., can be easily reached or read. Adequate electrical clearances are also to be provided from various line points of the transformer to earthed parts.

Type "ONAN" transformer depends entirely upon the surrounding air for carrying away the heat generated due to losses. For indoor installation, therefore, the room must be well ventilated so that the heated air can escape easily and be replaced by cool air. Air inlets and outlets should be of sufficient size and numbers to pass adequate air to cool the transformer. The inlets should be as near the floor as possible and outlets as high as the building allows. Where necessary, exhaust fans can be installed for the purpose. The transformer should always be separated from one another and from all walls and partitions to permit free circulation of air.

Where rollers are not fitted, level concrete plinth with bearing plates of sufficient size and strength can be adopted for outdoor transformers. The formation of rust, due to the presence of air and water in the space between the plinth and the base of the transformer should be prevented by use of rust preventive bituminous compound. Where rollers are fitted, suitable rails or tracks should be used and the wheels should be locked to prevent accidental movement of the transformer.

Bushing should be lifted properly as shown in Fig. 17.1. Mounting of stress shield must be properly checked as per drawing and placed accordingly. Lead should be pulled slowly so that joints should not give way.

17.4.2 Special Precautions for Installing 245 kV Re-entrant Type Bushing

The special feature of this bushing is the re-entrant control at the oil end which is brought about by special arrangement of condenser layers. Starting from the earth foil, the limiting area of foil ends in conformity with regular constructional principle form a core oriented away from the flange. At about two-third of the voltage between flange and tube, the foil ends are reversed and follow a course of

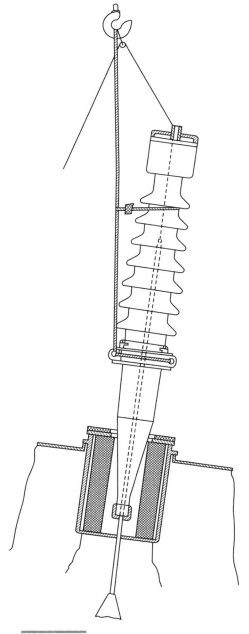

Figure 17.1 *Installation of bushing.*

core reverting towards the flange and goes as far as the conductor tube. The re-entrant control has the following advantages:

— The lower end is shorter than the conventional ends.
— Orifice of the lower end has a larger diameter and makes possible introduction of thick insulated cable.
— Stress shield is avoided.

It is important to ensure the concentricity and axial position of the insulated lead of transformer winding inside the re-entrant oil end of the bushing. Following precautions must be followed (Refer to Fig. 17.2):

— Ensure that the position of the transformer winding lead is correct axially and radially.

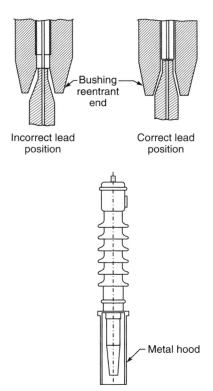

Figure 17.2 *Handling of 245 kV re-entrant type bushing.*

- Guide the lead while lowering the bushing on transformer tank so that it is concentric with the bushing orifice and angular gap all around is uniform.
- Check that lead is well insulated and in its final position the insulation is extended to the inside of the bushing tube.
- Do not allow the bushing weight to fall on the lead insulation while lowering the bushing into tank.

17.4.3 Oil Filling

Before filling with oil, transformer should be fitted with all accessories, such as valves, plugs, and made oil tight.

The oil which is to be filled in transformer must be tested for dielectric strength and water content and should be in line with the recommendations. If not, it must be filtered with stream line filters with built-in heaters and vacuum pumps for improving the quality of oil. During oil filling operation, it should be ensured that no air pockets are left in the tank and no dust or moisture enters the oil and it should be warmer than surrounding air.

For transformers dispatched gas filled, the filling of oil inside the tank should always be done under vacuum. While evacuating the transformer tank, care should be taken to ensure that bakelite cylinders, panels, etc., are not subjected to pressure. The vacuum should be maintained for a short time after the tank is filled with oil.

The vacuum pulling in the tank will avoid chances of air getting trapped and forming pockets. Oil filling will protect the windings against exposure to atmosphere at the time of erection of bushings, etc. Following precautions should be taken during this operation:

- Oil is easily contaminated. It is very important, when sampling the oil and filling the tank, to keep the oil free from contamination.
- All equipment used for handling the oil should be cleaned and flushed with clean transformer oil before use (the oil used for washing must be discarded). Particular attention should be paid to the cleanliness of bungs, valves and other parts where dirt or moisture tends to collect.
- For sampling, glass containers with glass stoppers are to be preferred over the metal type which are susceptible to contamination by dirt, etc. Cleanliness is essential as even

small amount of dirt and water will affect the accuracy of test results. Wax should not be used for sealing the oil sample bottles. However, the stopper can be covered by a pack of silicagel tied in a piece of cloth.
— Flexible steel hose is recommended for handling insulating oil. Some kind of synthetic rubber or PVC hoses are also suitable, but only those known to be satisfactory should be used. Ordinary rubber hose should not be used for this purpose, as oil dissolves the sulphur from the rubber and is thereby contaminated. Hose used for handing oil should be clean and free from loose rust or scales.
— Transformer must always be disconnected from the electricity supply system before the oil level in the tank is lowered.
— Oil must not be emptied near naked lights as the vapor released is inflammable.

17.4.4 Drying of Transformers

Drying of transformer is necessary in case insulation has absorbed moisture.

The process of drying out a transformer is one requiring care and good judgment. If the drying out process is improperly performed, great damage may result to the transformer insulation through overheating, etc. A properly dried out and correctly installed transformer is one of the most reliable electrical appliances. In no case should a transformer be left unattended during any part of the dry-out period; transformer should be carefully watched throughout the dry-out process and all observations should be recorded properly.

Drying of Core and Coils Using Oven

Where a suitable oven is available, the core and coils can be effectively dried in it by raising the temperature to a level not exceeding 80°C. A large volume of air should pass through the oven to remove moisture and vapors. Insulation resistance check will indicate when the coils are dry.

Core and coils can also be dried in its own tank in an oven. Transformer tank should be suitable for full vacuum (low vacuum for low voltage class). Full vacuum is kept in the tank and a temperature of about 75°C is maintained. Dry nitrogen is used for breathing the vacuum.

Drying by Short-circuit Method

The transformer can also be dried by heating the coils by short-circuiting the low voltage winding and supplying a reduced voltage at high-voltage terminals. Current should not exceed 70% of normal rated current and oil temperature should be of the order of 75°C. Winding temperature should in no case exceed 90°C. The winding temperature can be monitored by measuring winding resistance. This method is more effective in drying the insulation at site.

Drying out by Streamline Filter Machine (Ref. Fig. 17.3)

The most practical method of drying out is by circulation of hot oil through *streamline* filter machine incorporating oil heater and vacuum chamber. The vacuum pump of the filter machine should have the capacity of creating vacuum as high as possible but not less than 710 mm of mercury. Drying out process can be made faster by creating vacuum in the transformer tank by lagging the transformer tank to prevent loss of heat. The oil temperature in transformer should be of the order of 75°C.

It should be seen that the oil temperature at the filter machine in no case exceeds 85°C.

Drying process can be terminated when transformer oil characteristics are achieved within permissible limits and insulation resistance of winding shows a constant or rising trend.

17.5 Commissioning

Transformer must be healthy in all respects before energizing and, therefore, it should be thoroughly checked before commissioning. The following checks should be carried out.

17.5.1 Transformer

Measurements

Measurement of ratio, resistance, vector group and magnetizing current should be of the order of works' test results. BDV and water

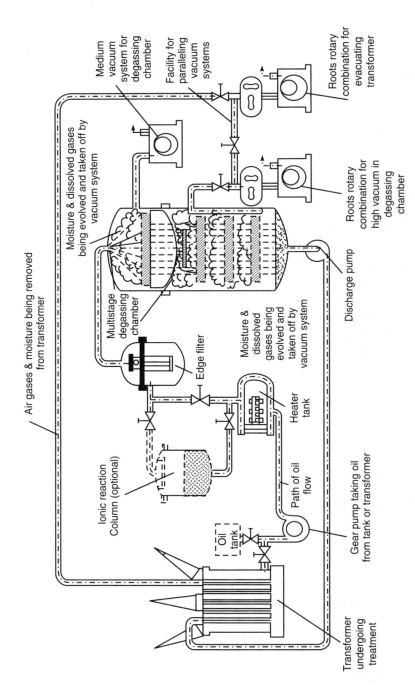

Figure 17.3 *General diagram for treatment of oil and conditioning of transformer.*

content of oil in transformer tank, tapchanger and cooler bank (the oil results should be within the permissible limits as mentioned in Appendix A1 of IEC 422) have also to be measured.

Observations

Bushing CT polarity, silicagel breather, earthing, bushing arching horn, valves, etc., must be checked for their correctness.

17.5.2 Protection

Buchholz relay, oil temperature, winding temperature, pressure relief device, magnetic oil level gauge, differential over current, earth fault, etc. Protection scheme should be checked for proper functioning.

17.5.3 Tapchanger

Manually, electrical (local and remote), parallel operations, IR value of motor and control wiring to be checked.

17.5.4 Marshalling Box

All control wiring should be checked.

When all the above checks have been done, a settling time of approximately 24 hr should be given to oil before charging the transformer. During this interval of settling, air should be released six hourly wherever possible. Voltage should be slowly built up to full level in around four to six hours wherever possible.

17.6 Maintenance

If a transformer is to give long and trouble-free service, it should receive a reasonable amount of attention and maintenance. The principal object of maintenance is to maintain the insulation in good condition. Moisture, dirt and excessive heat are the main cause of insulation deterioration. Maintenance consists of regular inspection, testing and reconditioning, wherever necessary.

Load voltage, load current, temperature of winding, oil and ambient should be recorded daily. Various fittings/accessories like silicagel breather, bushing, tapchanger, fan pump, oil in tank, tapchanger bushing, Buchholz relay, etc., should be checked periodically as per recommendation.

17.6.1 Reconditioning Procedure of Condenser Bushing (S.R.B.P. Type)

In case moisture is found in the bushing, the oil of the bushing is drained through drain plug. Two nylon tubes of suitable diameter are connected to an oil filtering plant, through suitable adaptation. The outlet tube from filter to be connected to the drain plug and the inlet tube of the filter to be submerged in oil at the top. The filtered hot oil should be circulated through bushing for three days by injecting it at the bottom and sucking from the top. This not only improves the oil quality but also dries out the bushing insulation. During the entire period of circulation, the bushing top must be covered so that no foreign matter can enter the bushing. After the reconditioning, the IR value of the bushing and dielectric strength of oil should be in line with the recommendations.

17.6.2 Analysis of Gases as a Means of Monitoring Transformers and Detecting Incipient Fault

Gases are produced in a transformer when transformer oil is subjected to high electrical or thermal stresses or due to breakdown of other insulation materials. These gases get dissolved in oil and if produced in substantial quantities, it may get collected in Buchholz relay. Analysis of these gases, together with the rate of their formation makes it possible to estimate the nature and seriousness of the fault.

The gases in Buchholz relay will, in general, be due to one of the following three main causes:

(a) Air introduced because of mechanical fault in the oil system or in some ancillary equipment.
(b) Gas produced by thermal or electrical breakdown of oil without damage to any solid insulants.
(c) Gas produced by a thermal or electrical breakdown of solid insulants which invariably includes the breakdown of oil.

It should, however, be kept in mind that one of the primary objects of fitting a Buchholz relay is to detect faults in their incipient stage and care should be exercised in interpreting the analysis results together with all other available data, to ensure that re-energizing the transformer does not cause an incipient fault to develop into a major fault.

In the event of low energy faults with only slight gas evolution, and in view of the fact that the oil solubility of the gases of decomposition may be very high, the detection of such gradual process is delayed, because the Buchholz relay does not respond until a sufficient volume of gas is collected. Further, because of the different solubility of the various gas constituents and the partial exchange with other gases dissolved in the oil, e.g., nitrogen or air, the composition of the gas mixture that is collected in the Buchholz relay differs from that at the point of origin. It is, therefore, hardly possible to obtain correct diagnosis of the nature of such an incipient fault by simply analyzing the gas collected in the Buchholz relay. A complete and reliable explanation of the faults can only be obtained by supplementing the gas analysis of the Buchholz relay by an analysis of the gases dissolved in the oil.

An analysis of the gas dissolved in the oil is possible by means of gas thromatography, which determines the individual constituents of such gas mixtures with a high degree of accuracy. IEC 599, *Interpretation of the analysis of gas in transformers and other oil filled electrical equipment in service,* may be referred to for this analysis. A comparison between the gases accumulated in the Buchholz relay and those dissolved in the oil can be very useful in diagnosing the nature and severity of the particular fault.

17.7 Dos for Power Transformer

1. Connect gas cylinder with automatic regulator if transformer is to be stored for long duration, in order to maintain positive pressure.
2. Fill the oil in the transformer at the earliest opportunity at site and follow storage instructions. It must be commissioned as soon as possible.

3. Open the equalizing valve between tank and OLTC diverter compartment, whenever provided, at the time of filling the oil in the tank and close the same during operation.
4. Clean the oil conservator thoroughly before erecting.
5. Check the pointers of all gauges for their free movement before erection.
6. Inspect the painting and if necessary do retouching.
7. If inspection covers are opened or any gasket joint is tightened, tighten the bolts evenly with the proper sequence to avoid uneven pressure.
8. Clean the Buchholz relay and check the operation of alarm and trip contacts.
9. Check the oil level in oil cup and ensure that the air passages are free in the breather. If oil is less, make up the oil level.
10. Check the oil in transformer and OLTC for dielectric strength and moisture content, and take suitable action for restoring the quality of oil.
11. Attend to leakages on the bushing immediately.
12. Check the diaphragm of the relief vent. If cracked or broken, replace it.
13. Remove the air from vent plug of the diverter switch before energizing the transformer.
14. Check the gear box oil level in the tapchanger. If less, top up with specified oil.
15. Check the OTI and WTI pockets and replenish the oil, if required.
16. Check the oil level in the diverter switch and if found less, top up with fresh oil.
17. Examine the diverter and selector contacts of tapchanger and if found burnt or worn out, replace the same.
18. Check and thoroughly investigate the transformer whenever any alarm or protection is operated.
19. Examine the bushings for dirt deposits and coats and clean them periodically.
20. Check the protection circuits periodically.
21. Check all bearings and operating mechanism of the tapchanger and lubricate them as per schedule.

Erection and Commissioning 453

22. Keep the valve connected between conservator of the tap-changer and its diverter compartment open, during transformer operation.
23. Check the silicagel charge. If it is found pink, regenerate or replace it with blue silicagel charge.

17.8 Don'ts for Power Transformers

1. Do not use low capacity lifting jacks on transformer for jacking.
2. Do not allow WTI, OTI temperature to exceed 75°C during dryout of transformer, and filter machine temperature beyond 85°C.
3. Do not re-energize the transformer, unless the Buchholz gas is analyzed.
4. Do not re-energize the transformer without conducting all pre-commissioning checks.
5. Do not energize the transformer, unless the off-circuit tap switch handle is in locked position.
6. Do not leave off-circuit tap switch handle unlocked.
7. Do not leave tertiary terminals unprotected outside the tank. Follow manufacturer's recommendations in this regard.
8. Do not leave marshalling box doors open. They must be locked.
9. Do not leave any connection loose.
10. Do not meddle with the protection circuits.
11. Do not leave maximum temperature indicating pointer behind the other pointer in OTI and WTI.
12. Do not change the settings of WTI and OTI alarm and trip frequently. The setting should be done as per the site conditions.
13. Do not allow oil level in the bushings to fall; they must immediately be topped up.
14. Do not allow conservator oil level to fall below one-fourth level.
15. Do not parallel transformers which do not fulfil the required conditions.
16. Do not switch off the heater in marshalling box except in summer.
17. Do not leave secondary terminals on an unloaded CT open.

18. Do not allow water pressure more than oil pressure in differential pressure gauge in OFWF cooled transformer.
19. Do not switch on water pump unless oil pump is switched on.
20. Do not leave ladder unlocked, when the transformer is energized.
21. Do not allow unauthorized entry near the transformer.
22. Do not overload the transformer other than the specified limits mentioned in national/international standards.
23. Do not allow inferior oil to continue in transformer.
24. Do not handle the off-circuit tap switch when the transformer is energized.

17.9 Dos and Don'ts for HV Condenser Bushings

17.9.1 Dos

1. Check the packing externally for possible transit damage before unpacking.
2. Do unpacking with care to avoid any direct blow on bushing or porcelain insulator.
3. Store the bushing in a shed or covered with tarpaulin to protect it from moisture and rains. If removed from the crate, keep it indoors with lower end protectives intact.
4. Handle the bushing with manila rope slings without any undue force on porcelain insulator.
5. Clean the porcelain insulator thoroughly.
6. Remove the wax tape protection on the oil end of the bushing (at the time of erection) and clean the surface with hot transformer oil.
7. Check the oil level and IR value of the bushing in vertical position only, taking care that the bushing is cleaned and no rope or sling, etc., is touching the terminal and ground.
8. Check the breakdown value (BDV) of oil taken from drain plug or siphoning from the bottom-most portion of bushing. This should not be less than the recommended value.
9. Check the IR value and tan δ value (if possible) with bushing in position on transformer with jumper connection removed.

Record these readings for reference and guidance for future measurements.
10. Check BDV of oil and IR value of each bushing periodically during maintenance shut down. These values should be comparable with the values recorded at the time of commissioning.
11. Ensure to allow the air to escape from central tube to the atmosphere while filling the transformer tank.
12. Maintain the log book records of periodical checks (i.e., tan δ and BDV of oil) up to date.

17.9.2 Don'ts

1. Do not unpack the bushing from the crate unless required to be mounted on the transformer.
2. Do not remove the waxed tape protection/metal protective hood from the oil end portion unless bushing is required for use.
3. Do not store the bushing outdoors without any protective covering.
4. Do not measure the IR value and tan δ value without thoroughly cleaning the porcelain and oil end portion.
5. Do not store the bushing without oil in porcelain.
6. Do not keep the top cap cover open for any longer time than required as it contaminates the oil.
7. Do not tighten the nuts and bolts in excess to stop any leakage, this could damage the cemented joints on porcelain.

CHAPTER 18

Transformer Protection

B.L. Rawat

Transformer being a vital equipment, its protection is equally important. The subject of transformer protection can be categorized under two major headings.

(a) Protection of the transformer against the effects of faults occurring on any part of the system beyond the transformer.
(b) Protection against the effects of faults arising in the transformer.

18.1 Protection against External Faults

18.1.1 Short-Circuit

Short-circuit across any two or all the three lines may occur in the system. The over currents produced because of such faults depend upon system MVA feeding the fault, the voltage which has been short-circuited and upon the impedance of the circuit up to the fault. System short-circuits produce a relatively intense rate of heating of the feeding transformer, as the copper-loss increases in proportion to the square of per-unit fault current. The duration of external short-circuit that a transformer shall sustain without damage, if the current is limited only by self-impedance, is 2 s as per IS : 2026. Large fault currents produce severe mechanical stresses in the transformer; the maximum stress occurs during the first cycle of asymmetrical fault current and so cannot be averted by automatic tripping of the circuit. The control of such stress is therefore a matter of transformer design. The transformers connected to a large power system have their windings very securely braced in order to minimize the effects of the mechanical forces to which they may be subjected, due to short-circuit. IS : 2026 specifies typical values of short-circuit impedance, for different ratings of transformers, to limit the short-circuit current.

18.1.2 High Voltage Disturbances

High voltage disturbances are of two kinds
- (a) Transient surge voltages
- (b) Power frequency voltages

(a) *Transient Surge Voltage*

High voltage high frequency surge may arise in the system due to any of the following:

- (i) Arcing grounds if neutral point is isolated
- (ii) Switching operation
- (iii) Atmospheric disturbances

These disturbances principally take the form of travelling waves having high amplitudes and steep wave fronts. On account of both their high amplitudes and frequencies these surges may, upon reaching the winding of transformer, break down the insulation between turns adjacent to line terminal, causing short-circuits between turns and producing extensive damage to the transformer winding. The effects of these surge voltages may, however, largely be minimized by designing the winding to withstand the application of a specified surge test voltage and then ensuring that this test voltage is not exceeded in service by the provision of a suitable surge diverter mounted adjacent to the transformer terminals.

All types of surge diverters aim at attaining the same results, viz. of shunting disturbing surges from the lines to earth to prevent their reaching the transformer. In essence, the different kinds of valve-type surge diverters employ several spark gaps in series with a non-linear resistor. These non-linear resistors offer a low resistance path to high voltage surge waves and hence these disturbances are discharged to earth through the diverter. As the surge voltage falls, the diverter resistance automatically increases and prevents the flow of power current to earth. Figure 18.1 shows schematically how the various dielectric paths of a three-phase transformer should be shunted by surge diverters in order to protect the transformer bushings and windings against surge voltages. A surge protective device should have the following qualities:

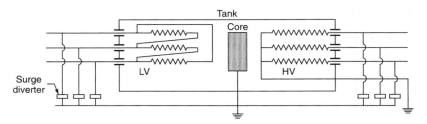

Figure 18.1 *Schematic diagram showing connection of surge diverters.*

(i) Rapid response to impulse over voltage
(ii) Independance of wave polarity
(iii) Non-linear characteristic
(iv) High thermal capacity
(v) High system follow current interrupting capacity
(vi) Consistent characteristic

Now, zinc oxide arrestors having high reliability and better characteristic are also available in the market.

(b) *Power Frequency Over-voltage*

Power frequency over-voltage causes an increase in the stress on the insulation and a proportionate increase in the working flux. The latter effect causes an increase in the iron-loss and a disproportionately large increase in magnetizing current. In addition, flux is diverted from the laminated core structure into steel structural parts. In particular, under conditions of over-excitation of the core, the core bolts which normally carry little flux, may be subjected to a large component of flux diverted from the saturated region of core alongside.

Under such conditions, the bolts may be rapidly heated to a temperature which destroys their own insulation and will damage the coil insulation if the condition continues.

Reduction of frequency has an effect, with regard to flux density, similar to that of over-voltage.

The overfluxing protection does not call for high speed tripping. Instantaneous operation is undesirable, as this would cause tripping on momentary system disturbances which can be born safely, but normal condition must be restored or the transformer must be isolated before significant damage is done to insulation structure. The fundamental equation for the generation of emf in a transformer can be arranged to give

$$\varphi = k\frac{E}{f}$$

It is necessary to detect a ratio E/f exceeding unity, E and f being expressed in per-unit values of rated quantities. An overfluxing relay whose E/f characteristic closely matches with that of the transformer should be used to give alarm and signal to correct the disturbance. If the condition persists for a long time, the transformer should be disconnected from the system to protect it from severe damage.

18.2 Protection against Internal Faults

Considering next the means to be adopted for protecting the transformer against the effect of faults arising in the transformer, the principal faults which occur are:

  (i) Breakdown to earth—either of winding or terminal gear
  (ii) Phase-to-phase fault
  (iii) Inter-turn fault
  (iv) Core faults

18.2.1 Earth Faults

(a) Star-connected winding with neutral point earthed through an impedance. The fault current for the fault shown in Fig. 18.2 is dependent on the value of earthing impedance and is also proportional to the distance of the fault from neutral point as voltage at the fault point will be directly proportional to this distance. The fault current in the primary winding will depend on transformation ratio between primary winding the short-circuited turns, which varies with the position of fault in the winding.

From Fig. 18.2, it may be noted that the fault in the lower third of the winding produces very little current through the primary terminals.

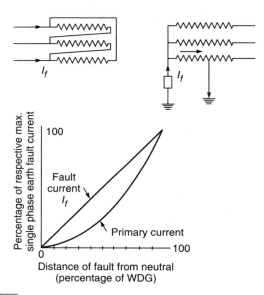

Figure 18.2 *Earth fault current in resistance earthed star winding.*

(b) Star-connected winding with neutral point solidly earthed. In this case, fault current is controlled by the leakage reactance of the winding which varies in a complex manner with the position of fault on the winding. As in the earlier case, the voltage available for fault current varies with the position of fault on the winding. It is seen that the reactance decreases very rapidly for the fault point approaching the neutral and hence the fault current is highest for a fault near the neutral end of the winding. The variation of current with fault position is shown in Fig. 18.3.

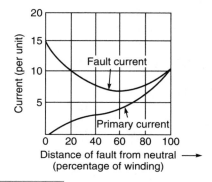

Figure 18.3 *Earth fault current in case of solidly earthed neutral.*

From Fig. 18.3 it can be noticed that the fault current magnitude stays very high throughout the winding. Also, after transformation ratio the input current curve remains at a substantial level for faults at most points along the winding.

(c) Restricted earth fault protection. A simple over-current and earth fault system will not give good protection cover for a star-connected primary winding, particularly if the neutral is earthed through an impedance. The restricted earth fault protection schematic shown in Fig. 18.4 improve the degree of protection very much. This scheme is operative for faults on the star winding of the transformer. The system will remain stable for all faults outside the zone. As whole fault current is measured, a good degree of protection of the winding is achieved even in the case of neutral being earthed through impedance. In the case of solidly earthed neutral, fault current remains at a very high value even to the last turn of the winding and hence complete cover for earth fault is obtained with restricted earth fault protection scheme.

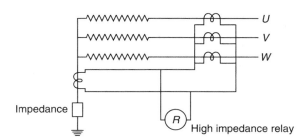

Figure 18.4 *Restricted earth fault protection for a star-connected winding.*

18.2.2 Phase-to-phase Fault

Phase-to-phase faults in the transformer are rare. If such a fault does occur, it will give rise to substantial current to operate instantaneous over-current relay on the primary side as well as the differential relay.

18.2.3 Inter-turn Fault

A high voltage transformer connected to an overhead transmission system is very likely to be subjected to steep fronted impulse voltages. A line surge which may be of a magnitude several times the rated system voltage will concentrate on the end turns of the winding, because of the high equivalent frequency of the surge front. Also LV winding is stressed because of the transferred surge voltage. It is reported that a very high percentage of all transformer failures arise from faults between turns. Inter-turn fault may also occur because of mechanical forces on the winding due to external short-circuit. Though there may be high short-circuit current between few turns loop, the terminal current will be very small because of the high ratio of transformation between the whole winding and the short-circuited turns. If this turn-to-turn fault is not detected in the earliest stage, the subsequent progress may destroy the evidence of the true cause.

(a) *Differential Protection*

A differential scheme can be arranged to cover the complete transformer protection. This is possible because of the high efficiency of transformer operation and the close equivalence of ampere-turns balance on both primary and secondary windings. The rated currents of primary and secondary windings differ in inverse ratio to the corresponding voltages. Therefore, current transformers should have their primary rating to match the rated currents of the transformer windings to which they are applied. To correct phase shift of current because of star/delta-connection of transformer winding, the current transformers secondaries should be connected in delta and star as shown in Fig. 18.5.

For correcting voltage variation, almost all transformers are provided with tapping, which in turn change the ratio from the mean. This will create an imbalance proportional to the ratio change. At maximum through-fault current, the spill output produced by the small percentage unbalance may be substantial. Therefore, differen-

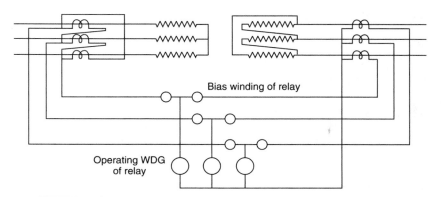

Figure 18.5 *Schematic diagram of differential protection scheme.*

tial protection should be provided with a proportional bias of an amount which exceeds in effect the maximum ratio deviation. This stabilizes the protection under through-fault condition while still permitting the system to have good basic sensitivity.

(b) Magnetizing Inrush Current Phenomenon

The magnetizing inrush phenomenon produces current input to the primary winding, which has no equivalent on the secondary side. The whole of the magnetizing current appears, therefore, as unbalances. The normal bias for the inrush current is not effective and increase of protection setting to a value (which would avoid operation of relay) would make the protection of little value.

As the magnetizing inrush current phenomenon is transient, stability can be achieved by providing a small time delay. This can be achieved by various means. A kick fuse can be connected as shunt to an instantaneous relay. This fuse is so rated that it carries inrush current without blowing. However, for internal fault the fuse blows and permits the relay to operate. An induction pattern relay may have both a suitable time characteristic and also a through bias feature. Two induction electromagnets operate on a single disc to produce opposing torque.

In case of severe inrush current, the set time delay might be insufficient to give stability. Also, to minimize the damage to important transformers, it may be essential to clear the fault without delay. For these cases, another solution to the inrush phenomenon must be found. It can be noticed that the waveform of inrush current and zone fault current differs greatly. This distortion of waveform can be used to distinguish between the conditions. The differential current is passed through a filter which extracts the second harmonic current. This component is then applied to produce a restraining quantity sufficient to overcome the operating tendency due to whole of the inrush current which flows in the operating circuit.

(c) Combined Differential and Restricted Earth Fault Scheme

A set of additional phase correcting auxiliary current transformers are required as shown in Fig. 18.6.

18.2.4 Core Faults

If any portion of the core insulation becomes defective or the laminated structure of the core is bridged by any conducting material which can permit sufficient eddy current to flow, it will cause serious overheating. The insulated core bolts are used for tightening the core. If

464 Transformers

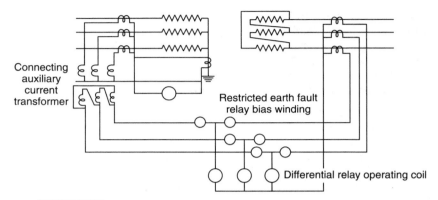

Figure 18.6 *Schematic diagram of differential and restricted earth fault scheme.*

the insulation of these bolts fails and provided easy path for eddy current, this will lead to over-heating. This additional core-loss, although it causes severe local heating, will not produce a noticeable change in the input current and cannot be detected by the normal electrical protection. However, it is desirable to detect over-heating condition before a major fault has been created. Excessive heating of core will breakdown transformer oil with evolution of gas. This gas rises to the conservator.

(a) All faults in transformer core and windings result in the localized heating and breakdown of oil. When the fault is of very minor type such as hot joint, gas is released slowly and rises towards conservator. A major fault where severe arcing takes place, causes rapid release of large volume of gas and oil vapor. This violent evolution of gas and oil vapor does not have time to escape and instead builds up pressure and bodily displaces the oil, causing surge of oil to pass up the relief pipe to the conservator. Recognition of the above action by Buchholz, and the limitation of other means of detecting certain types of fault, led to the development of a protective device known by his name. Construction of Buchholz relay has been described in Chapter 11. Two electrical contacts are available in the Buchholz relay, one for alarm and the other for trip. When the generation of gas is slow, the gas while moving towards the conservator gets trapped in the relay and displaces the oil. After a certain amount of oil is displaced the alarm contact is made and gives signal in the control room necessitating further investigation.

A violent action described above causes a surge of oil which makes the trip contact and the transformer is isolated from the circuit.

Because of its universal response to faults within the transformer, some of which are difficult to detect by other means, the Buchholz relay is invaluable.

(b) Because of heavy arcing inside transformer, if excessive pressure is generated, it is released through the pressure relief device. Also one electrical contact on the device is made which trips the circuit breakers and isolated the transformer from electrical circuit. Construction of the device has been described in Chapter 11.

(c) The overload or persistent fault increases oil temperature or winding temperature or both. If the temperature increases beyond safe working limit, the electrical contacts provided on these devices are made. When first contact is made it gives alarm signal in the control room necessitating investigation. If temperature rises further, another contact is made which trips the transformer. Construction of these devices have been described in Chapter 11.

REFERENCES

1. Austen Stigant, S., and A.C. Franklin, *The J and P Transformer Book*, Newness-Butterworth, London.
2. *Protective Relay Application and Guide*, General Electric Co. (U.K.).

CHAPTER **19**

Reactors

C.M. Shrivastava
S.K. Mahajan

Reactors find a number of applications in the transmission and distribution network as well as industrial plants. They are usually classified according to duty application, viz. current limiting, neutral earthing, shunt, damping, tuning or filter, arc-suppression, smoothing, etc. These reactors have typical characteristic requirements and call for different constructions, viz. with air core or with gapped iron core, with or without magnetic shield, for indoor or outdoor installation, for fixed and variable reactance, with or without taps, etc.

19.1 Series Reactors

These reactors are intended for series connection in a system, either for limiting the current under system fault condition or load sharing in parallel circuits.

19.1.1 Types

Series reactors are classified by their location or application in a system. Some of the more common types are described hereafter.

(i) *Generator Line Reactor*

This is usually used with a generator to reduce stresses under 3-phase short-circuit.

(ii) *Feeder Reactor*

This is located on feeders from operating stations or on sub-station buses and is used to minimize the resultant effects of a short-circuit on other parts of the system. This application often allows a lighter

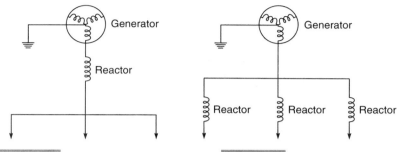

Figure 19.1 *Generator reactor.*   **Figure 19.2** *Feeder reactor.*

construction of feeder circuit which can result in appreciable cost savings due to their association with circuit breakers of smaller capacities.

(iii) *Duplex Reactor*

It is a center tapped reactor used as an effective device in dividing current source and limiting the magnitude of fault current.

It is usually applicable where:

(a) There is a heavy concentration of load as in a power house.
(b) Local generation is present.
(c) The utilization and the generation voltages are the same.
(d) There is a large amount of load distribution at the same voltage.

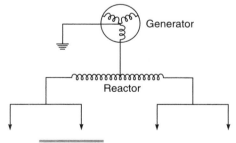

Figure 19.3 *Duplex reactor.*

(iv) *Tie Line Reactor*

It provides an easier method of expanding a system at the existing voltage. It has the following advantages:

(a) Existing switchgear can be used with a little modification.
(b) It is a relatively low cost means of expanding a system.

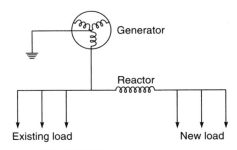

Figure 19.4 *Tie line reactor.*

(v) Synchronizing Reactor

This reactor provides a method for expanding a system when the utilization voltage and generated voltage are the same and it is not feasible to go to higher voltage. This is done by sectionalizing the existing bus and tying all the bus sections through reactors to a common bus. By this method interrupting duty is greatly reduced on the existing breaker and new loads and additional incoming capacity may be added. This arrangement is frequently found in industrial power plants on a very large system.

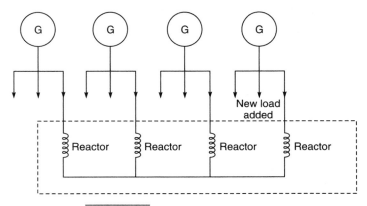

Figure 19.5 *Synchronizing reactor.*

(vi) Motor Starting Reactor

This is used when it is necessary to reduce the starting current of AC motors. This decreases stresses on the motor and reduces the system disturbances.

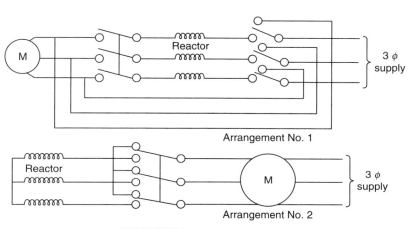

Figure 19.6 *Starting reactor.*

(vii) *Reactor for Capacitor Bank*

This reactor is used to limit high transient inrush current flowing into the capacitor bank when capacitor bank is switched on. This also helps in suppressing harmonics present in the system. This is also called damping reactor. This is connected in series with the capacitors and is rated for highest inrush current and continuous current.

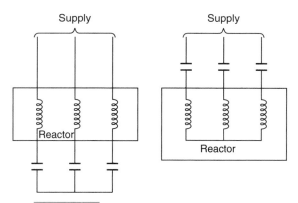

Figure 19.7 *Reactor with capacitor bank.*

(viii) *Smoothing Reactor*

This is employed to filter out all harmonics present in DC power system. This can be both HV DC smoothing reactor and low

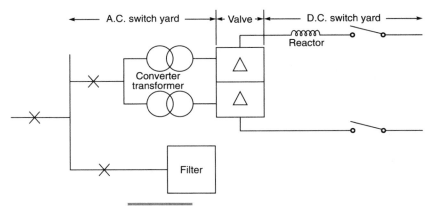

Figure 19.8 *Smoothing reactor.*

voltage DC smoothing reactor with large superimposed harmonic components.

(ix) *Line Trap*

This is used on power lines having carrier channels for equipment control and communication purposes and as such is a resonated inductance coil. It helps in blocking out unwanted frequencies and noise from the carrier channels.

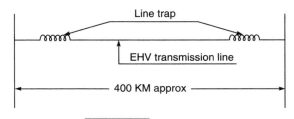

Figure 19.9 *Line trap.*

19.1.2 Calculation Data

If it is desired to reduce fault from Z to Y MVA by connecting the reactor in series

$$X_i = U^2 \left(\frac{1}{Y} - \frac{1}{Z} \right)$$

$$I = \frac{A \times 1000}{\sqrt{3}U}$$

$$I_S = \frac{Y \times 1000}{\sqrt{3}U}$$

$$Q = I^2 \times X_1 \times 10^{-3}$$

where U = System voltage (kV)
 A = System throughput power (MVA)
 I = Reactor current (amperes)
 X_i = Reactor reactance (ohms)
 Q = Rating (kVAR per phase)
 I_s = Short time current (kilo-amperes)

19.1.3 General Consideration/Characteristics

(i) *Linear Characteristic*

The reactor should have linear characteristic which means a straight line relationship between its current and voltage. Such a characteristic helps in limiting current under system fault conditions. The advantages of these characteristics are mentioned below:

(a) Expensive equipments are protected from failure due to excessive mechanical stresses and heating caused by short-circuits.
(b) Service on other parts of the system is not interrupted during a fault, since the main bus voltage is not appreciably reduced.
(c) There is a reduction in the current interruption duty of circuit breaker, thereby permitting the installation of circuit breaker of reduced rating.

However, it may be noted that in case of starting reactor it is not necessary to have a linear characteristic as it is required for shorter duration and not for limiting the short-circuit current.

(ii) *Ability to Withstand Short-Circuit*

Since the reactor is basically required to limit short-circuit current, mechanical forces (radial and compressive forces) are taken care of during design by selecting proper parameters of reactor and by proper clamping of the winding. Reactor must also be able to withstand short-circuit current from thermal point of view. The maximum permissible value of temperature of the winding during a short-circuit is given below as per IS : 2026.

Type	Class of Insulation	Temperature°C	
		Copper	Aluminum
Oil	A	250	200
Dry	A	180	180
Dry	E	250	200
Dry	B	350	200
Dry	F and H	350	—

(iii) *Fully Insulated Winding*

Series reactor is connected in series in the system and therefore both the ends should be suitable to withstand lightning impulse voltage. This can be achieved by having fully insulated winding.

(iv) *Other Requirements*

Rated continuous current and rated short-time currents are usually specified for such reactors. The rated impedance, in ohms per phase, for such a reactor, usually pertains to short-time current at rated frequency. Rated impedance in such a case is determined from the oscillogram records of the steady-state value of test voltage and current during demonstration of dynamic ability in accordance with IS : 2026 (Part I)-1977. This is classified as a special test. The routine test for impedance measurement is done at continuous current, if applicable. In case of reactor without magnetic shield, this test also verifies the rated impedance (at short-time high current). Measurement of loss applies only on reactor where continuous current is specified.

19.1.4 Construction Features

Series reactors are manufactured in various ways, some of them are described below:

(i) *Oil Immersed Coreless Reactor*

Such a reactor can be manufactured for either indoor or outdoor services and in general for oil immersed types, the reactor and its external appearance, fittings and accessories closely follow established power transformer practice. A high proportion of the magnetic flux due to the current in a reactor winding passes through the coil itself. This flux would produce local eddy losses particularly at the coil ends and could possibly cause local hot spots. To obviate the above condition it is preferable to use thin stranded conductor and frequent transposition in the coils. Transposed cables are also sometimes preferred. Due to heavy concentration of flux in and around the coil, non-magnetic materials are used except main tank body. To achieve constant ohmic reactance during short-circuit conditions, usually coreless construction with non-magnetic shielding of winding is provided. The non-magnetic shield (generally made of aluminum or copper) is in the form of cylinder and located around the winding. Eddy currents are induced in the shield, which not only help in reducing the tank losses but also reduced any tendency for the coil to bulge out during short-circuit.

(ii) *Dry Type Coreless Reactor*

Dry type reactor is usually cooled by natural ventilation but can also be designed with forced air. Free circulation of air must be maintained to provide satisfactory heat transfer.

Steel structures such as I beams, channels, plates and other metallic members either exposed or hidden should also be kept away from the reactor and also they should not form closed circuit.

Winding conductor consists of one or more cables, depending upon the current capacity, wound in radial layers. Cross-overs between layers are made of minimize the eddy losses. Aluminum conductor is normally used. Fiber glass insulation, the density of which plays an important part in the overall rigid construction of the reactor, is adopted normally in the construction.

Porcelain insulators are used for achieving sufficient electrical clearance to ground.

It is desirable that some safety enclosure which protects the personnel and equipment should be put around the reactor. The 3-phase unit can be made up of 3 single phase coils either stacked or arranged in one and the same plane. It is non-inflammable and owing to its design, this equipment requires little supervision during service. Maintenance is limited to clearing only.

19.2 Shunt Reactor

It extra high voltage networks, capacitive generation often creates problem during operation at low loads and during switching operations. The severity of such problems increases with higher system voltage, with extension of network and with longer line sections. In most of the cases shunt reactors often represent an economically and technically sound means of compensating part of the capacitive generation.

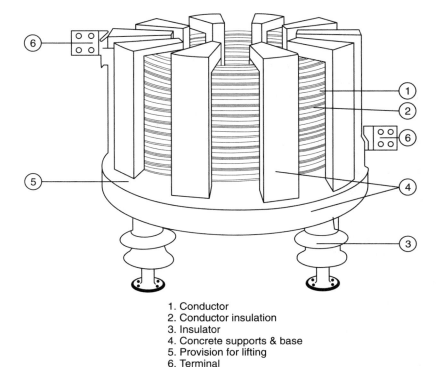

Figure 19.10 *Dry type current limiting reactor.*

1. Conductor
2. Conductor insulation
3. Insulator
4. Concrete supports & base
5. Provision for lifting
6. Terminal

19.2.1 Types

The shunt reactor can be classified as follows:

(a) *From the Service Point of View*

— Normally in service
— Switched in, switched off type

If reactor is kept normally in service, during peak load time the excitation of generator can be increased and consequently the stability of the system. Under service fault conditions the reactor may be switched off as a further measure to improve the margins of stability. This disconnection momentarily increases the voltage of the network. The use of shunt reactors in normal services may also result in poor voltage levels in the underlying systems and increase system losses. Depending upon loading condition, this may also necessitate installation of static compensator banks for operation during peak load condition. Switched-in reactor is normally used when system voltages are required to be controlled. The amount of reactance to be switched in at a time must be limited to suitable value otherwise it may result in objectionable voltage fluctuations in the entire system. For switching in and switching off the shunt reactor, protective gears should be more reliable. The switched in, switched out type of reactor is sometimes connected to a low voltage tertiary winding of a large transformer. The tertiary connected low voltage reactor is obviously cheaper. But, it causes extra losses in the transformer. Its rated voltage must be carefully selected because of the large voltage drop in the series reactance of the transformer between high voltage and tertiary winding.

(b) *From the Construction Point of View*

— Coreless (magnetically shielded)
— Gapped core
(Refer Clause No. 19.2.3)

19.2.2 General Consideration/Characteristics

Following characteristics are required to be considered:

(a) *Highest System Voltage and Linearity*

Normally the rated voltage for a reactor is defined on the highest system voltage unless otherwise stated. Most of the specifications require that the impedance of the reactor shall be linear to a certain level above rated voltage.

The magnetization curve of the reactor is a relation between crest values of current and flux. In order to avoid harmonic current generation under system over voltage conditions or the risk of non-linear ferroresonance of heavy inrush currents, shunt reactor should have constant impedance up to about 1.3 times rated voltage. Sometimes, constant impedance up to 1.5 times is specified. Furthermore, the impedance should be accurately balanced between phases of 3-phase reactor. The acceptable tolerance on the current of any single reactor at rated voltage is, according to IEC Pub. 289, ±5 percent of the rated current. Any one phase may deviate from the mean of all three phases by ±2 percent.

(b) *Losses*

The total losses in a shunt reactor from economical reasons should be as low as possible. Low loss design with natural cooling generally works out to be more economical. The tan delta figure of large reactors is in the range of 0.30 percent down to 0.15 percent at the highest capitilisation.

(c) Zero Phase Sequence Impedance

The requirement of zero phase sequence impedance is dependent upon the system conditions. If this is 90 to 100% of positive sequence impedance, coreless design with magnetically shielded or 5 limbed gapped core reactor is most suitable. If this value is not very much important or a value of about 50 to 60% of positive sequence impedance is permissible, the 3 limbed gapped core reactor may be used.

(d) Single-phase or Three-phase Units

Large rating reactor sometimes poses problems of testing. To overcome the above problem, banks of 3 single-phase units are most suitable. Further, the banks of three single-phase units have the following advantages:

— Zero phase sequence impedance is equal to positive sequence impedance.
— Spare unit requirement will be only one single-phase reactor instead of a 3 phase reactor.

Three- phase reactors are manufactured for system voltages of 400 kV and below. It is technically feasible to extend the range up to 500 and 800 kV, but this has not been done yet. The reactors in 800 kV class are usually banks of single-phase units installed with a spare reactor available in the station.

19.2.3 Construction

(a) Core-less

In a core-less reactor there is no magnetic material inside the coil and the dimensions of the winding are identical to those of the air gap volume, which imposes restrictions on the geometrical proportions. The winding must be made wide and flat, so that the magnetic flux path length is limited. The flux density in practice is much lower than in the gapped core design and the total volume is larger for a given reactive power rating. Flux with full intensity penetrates the inner turns of the winding and gives rise to relatively high eddy current losses. The flux density then sinks gradually towards the periphery of the coil, where it approaches zero. The average circumference of the coil is appreciably larger than for the gapped core reactor and the winding losses tend to be higher.

Core-less designs have a variety of alternative constructional arrangements.

In the USA, there has been frequent use of open, dry type coils (previously of concrete type). These are applied at low system voltages for the compensation of cable systems. The design is practically the same as for a current limiting reactor but it is larger in dimensions.

Another design consists of a cylindrical coil with vertical axis, placed in a cylindrical oil tank. The tank is lined with strips of core steel which serve as magnetic shielding for the returning flux outside the coil. The whole coil volume is "air gap."

Two cylindrical coils placed side by side between an upper and a lower yoke of core steel form a single phase reactor where the flux goes upward in one coil and down in the other. The coil volumes are air gap. Three coils side by side in the same manner make a 3 phase reactor. A compact non-circular pancake stack, surrounded by a rigid frame of core steel forms a type of reactor which is often referred to as "shell form," because of its similarity in shape and coil design with a shell form transformer. The principle of shell type reactors gives the possibility of arranging each phase in its own magnetic box.

Another type of construction without yokes, is a triangular conformation of coils with the axis of each coil at the corners of a triangle (Fig. 19.11). There is no yoke. The top and bottom of the tank are screened from the coil flux by plates of high conductivity material. Magnetic shields consisting of core pockets of core steel laminations, all around the tank, restrict the flux linkage with the tank thus reducing stray losses and heating of tank.

(b) *Gapped Core Reactor*

The achieve high impedance of reactor in a core type construction gaps having suitable size are inserted in the magnetic circuit. The required effective length of the magnetic field is mainly dependent on the dimension of the distributed gaps in the iron core and is largely independent of winding axial height. The magnetic field of gapped core is controlled by means of gaps. Compared with magnetically shielded coreless reactors, gapped core reactors can be operated at higher flux density, the selection of which is dependent upon the requirement of linear characteristic. Core elements are constructed by using CRGO laminations. Various types of assembly

478 *Transformers*

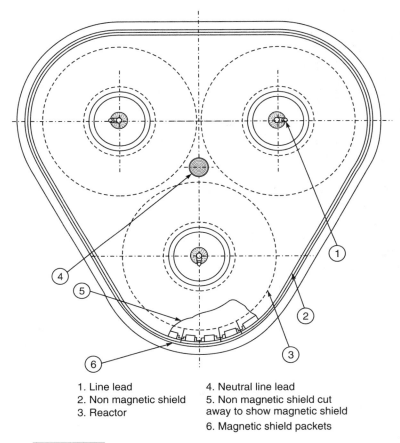

1. Line lead
2. Non magnetic shield
3. Reactor
4. Neutral line lead
5. Non magnetic shield cut away to show magnetic shield
6. Magnetic shield packets

Figure 19.11 *Sectional arrangement of coreless shunt reactor.*

of laminations are possible. Conventional type (transformer core section), involute disc type and radially arranged moulded types are normally employed to make the core elements. In modern reactors, moulded type radial core sections moulded in epoxy resin to prevent movement between individual laminations are commonly used. The radial laminations prevent fringing flux from entering flat surfaces of core steel, which would result in eddy current overheating and hot spots.

Stiff material having very high dimensional stability under pressure and high temperature conditions are used, to form gaps between core elements. The core segments are accurately stacked and cemented/glued together to make a solid core limb column. Top and

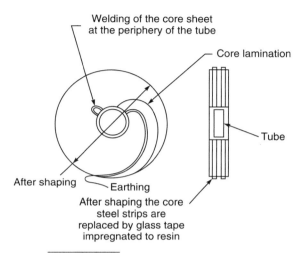

Figure 19.12 *Involute disc type core.*

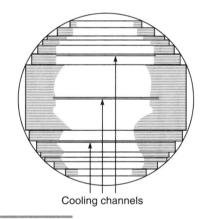

Figure 19.13 *Conventional type core.*

bottom yokes are butt or miter jointed with legs. The core elements are tightened under high pressure by tie rods and clamping beams which maintain the pressure on core elements.

19.2.4 Noise and Vibration Problem

The magnetic field creates pulsating forces across air gaps which amount to tens of tons. The reactor cores should, therefore, be very stiff to eliminate objectionable vibrations. The spacers supporting the

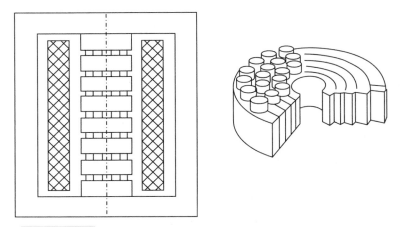

Figure 19.14 *Gapped core with radially laminated core segment.*

air gaps should be of stiff material (usually ceramic blocks or stone spacers) with a high modulus of elasticity so that their dimensional stability is maintained and the core does not become loose. Besides the gapped legs, care is also taken in the design of core frame. The fundamental mechanical resonance mode o++f the reactor frame must be designed to fall away from twice the power frequency.

The vibrations of active part are transferred to the tank through mechanical contact at the bottom and also through the oil. Sometimes, antivibration pads are used between active part and tank bottom. The tank walls set up a complex pattern of vibrations, and there are possibilities for resonance of individual panels between stiffener beams.

A large vibration amplitude in the middle of a large panel of thin sheet may be less objectionable than a small amplitude on thicker material. The resulting dynamic stress in tank and cover should be limited to less than 20 N/mm^2. Maximum vibration (double amplitude) on tank wall at rated voltage should not exceed 0.2 mm.

Customer specifications also frequently specify sound levels and the values are selected in analogy with those specified for transformers with same MVA rating and BIL. NEMA TR-1 recommended levels for transformers are generally applicable.

In the early days of shunt reactor design the noise and vibration problems caused great concern to the users and manufacturers. Sound level proved to be intolerable, and tanks developed oil leaks or even cracked open due to vibration fatigue along weld seams. In modern designs, both gapped core and coreless with magnetic shields, these problems have been well sorted out. High degree of dimensional accuracy is required while manufacturing the core to obtain reactance and phase currents within permissible limits.

19.2.5 Calculation Data

A reactor absorbs magnetic field energy. The energy per unit volume, or energy density in the field is

$$\frac{B \cdot H}{2} = \frac{B^2}{2\mu\mu_0}$$

The energy density shall be low, in core steel with high permeability. The energy storage is concentrated in volumes with $\mu = 1$, i.e., in air gaps. In a gapped core reactor, this storage volume consists of gaps which are introduced in the magnetic circuit and in a coreless or "shell form" reactor, the whole volume inside the magnetic shield represents the energy storage.

Assume for simplicity, homogeneous field in the total volume. The peak value of energy storage in the reactor is then,

$$W = \frac{\hat{B}^2}{2\mu_0} \text{ (Gap volume)}$$

Where $\hat{B}$ = peak value of flux density at current maximum.
In terms of circuit elements

$$W = \frac{L\hat{I}^2}{2} = L I^2$$

Where I = rms value of current
Rated reactive power

$$Q = \omega L I^2 = \omega W = \omega \cdot \frac{\hat{B}^2}{2\mu_0} \text{ [gap volume]}$$

The gapped core design is based on high flux density in a relatively small gap volume.

Also, Field energy per unit volume

$$P = \frac{B^2}{2\mu_0} = \text{Magnetic pulling force}$$

19.3 Neutral Earthing Reactor

Modern systems normally operate with their neutral points grounded. Neutral earthing reactor is one of the means to ground the neutral. There is no idea in purposely increasing the grounding reactance of the system beyond that required to keep currents within non-destructive range, except of course for the special case of ground fault neutralizers used in resonant grounding.

In the transmission systems which require single pole opening and reclosing of the EHV lines from the consideration of transient stability, successful single pole reclosing requires that extinction of secondary arc and the deionization of arc path in faulty phase should occur before reclosing is effected. In this case it may be necessary to reduce secondary arc current and thus may call for compensation of phase to phase and phase to ground capacitance of the line. In EHV system this is achieved by installing a single phase shunt reactor (called neutral earthing reactor) between neutral point of EHV reactor and the earth. Studies carried out have indicated that for 400 kV lines, basic insulation level of 550 kVp may be adequate for the line terminal of neutral reactor.

The neutral earthing reactor is classified under the category of current. For this reactor, no rated continuous current is applicable, unless otherwise specified. The reactor should withstand without undue heating or excessive mechanical stresses when short time current is carried for a specified duration which is usually 10 sec. Measurement of loss applies only when a continuous current is specified. Induced over voltage withstand test purported to be routine test is to be carried out at twice the voltage occurring across the winding at rated short time current. In lieu of this test, lightning impulse test is conducted and the impulse is applied to the terminal which is to be connected to transformer or shunt reactor neutral, whereas the other terminal is earthed. The duration of the front time in such a case is permitted up to 13 microseconds.

19.4 Tuning for Filter Reactors

Tuning reactors have been defined as those reactors in a.c. systems which are connected with capacitors to tuned filter circuits with resonance in the audio frequency range for reducing, blocking or filtering harmonics or communication frequencies. These reactors can be connected either in parallel configuration in the system or in series configuration and can be oil immersed or dry type. Usually these reactors are designed with means for adjusting the inductance value within a limited range by means of tappings or by movement of core and/or coils. Tuning reactors have rated quantities viz., continuous current, voltage, inductance, Q factor at tuning frequencies.

19.5 Arc Suppression Reactors

Arc suppression reactors are single-phase reactors used to compensate for the capacitive current occurring in the case of line-to-earth faults in a system with an insulated neutral. They are connected between the neutral of a power transformer or an earthing transformer and earth in a three-phase system. Sometimes a low voltage secondary winding is provided for connection of a loading resistor or for measuring purpose. Arc suppression reactors are usually specified with their inductance variable either in steps or continuously over a specified range to permit tuning with the network capacitance. Adjustment of reactance can be accomplished in one of the following ways:

(a) by adding additional sections of winding in finite steps with an off-load or on-load tapchanger.
(b) by adjusting the air gap of the magnetic circuit by mechanical means.

Compared to a neutral grounding reactor, arc suppression coils have closer tolerances of rated current and currents at other adjustments. Further a higher duration of rated current is envisaged and temperature these limits are specified for continuous, 2 h and 30 minutes durations.

19.6 Earthing Transformers (Neutral Couplers)

Earthing transformers are three-phase transformers or reactors and are used to provide an artificial neutral for earthing of a system.

Earthing transformers are usually connected in zigzag. Sometimes, a secondary (low voltage) winding having a continuous rated power for station auxiliary supply is provided.

19.7 Standards on Reactors

Indian Standard 5553 "Reactors" is the governing standard on reactors. This is in line with IEC 60289 (1988) reactors. It is comprised of 8 parts instead of 3 parts of 1970 as follows:

    Part I   : 1989 – General
    Part II   : 1990 – Shunt reactors
    Part III : 1990 – Current limiting reactors and Neutral earthing reactors
    Part IV : 1989 – Damping reactors
    Part V   : 1989 – Tuning or Filter reactors
    Part VI : 1990 – Earthing transformers (Neutral Couplers)
    Part VII : 1990 – Arc suppression reactors
    Part VIII : 1990 – Smoothing reactors.

Almost entire range of important reactors are now covered. Special purpose reactors like high frequency line taps or reactors mounted on rolling stock are, however, not covered.

First revision of IS : 2026 (Part I to IV) is a necessary adjunct to IS : 5553. References have been made to this standard at several places specially for temperature rises and ambient conditions, dielectric requirements and testing methods.

19.7.1 Transformer Standards

All transformer standards wherever applicable are also applied to reactors. A few to mention are:

(i) IS : 335–1993 : Specification for New Insulating Oil (Fourth Revision)

(ii) IS : 1866–2000 : Code of Practice for Electrical Maintenance and Supervision of Mineral Insulating Oil in Equipment (Third Revision)

(iii) IS : 2099–1986 : Specification for Bushings for Alternating voltages above 1000 volts: (Second Revision)

19.7.2 CEA Report

The CEA report of "The committee for standardization of the parameters and specifications of major items of 400 kV substation equipment and transmission line materials" has standardized the specifications of 420 kV shunt reactors. Presently, requirement of 50, 63 and 80 MVAR reactors in Indian systems have been envisaged and, therefore, only these ratings have been standardized. 1300 kVp lightning and 1050 kVp switching impulse withstand levels have been specified for these reactors. The neutral terminal of these star-connected reactors is earthed directly when 3 pole opening and reclosing of 400 kV transmission line is practiced. In case of single-pole opening and auto-reclosing of lines, a neutral reactor is used for earthing the neutral terminal. In such a case a BIL of 550 kVp is considered necessary for the neutral terminal of shunt reactor.

The CEA report specifies 1.5 linearity for the V/I characteristic. X_0/X_1 ratio has been specified to be between 0.9 to 1.0. A maximum value of vibrations has been fixed as 200 microns anywhere on the tank surface and the resulting permissible stress is 2 kg/mm^2.

CHAPTER 20

Traction Transformers

J.M. Malik

The traction transformers require special considerations for their design due to limitation of space availability and problems due to vibration. They have to be designed and manufactured to stringent specifications, so as to withstand heavy stresses in this type of applications.

20.1 Types of Traction Transformers

The following types of traction transformers are in use with the electric locomotives of the Railways of the various countries.

20.1.1 Tapchanging Transformers for Low-voltage Control

The voltage is adjusted on the secondary side by a tapchanger or tapping contactors. These types of traction transformers are generally used under the coach of Electrical Multiple Units with diode rectifiers. The schematic diagram for the arrangement is given in Fig. 20.1.

20.1.2 Tapped Transformers for High Voltage Control

The variable voltage is produced on the HV side of an auto-transformer by a tapchanger and is conveyed to a main transformer with a fixed transformation ratio. This variant is used in powerful locomotives with diode rectifiers. The schematic diagram for the arrangement is given in Fig. 20.2.

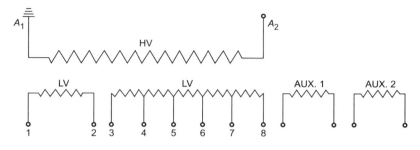

Figure 20.1 *Schematic diagram of A.C. Electrical Multiple Unit transformer.*

20.1.3 Rectifier Transformers for Pulse-Controlled Convertors

Main convertors are fed through several secondary windings with a fixed transformation ratio, e.g., six-pulse, four-quadrant convertors. The schematic diagram for the arrangement is given in Fig. 20.3.

Here a polygon equivalent circuit diagram is illustrated to show how a transformer with four secondary winding systems may be connected.

The traction transformers for 25 kV AC locomotives using on-load tapchanger for voltage control and traction transformers for thyristor controlled locomotives are described in this chapter in detail.

20.2 Special Considerations

As distinct from an ordinary power transformer, the traction transformer has to withstand frequent mechanical and electrical shocks and cater for overloads under adverse conditions. The traction transformers have to withstand mechanical shocks in line with IEC-310, which lays down a value of $3g$ in the direction of motion, $2g$ in the transverse direction and $1g$ in the vertical direction, where g is the value of acceleration due to gravity.

The following operating conditions must be taken into account, particularly in design:

— Fluctuation of the contact wire voltage.
— Different short-circuit capacities of supply systems.
— High closing rate due to bouncing by the pantograph.

488 *Transformers*

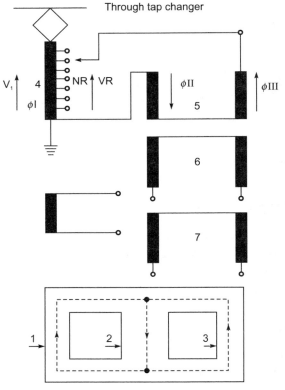

1, 2, 3. Limbs
4. Regulating auto-transformer with steps
5. Primary winding of the main transformer
6. Secondary winding 1 of the main transformer
7. Secondary winding 2 of the main transformer
8. Auxiliary winding

Figure 20.2 *Basic circuit diagram.*

—Special climate conditions, such as when operating in long tunnels or tropical areas.

The transformer windings should also be suitably designed to withstand electromagnetic forces due to flash-overs in the rectifiers/traction motors or electrical stresses due to the voltage surges travelling along the exposed 25 kV overhead conductor system.

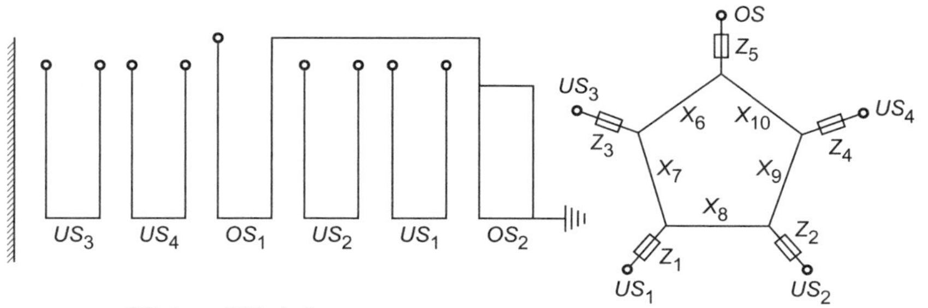

Figure 20.3 *Schematic diagram of four-quadrant convertors.*

- OS_1-Inner HV winding
- OS_2-Outer HV winding
- US_{1-4}-Four secondaries
- Z_1-Z_5, X_6-X_{10}-Impedances of equivalent circuit diagram

Any weakness in the insulation system or defective layout of the windings could be fatal to the transformer. The voltage variations are unusually large in traction duty and form an important consideration in the design of equipment. Indian Railways specify the following values for this purpose.

Normal for traction	22.5 to 27.5 kV
Minimum for traction	19 kV
Minimum for auxiliaries	17.5 kV

Limitation due to locomotive design and other related conditions are as under.

— Low weight
— Location of fixing
— Grading (the minimum possible number of steps in tractive effort) and the high switching rate of tapchangers
— Auxiliary winding for supplying power to auxiliary equipment and train lighting, etc.

20.3 Design and Constructional Features

20.3.1 Basic Design Concept

The basic circuit diagram of Fig. 20.2 shows the individual windings and their arrangement on the three cores. The main transformer (limb 2 and 3) is fed with variable voltage by the regulating auto-transformer (limb 1).

(a) Regulating Auto-transformer and Main Transformer

The entire winding of regulating auto-transformer is divided into individual groups which are suitably interleaved. The initial 80% steps are of the same magnitude and consist of one double section each, while the last 20% steps are somewhat smaller in magnitude. As the currents on the series and parallel connection of traction motors change from tap to tap, the transformer is designed for continuous operation at each step with a constant secondary current. The same limb 1 carries auxiliary winding also.

Both limbs 2 and 3 of main transformer are identical except the core width. The primary winding of each core consists of two groups connected in series. The low voltage groups are symmetrically positioned in relation to the primary groups. The two secondary branches have the same short-circuit voltage and are closely coupled to each other. To obtain the optimum conductor cross-section, each low-voltage winding has a number of conductors in parallel.

(b) Flux Relationship in Core and Iron-Loss

The flux ϕ_I in the limb 1 is proportional to the supply voltage and independent of the position of the tapchanging switch. The fluxes ϕ_{II} and ϕ_{III} in the two other limbs depend on the winding ratio NR/NP and are under.

$\therefore$
$$\phi_{II} = \frac{\phi_I}{2}\left(\frac{NR}{NP} + 1\right)$$

$\therefore$
$$\phi_{III} = \frac{\phi_I}{2}\left(\frac{NR}{NP} - 1\right)$$

where
V_I = incoming supply voltage
V_R = regulating output voltage
N_1 = series + common winding turns
NR = regulating winding turns
NP = primary winding turns

So the number of turns on one hand and the core cross-section on the other are selected so that the flux in all three cores is approximately equal on traction operation steps, which correspond to primary voltage of 22.5 to 27.5 kV.

(c) *Copper-Losses in Relation to Tapchanging Switch Position*

The copper-losses of the regulating transformer vary widely with the tapchanging switch position, while the losses of the main transformer remain constant, since the secondary current remains constant.

With a constant secondary current, the main transformer impedance voltage varies with step number. This means, that the voltage is lowest at the highest step and increases when the steps are lowered. To ensure that the overall impedance voltage is as constant a characteristic as possible, it is essential to ensure that the regulating transformer impedance voltage to be added does not vary much.

20.3.2 Constructional Details and Other Features

(a) *Core and Coils*

Like any other power transformer, the traction transformer core is made from cold rolled grain oriented steel. The core has a three-limbed construction.

Due to height limitation in electric locomotives, the core is placed horizontally. It also requires careful tightening of the core, end frames and cores together, to form a rigid assembly; otherwise there is a possibility of sagging of the core limbs. The windings used in traction transformer are of pancake construction.

The regulating winding and auxiliary winding are placed on limb 1, and limbs 2 and 3 are provided with HV and LV windings. The conductors used are of paper-covered high conductivity strips. As the number of tapping leads are large they are carefully insulated and anchored to prevent movement under mechanical forces. The entire core and coil assembly is suspended from the tank cover and held in position by steel pins at the bottom to prevent shifting of the transformer.

(b) *Insulation*

Bulk of the insulation is precompressed pressboard, which does not chip or get crushed under shock loads. The entire coil assembly is carefully shrunk under predetermined clamping pressure and controlled heat. As a further safeguard against any minor shrinkages in service, spring-loaded pressure pads are employed which exert a steady pressure on all the windings through a sturdy steel plate. As the space is at a premium, the precompressed inter-section insulation is machined to close tolerances to avoid coil build-ups. Insulation angle rings and angle barriers are placed between HV and LV windings. They are of a moulded construction, which brings compactness and mechanical strength to the coil assembly. The insulation blocks are carefully dimensioned and positioned between the windings, so as to ensure adequate strength to the coils and effective cooling. All bolts and nuts are properly secured to prevent loosening in service. All leads are suitably clamped to preclude any possibility of a short-circuit due to vibrations. There is an arrangement of lead connections to bus bars for LV terminations and leads for tapchanger.

(c) *Tank and Accessories*

The mild steel tank, oil conservator and cooling pipe-work are fabricated to close tolerances to facilitate their mounting in the limited space in the locomotive. In their design, care is exercised to keep their weight down to a bare minimum but at the same time ensuring sufficient strength against mechanical shocks. The tank rim and cover are carefully matched to ensure oil tightness under the worst operating conditions encountered during acceleration or retardation of the locomotive, which bring in additional oil pressure at the joints. The covered suspended transformer is provided with special screws to adjust the height of the assembly, so that in fully tightened position it sits evenly on base pads.

(d) *Fittings*

To bring in the desired compactness, the coolers are always of the *forced-oil forced-air type*. The rate of oil flow and position of inlet and outlet of oil from the transformer are governed by considerations of optimum cooling. The oil pump is of a glandless-type construction and is placed near the oil outlet. The oil passes through a cooler consisting of finned tubes, over which the air is forced by a blower. The oil from the cooler enters the transformer at the bottom and then rises to the top near the outlet. To facilitate servicing and replacement of the major components in the cooling circuit, valves are introduced at the appropriate places. The tapchanger with the conservator at the top and in the center is the blower motor for forced-air cooling.

The traction transformer has an in-built system of protection for maximum safety in service. This is achieved by a number of protective devices such as oil and air flow indicators in the cooling circuits. The separately mounted CT in the primary circuit ensures protection against excessive overloads and short-circuits. The primary side is protected from the voltage surges by means of arcing horns placed on the roof of the locomotive. The auxiliary windings and associated equipment which are subjected to sudden rises in voltages are protected by surge condensers connected across the winding. The breather which prevents the ingress of moisture is also of a special construction. The breather has a special shaped nozzle at the entrance to filter out the heavy particles.

The transformer is provided with a condenser bushing for connection to 25 kV end. All other HV, LV, auxiliary bushings are porcelain type and are mounted on top of the tank cover.

20.4 Traction Transformer for Thyristor Controlled Locomotives

20.4.1 Introduction

The thyristorized controlled locomotive has the following advantages, due to elimination of the tapchanger as well as of the silicon diodes.

— Smooth acceleration and retardation by stepless control.
— Full regenerative braking capability.
— Better adhesion and high speed detection and correction of wheel slip and wheel skid.
— Facility of automatic current and speed control.
— Absence of wearing parts and consequent saving in maintenance costs.

20.4.2 Main Feature

The transformer is provided with a number of traction windings for connection to thyristor system, an auxiliary winding for supply to auxiliary machines and excitation winding to feed the traction motors fields during rheostatic braking. The transformer design is suitable for taking external connection of traction windings in series or series–parallel combination. The transformer is provided with two primary windings on two-limbed core and connected in parallel with corresponding secondaries suitably magnetically linked, in order to have

494 *Transformers*

minimum mutual coupling between the windings. The schematic diagram of the transformer is shown in Fig. 20.4.

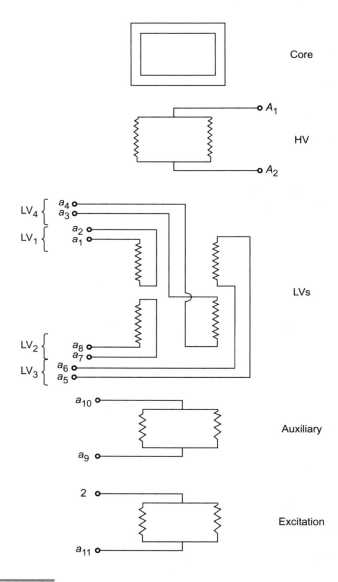

Figure 20.4 *Schematic wiring diagram of thyristor controlled traction transformer.*

The construction of the transformer is generally the same as that of the tapchanger controlled transformer, except that no regulating winding is provided and core is two-limbed construction. The construction of the transformer is such as to keep its weight and dimensions as low as possible. In the design and construction of this traction transformer, care is taken to incorporate all the salient design and constructional features of traction transformer with on-load tapchangers as described in Secs. 20.3 and 20.4.

REFERENCES

1. Ecknaver, E., "Traction Transformers for the Indian Railways," *Bulletin Oerlikon*, No. 361.
2. Gupta, D.P., R.K. Shukla and J.M. Malik, "Salient Features of Traction Transformer," *Electrical India*, June 19/3.
3. Bohli. W.U., H.M. Deng and W. Muller, "Transformers and Smoothing Reactors for AC Traction Vehicles," *Brown Boveri Review*, Publication No. CH–B 0540 E.

CHAPTER **21**

Rectifier Transformers

J.S. Sastry

Chemical plants, aluminum plants, etc., where electrolytic processes are adopted, require electrical energy in the form of direct current. In these industries, the direct current requirement is so large that it is considered to be one of the basic inputs for the production of caustic soda, aluminum, etc. For example, about 3600 kWh of energy is required for the production of one ton of caustic soda and about 17000 kWh for aluminum.

Basically, the different methods for converting ac into dc are:

(a) Motor generator set
(b) Synchronous convertor
(c) Mercury arc-rectifiers
(d) Rectifier transformer along with rectifier units using silicon diodes

The advantages of using a rectifier transformer along with rectifier unit is that it can be connected directly to the supply lines. This being a static unit, the efficiency is higher and enjoys all the advantages of static unit. The third generation of electrochemical plants have capacities as high as 250 kA. The present trend calls for single unit of rectiformers having current rating as high as 100 kA and more. Design and manufacture of such units pose many difficult problems and place several constraints for the designer, which require special attention.

21.1 Comparison between Rectifier Transformer and Power Transformer

Though outwardly a rectifier transformer may appear to be similar to a power or distribution transformer, it actually differs in many respects. A power transformer is used to step the voltage up or down for the transmission of the power, whereas a rectifier transformer which steps down the voltage, is used in conjunction with rectifier unit for conversion of ac power into dc.

The salient features which make a rectifier transformer unique when compared with power transformers are given here under:

(a) The general expression for direct current I_d in the resistance load "R" connected to a transformer with m-secondary phases and phase connected to a diode, assuming negligible diode drop and transformer impedance, can be written as

$$I_d = \frac{m}{2\pi} \int_{-\pi/m}^{+\pi/m} \frac{E_m \cos \omega t}{R} d\omega t$$

$$= \frac{E_m}{R} \cdot \frac{m}{\pi} \cdot \sin \frac{\pi}{m}$$

where, E_m is peak phase to neutral voltage of the transformer. This is because each diode conducts for an interval of 360/m degree or $2\pi/m$ radians per cycle. Similarly, the rms current per diode is

$$I_{rms} = \sqrt{\frac{1}{2\pi} \int_{-\pi/m}^{+\pi/m} \frac{E_m^2 \cos^2 \omega t}{R^2} d\omega t}$$

$$= \frac{E_m}{R} \sqrt{\frac{1}{2\pi} \left[\frac{\pi}{m} + \sin \frac{\pi}{m} \cdot \cos \frac{\pi}{m} \right]}$$

and the rms value of the current in the load having m pulses* per second is $\sqrt{m}$ times I_{rms}. The ripple value, defined by the expression

$$\text{Ripple} = \sqrt{\left[\frac{I_{rms}}{I_d}\right]^2 - 1}$$

The ripple decreases rapidly with increase in the number of pulses or phases and hence the distortion of primary current

---

* Since waves are gradually changing alternating currents or voltages, each cycle consists of a positive and a negative half, and the frequency refers to how many cycles are completed in each second. In contrast a pulse is an abruptly changing voltage or current which may or may not repeat itself.

waveform is less and this is important since harmonics introduced in the primary windings may cause interference in adjacent telephone lines. This also raises the dc output voltage and increases the load current. A rectifier transformer may, therefore, have more than one LV winding and sometimes it may have as many as four LV windings, unlike an ordinary power transformer. Typical computed values of ripple are given in Table 21.1.

➤ **Table 21.1** Computed Values of Ripple

m	2	3	6	12
Ripple frequency for 50 Hz	100	150	300	600
Ripple	0.47	0.17	0.04	0.014

(b) Transformers which supply rectifiers do not carry sinusoidal current in their windings or may carry currents in various coils only over a portion of a cycle. The harmonic component of these distorted waves contribute to transformer heating but produce no useful dc output. To provide for these harmonic currents, a transformer rating in watts, greater than the rated dc power output, is required. Hence the Volt–Ampere rating of the primary and secondary windings for rectifier transformers may be different.

(c) The rectifier units connected to the rectifier transformer will supply dc current for electrolytic processes as stated earlier. This direct current has to be constant at all voltages. Hence a rectifier transformer is basically a constant current transformer and consequently the KVA rating changes with output voltage.

(d) The LVs of a rectifier transformer carry very high current. So, heavy bus-bars will have to be used for the connections between the windings and the terminals. The size and the disposition of bus-bars carrying these high currents need special attention so as to reduce the stray losses and also the reactive drop.

(e) Other associated equipment of rectifier transformers are also housed in the same tank. These may be:

(i) A regulating transformer which works on auto-transformer principles for voltage control.
(ii) Interphase transformer, if two star-connected rectifier groups are to be paralleled.
(iii) A transductor unit for smooth voltage control on the secondary side.

21.2 Rectifier Circuits

The main parameters of the rectifier transformer, viz. KVA rating of the primary and secondary windings, currents and voltages, number of secondary phases, etc., depend on the rectification circuits adopted, which in turn are based on the requirements of the direct current, voltage, ripple value, etc. Since the electrical parameters of a rectifier transformer and its associated equipment have a direct bearing on the circuit adopted, it is necessary here to briefly touch upon some of the most commonly used circuits.

21.2.1 Six-Pulse System

(a) *Double star-connection with Interphase Transformer*

This gives a symmetrical load to the three-phase system. The current transfer from one phase to the other occurs as soon as the potential of the succeeding phase rises above that of the preceding phase. Hence, the phase with the higher positive (or negative) potential (depending on the polarity of the diodes) carries the load current and determines the direct voltages.

Two three-phase connections are frequently paralleled through an interphase transformer in order to double the load current and to obtain six-pulse performance. This requires 180° phase shifted transformer secondaries, i.e., one secondary connected in star and the other inverted star as shown in Fig. 21.1. Each secondary winding conducts during one-sixth of a cycle. Reflected to the primary side, the secondary currents in the two windings on each limb are opposite and 180° displaced and they can be perfectly balanced by symmetrical primary current, regardless of the primary connection. Since the primary side carries the current pulses of both the secondary stars, the primary windings are effectively used

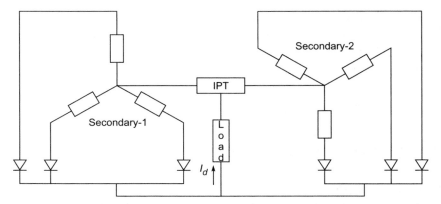

Figure 21.1 *Six-pulse double star-connection with interphase transformer.*

but the secondary rating is high as each secondary winding conducts only for one-sixth of a cycle.

(b) *Bridge Connection*

In the connection described in (a) above, the transformer secondary windings are idle most of the time, resulting in a high transformer rating compared with the converted power. These windings can be put to use more efficiently, if they feed bridge connected rectifier circuit. This is accomplished by combining two three-pulse commutating groups as illustrated in Fig. 21.2. A load connected between points a and b is supplied with the total of two voltages when viewed as separate groups. Since the two commutating groups operate with opposite polarity, their ripple voltages are displaced and the total voltage shows six-pulse ripple.

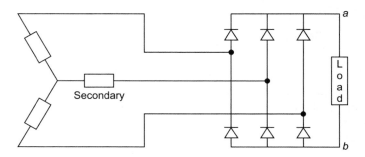

Figure 21.2 *Six-pulse bridge connection.*

However, interphase transformer connection is competitive over the three-phase bridge connection in certain voltage-current range, despite the relatively high transformer rating.

21.2.2 12-Pulse System

As already discussed in Sec. 21.1 (a), the lower the ripple of the direct voltage and the distortion of the alternating line currents, the higher is the pulse number. This is of minor importance for small power supplies where simplicity is the first requirement, but it is the leading factor when selecting the connection for a large rectifier installation which usually consists of several independently commutating units that are phase-shifted in such a way that the total pulse number is 12 or even higher. There is considerable reduction in the input power and also the ripple content, if the pulse number is increased from 6 to 12, but not much more if it is further increased. One of the reasons for occasionally employing a higher pulse number is to reduce interference with other electrical systems, e.g., communication systems, caused by harmonics in the alternating current.

(a) *Quadruple Star-Connection with Interphase Transformer*

In this type of connection shown in Fig. 21.3, phase-shift of 30° is achieved by providing two transformers, viz. one with the star primary and the other with delta primary, each transformer having two star-connected secondary windings, the neutrals of which are connected to their associated interphase transformers. This would require two rectifier cubicles placed on their side of the transformer tank. It is important to note that the two cubicles are of opposite polarities, so that the two fluxes oppose each other and minimize dc magnetization of the transformer tank. The phase-shift in the primary windings can also be achieved by using zig-zag (interconnected star) windings with +15° (leading) phase-shift on one transformer and −15° (lagging) on the other, making it a total of 30°.

(b) *Bridge Connection*

A transformer with a primary star and secondary delta winding or vice versa, causes a phase-shift of 30°.

Hence two complementary six-pulse systems can be built if one is given a star-star or a delta-delta transformer and the other a star-delta or a delta-star connected transformer. Even one transformer is sufficient, if it has two secondary windings, one connected in star and the

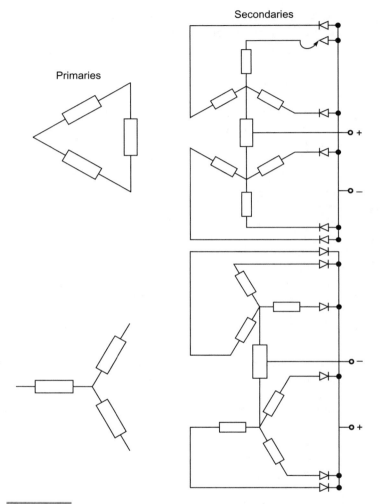

Figure 21.3 *Quadruple star-connection with interphase transformer.*

other in delta, provided that the rectifier circuits are bridge connected as shown in Fig. 21.4. The two six-pulse systems can either be connected in series or in parallel.

21.2.3 Transformer Utility Factor

As already stated, the windings of the transformer which supply to rectifier units, carry currents only over a portion of a cycle. The

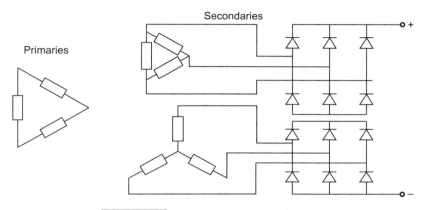

Figure 21.4 *12-pulse bridge connection.*

harmonic components of those distorted waves contribute to transformer heating only, without any additional output. To provide for these harmonics, the transformer equivalent kVA rating is increased. The utility factor of a rectifier transformer winding may be defined as the ratio of the dc power output to the equivalent volt-ampere carried. It is a function of the waveform and the ratio of rms to average voltage. Table 21.2 gives the performance data for the circuits described in this chapter.

21.3 Design Features of Rectifier Transformers

Depending on the circuit adopted, the tank of the rectifier transformer also houses any or all of the associated equipment listed below:

(a) Regulating transformer — one
(b) Interphase transformer — one or two
(c) Transductor — one or two sets

A typical electrical scheme for 12-pulse system is shown in Fig. 21.5 for describing the important features of the above equipment.

21.3.1 Regulating Transformer

The technology used for the electrolytic and smelting processes requires variable voltage. The ratio of the minimum to maximum

> **Table 21.2** Performance Data for Different Circuits

No. of sec. phases (m)	Bridge connection		Interphase connection	Transformer
	6	12	6	12
Conduction angle	60	30	64	30
E_{rms}/phase secondary	$E_{do}/2.70$	$E_{do}/2.34$	$E_{do}/1.17$	$E_{do}/1.17$
I_{rms} of secondary	$0.577\ I_d$	$0.408\ I_d$	$0.289\ I_d$	$0.144\ I_d$
P_{eq} of primary	$1.28\ P_{do}$	$1.01\ P_{do}$	$1.05\ P_{do}$	$1.05\ P_{do}$
P_{eq} of secondary	$1.28\ P_{do}$	$1.05\ P_{do}$	$1.48\ P_{do}$	$1.48\ P_{do}$
P_{total}	$1.05\ P_{do}$	$1.01\ P_{do}$	$1.05\ P_{do}$	$1.01\ P_{do}$
PUF	0.78	0.99	0.95	0.95
SUF	0.78	0.95	0.625	0.625
Ripple value	4%	1.4%	4%	1.4%

The various symbols used in Table 21.2 represent the following:

E_{do} = Average value of theoretical direct voltage at no-load.
I_d = Average value of direct current in load circuit.
$P_{do} = E_{do} \times I_d$
E_{rms} = RMS value per phase of the primary or secondary alternating voltage at no-load.
I_{rms} = RMS value of the alternating current.
P_{eq} = Equivalent rms volt-ampere rating of the transformer based on the supply frequency.
P_{total} = Total volt-ampere input power (rms value), neglecting transformer excitation current.
PUF = Primary utility factor.
SUF = Secondary utility factor.

secondary voltage is sometimes as high as 1 : 20. The realization of this voltage regulation demands special considerations. The low-voltage winding of the rectifier transformer is never provided with tappings due to the obvious technical reasons. Hence, the desired secondary voltage regulation is brought about by tappings arranged on the high voltage side. This can be done in two ways.

(a) By providing taps on the primary of the rectifier transformer. This is not always desirable due to technical and economical considerations even in case of a six-pulse circuit, when the voltage regulation is high. For 12-pulse circuits where primary phase snift is done, this would be more complicated and uneconomical.

Rectifier Transformers

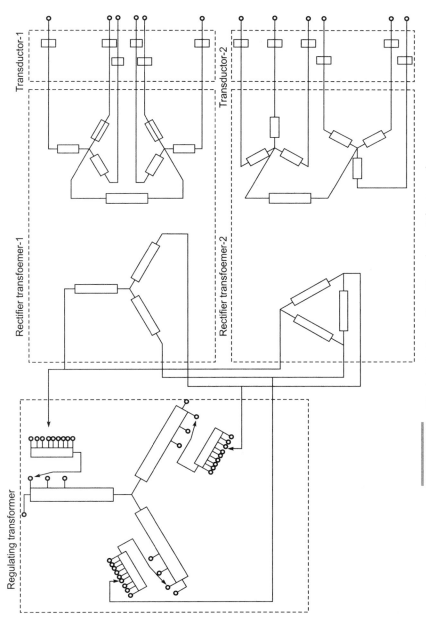

Figure 21.5 Electrical scheme for a 12-pulse rectifier transformer.

(b) Feeding the primary of the rectifier transformer through a separate auto-regulating transformer, which is a widely accepted practice. The rectifier transformer is a constant-current transformer; the output current of the regulating transformer also is constant. Hence the KVA of the regulating transformer changes with output voltage.

The voltage variation in the output voltage of the regulating transformer can be achieved with the help of a regulating or tapped winding. The tapped winding, besides serving the purpose of changing the output voltage within the given range, also maintains the required output voltage when there is any variation in the input supply voltage. For this purpose, a few taps on the regulating winding are intended to cater for the input supply voltage variation.

The output voltage of the regulating transformer is decided on the parameters of the available on-load tapchangers, viz. the maximum continuous current, voltage per step, voltage across the range and the impulse level. Electrical location of the tap windings is largely a choice of the designer and depends on the suitability of the on-load tapchangers. The tapping arrangement can be either linear or reversing or coarse-fine type. Thus, it is seen that the selection of the on-load tapchanger plays a vital role in the design of a regulating transformer. The design philosophy of the regulating transformer is very much similar to that of a normal auto-transformer.

21.3.2 Rectifier Transformer

This is also called the main transformer, as the output terminals of this transformer are connected to the rectifier units. This is essentially a stepdown transformer, the secondary voltage of which corresponds to the output dc voltage. The HV terminals of this transformer are connected to the output terminals of the regulating transformer, if any, or directly to the supply lines in the absence of a regulating transformer.

(a) *HV Winding*

The HV windings of this transformer are usually disc type and are connected in either star or delta or zig-zag, which is described below.

It was mentioned that the phase shift between the input and output voltage vectors can be achieved by zig-zag connection. This can be explained with the help of Fig. 21.6(a) and (b).

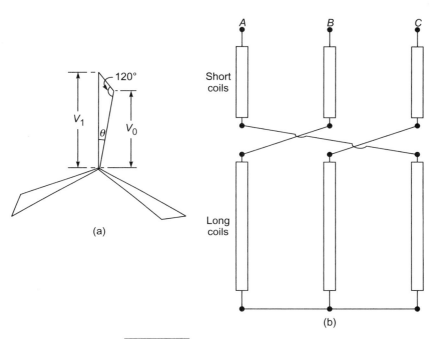

Figure 21.6 *Zig-zag connection.*

We know that the three phases in a three-phase supply are 120° apart. Now referring to the winding connections in Fig. 21.6(b) in which two coils—one long and the other short, located on two different limbs are connected in series. From the corresponding vector diagram shown in Fig. 21.6(a), it would be noticed that the output vector V_o lags the input vector V_i or in other words, there is a phase shift of $-\theta°$ between the input and the output vectors. A positive phase shift can be obtained by slightly modifying the winding connections. The phase shift angle θ depends on the ratio of the turns in the short and long coils, which can easily be calculated. Table 21.3 gives the voltages of short and long coils as a percentage of the input voltage for different phase shift angles normally in use.

(b) LV Winding

The secondary voltage of the rectifier transformer is very low and of the order of a few hundred volts. Obviously, the turns in the secondary winding can be as few as two or three. On the other hand, current carried by these windings is very high. For this reason a special winding called *half* and *sections* (H and S) is used. This resembles the disc

> **Table 21.3** Variation of Phase Shift Angle

θ in degree	Short coil voltage in %	Long coil voltage in %	Incoming voltage in %
3.75	7.552	96.010	100
7.5	15.072	91.609	100
10	20.051	88.455	100
11.25	22.527	86.815	100
15	29.886	81.650	100
20	39.493	74.220	100
22.5	44.188	70.294	100
30	50.000	50.000	100

winding to the extent that each coil is wound in two discs accommodating the total number of turns. A number of coils of this type are connected in parallel by bus-bars for sharing the total current. The advantage of this coil is that odd number of turns can be accommodated in two discs of the coil without any loss of space, which cannot be done with the normal disc-winding technique. But the condition for this is that there should be even number of parallel conductors.

As an example, a coil with three turns and two conductors in parallel is shown schematically in Figs. 21.7 and 21.8.

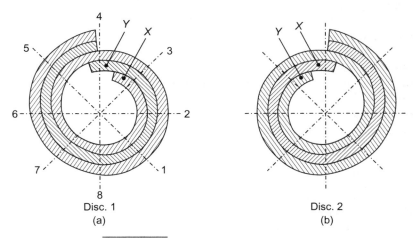

Figure 21.7 *Discs of half and section.*

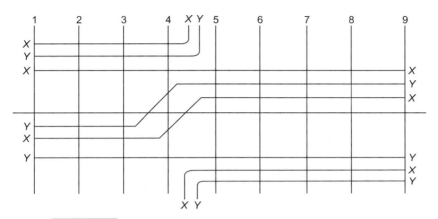

Figure 21.8 *Development of half and section winding discs.*

(c) Interleaving of LV Windings

The two groups of coils of both the secondaries are axially interleaved as shown in Fig. 21.9. The interleaving is done to ensure that the impedance between the two secondaries is minimum and the impedance of the secondaries with respect to the primary windings is the same so that all the coils of both the secondaries share the current equally. The ends of the bus-bars are then connected in star or delta as the case may be.

(d) Coil Disposition

Since the secondary coils carry heavy currents and the coils are connected in parallel by means of bus-bars, it is essential that the secondary coils are placed outermost for ease of connection. Consequently, the HV coil is placed concentrically over the core, over which the LV coils are placed.

(e) Bus-bars

In addition to carrying LV currents, bus-bars are also used for neutral formation, neutrals to interphase transformer connection and for connecting the LV bus-bars to the transductors. The design of the bus-bars and the arrangement has to be keeping the following important considerations in view.

(i) Skin effect. A reasonably accurate determination of the current rating of the conductors carrying direct current is possible from the theoretical considerations, but with heavy alternating currents the

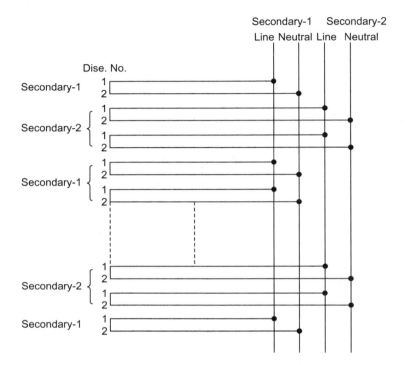

Figure 21.9 *Axial disposition of secondary windings.*

current density is not uniform due to the skin effect and the apparent ac resistance is always greater than that measured by dc methods. In the case of flat bus-bars, skin effect results in a concentration of current at the edges of the bus-bars. With the larger sizes of the conductors for a given cross-sectional area, skin effect in a thin flat bar or strip is usually less than in a circular rod but greater than in a tube. It is dependent, among other things on the ratio of the width to the thickness of the bar and increases as the thickness of the bar increases. The curves in Fig. 21.10 give approximate skin effect ratios for flat bars of various cross-sectional areas and proportions. The skin effect ratio is the ratio of the apparent ac resistance of a conductor (R') to its dc resistance (R).

(ii) Proximity effect. In the foregoing consideration of the skin effect, it has been assumed that the conductor is isolated and at such a distance from the return conductor that the effect of the current in it can be neglected. When conductors are close together, as is often the case, a

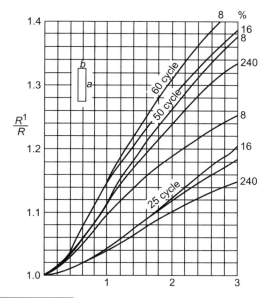

Figure 21.10 *Skin effect ratios for flat bus-bars.*

further distortion of current density results from the interaction of magnetic fields of the other conductors. The magnetic flux of one conductor may produce emf by induction in any other conductor that is close enough. If two such conductors carry currents in opposite directions, their electromagnetic fields are opposed to one another and tend to force one another apart. There is, therefore, a decrease of the flux linkages around the adjacent parts of the conductors and an increase around the more remote parts, which results in a concentration of current in the adjacent parts where the opposing emf is minimum. If the currents in conductors are in the same direction, the action is reversed and they tend to crowd into the more remote parts of the conductor. This is called the proximity effect. In some cases, however, proximity effect may tend to neutralize the skin effect and produce a better distribution point, as in the case of strip conductors arranged with their flat sides towards one another. If such conductors are arranged edgewise to one another, the proximity effect increases the skin effect.

(iii) Bus-bar reactance: An accurate knowledge of bus-bar reactance is necessary, as this would directly add to the transformer reactance for estimating the regulation drop under fully loaded conditions. The bus-bar reactance is calculated in two parts. The first part consists of "Go"

and "Return" path, i.e., up to the point where the line and neutral bus-bars (risers) run together. The second part is for single-phase path, i.e., from the point where neutral is made to the point of connection to the line terminals. In calculating this reactance, it is assumed that in the case of laminated bus-bars connected in parallel, provided the spacing between the individual bars is small, the reactance may be taken to be the same as that of a single conductor having the same overall dimensions as the group.

1. *Go and return path.* Referring to Fig. 21.11, let the values of a, b and d be as shown. The ratios (b/a) and (d/a) are calculated and the corresponding D_m, geometric mean distance between two paths is read off from the curve given in Fig. 21.12. Then D_s, the geometric mean distance* of each path from itself is calculated from $D_s = 0.2235 \times (a + b)$.

   Then associated inductance L_a is calculated by equation

   $$L_a = 0.002 \times l \times \log (D_m D_s) \, \mu H$$

   where l is the length of "Go" and "Return" path. Hence, the associated inductance L_a included both self and mutual effects, i.e., $L_a = L - M$. From this inductance, reactance can easily be calculated.

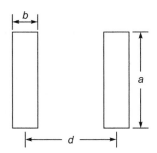

Figure 21.11 *"Go" and "return" bus-bars.*

2. *Single-phase path.* As before, the value of D_s is calculated. Then inductance L is given by $L = 0.002 \times l \times$

---

* Bus-bar sections are not filaments of negligible thicknesses, leaving no doubt as to the distance between one bus-bar and another, in fact they are large finite sections. The distance between two points may be different from that between any other pair and hence the concept of geometrical mean distance D_m.

Rectifier Transformers 513

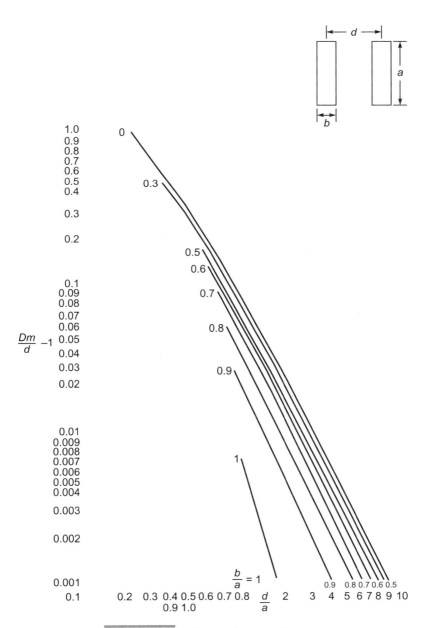

Figure 21.12 *Geometric mean distance.*

$\left(\log \dfrac{2l}{D_s} - l + \dfrac{D_s}{L}\right)\mu H$ where l is the average length of the single-phase path. From the above inductance, the reactive drop for the single-phase path can again be calculated.

The sum of the above two paths gives the reactive drop introduced by the bus-bars.

21.3.3 Interphase Transformer

A number of batteries can be paralleled only when their terminal voltages are equal, as otherwise there will be internal circulating current. Parallel operation of two rectifier systems is more complicated as their direct voltages are fluctuating. Such systems can be connected in parallel without any circulating current, if their direct voltages are equal at any instant, i.e., if the average values are equal and ripple voltages coincide. Usually, however, it is desired that the ripple voltage instead of coinciding, be so displaced that the combination results in a system with higher pulse number. The parallel connection, then, must be made in such a manner that it does not affect the operation of the individual group. Through an interphase transformer, two rectifier systems with displaced ripple voltages are paralleled. It absorbs, at any instant, the difference between the direct voltages of the individual systems and must be designed for the time integral of this voltage.

Figure 21.13 shows the direct currents through an interphase transformer. This transformer absorbs at any instant, the voltage difference between the individual groups and thus maintains independent operation of these groups.

With respect to the voltage difference to be absorbed, the two windings of the interphase transformer of Fig. 21.13 are in series connection. Thus, the voltage difference can be balanced by the emf induced in these windings—just as in a normal transformer. The voltage impressed on the primary windings is balanced by the induced emf. However, inducing the balancing emf needs a changing magnetic flux and exciting ampere-turns, which is the difference of the direct currents to be combined (since these currents pass in opposite direction through the window of the interphase transformer).

If these currents are well balanced, the core will not be driven into saturation by the dc ampere-turns, even without an air gap. The time integral of the voltage to be absorbed by the interphase transformer is

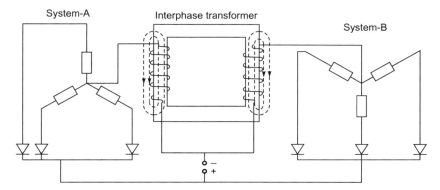

Figure 21.13 *Paths of direct currents in interphase transformer.*

a function of the dc voltage, of the operating conditions (angles of phase control and commutation) and of pulse number. For regular transformer, operating at 50 Hz, the maximum flux density is chosen in the vicinity of 1.6 tesla. For an interphase transformer, a lower value, say 0.9 tesla, is taken, since the magnetic flux is alternating with three times the supply frequency, if the two three-pulse systems are combined.

Design Features

An interphase transformer can be either a wound type or a bar-mounted type depending on the voltage to be absorbed and the direct current rating.

(a) Wound type interphase transformer. The wound type interphase transformer is like a single-phase transformer with windings on the two legs of the core.

The total number of turns in each windings is divided equally and accommodated on the two limbs. The connection diagram is as shown in Fig. 21.14. The terminals N_1 and N_2 are connected to the respective neutrals of the two secondaries of the rectifier transformer and N forms the negative terminal of the dc supply. Since the windings carry large currents and if the number of turns are few, H and S coils are used.

(b) Bar-mounted type of interphase transformer. In this type of interphase transformer, the bus-bars from the neutrals pass through rectangular cores forming a single turn winding. The arrangement is shown in Fig. 21.15. As before N_1 and N_2 are connected to the two neutrals and N to the dc negative terminal.

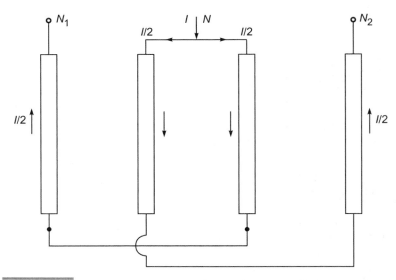

Figure 21.14 *Schematic diagram of wound type interphase transformer.*

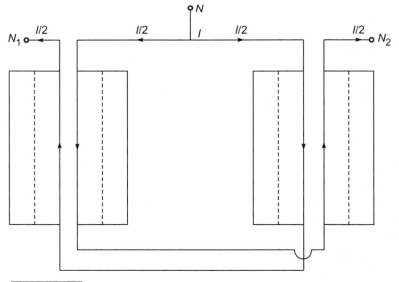

Figure 21.15 *Schematic diagram of bar-mounted type interphase transformer.*

21.4 Transductors

A transductor is a device consisting of one or more ferromagnetic cores with windings, by means of which an alternating voltage or current can be varied by a direct voltage or current utilizing the saturation phenomenon of the core material. Coarse control of the rectifier output voltage is achieved by changing the tapchanger position, i.e., increase or decrease of the output voltage can be achieved in steps by the tapchanger. The function of the transductor in a rectifier system is to smoothly control the output voltage between any two tapchanger positions. The transductor is usually designed to cover a range of two steps of voltage of the tapchanger, so that the frequency of operation of the tapchanger is reduced.

It is well known that the magnetization produced in an iron core by passing current through a coil wound round it is not constant, but depends on the magnitude of the current. This property can be used to control the power in a circuit in a convenient and flexible manner.

With increase in current, the magnetization increases rapidly at first and is roughly proportional to the current but soon the magnetization curve begins to flatten out. Increasing the current still produces increased magnetization but not in the same proportion and finally a point is reached when further increase in the current produces no change in the magnetic field intensity.

In this condition, the iron is said to be saturated. If an alternating voltage is applied across the coil, the current which flows, neglecting resistance, is determined by the inductive reactance, i.e., the product of the angular frequency of the current and the inductance of the coil. The inductance of the coil is proportional to the degree of magnetization. In other words, current which will flow is dependent on the state of iron core. Suppose, therefore, we place two coils on the core as shown in Fig. 21.16. One is a control winding connected to a source of dc which serves to determine the state of the iron core. But the other is connected to a source of ac supply. The circuit is completed through a load.

The load current will be determined by the combined effect of the coil L and the load R. If the inductance of the coil is small, all of the applied voltage appears across the load. On the other hand, if the inductance is high, most of the applied voltage will be expended in over-coming the impedance, so that the current through the load

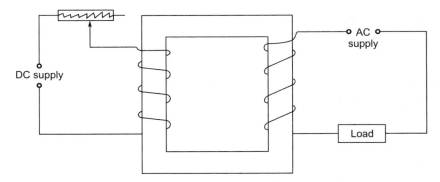

Figure 21.16 *Transductor schematic diagram.*

becomes very small. But we have seen that the inductance of the coil L depends on the state of iron which can be controlled by current in the dc winding. Hence, by suitable adjustment of the direct control current, the power in the load can be regulated within wide limits.

In normal practice, there will be two dc windings—called bias winding and control winding. The current through the bias winding is preset at a particular level and the current through the control winding is smoothly varied.

The usual form of transductor core is toroidal, through which the secondary bus-bars of the rectifier transformer are threaded as shown in Fig. 21.17.

The control and bias windings are in the form of copper rods, also threaded through the ring core. The ring cores are made of CRGO steel. A number of such cores, depending on the voltage to be absorbed or released (or simply the range of the transductors) are clamped between two permawood boards by means of insulated bolts. There will be as many stacks of the transductor as the number of the secondary phases.

21.5 Constructional Features of Rectifier Transformers

21.5.1 Types of Construction

Use of silicon rectifiers along with rectifier transformers as a heavy dc power source for the electrolytic industry has come to be widespread. The relative placement of these two units is of considerable importance from the economics and operational point of view. This is due to the

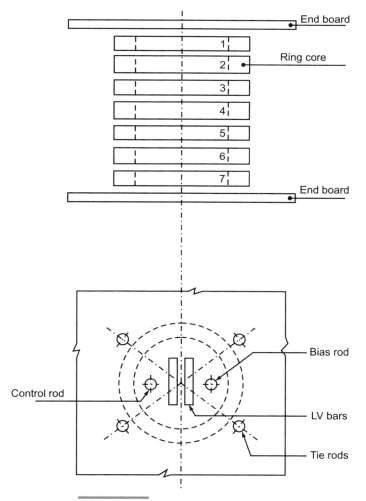

Figure 21.17 *Transductor assembly.*

fact that enormous conductors carrying heavy currents are necessary between rectifier units and the transformer. The length of the connectors has the following implications:

(a) The cost of the conductors and their installation account for a considerable part of the total construction cost of the equipment.
(b) The current carried by this conductor is of rectangular waveform and consequently eddy loss is inevitable. This,

together with high current lowers the overall efficiency of the equipment.

(c) Inductive voltage drop due to the heavy currents lowers the power factor, and also the regulation becomes poorer.

(d) Strong magnetic field around the conductor carrying currents causes inductive heating. For this reason, the conductor's support frame must be made of a non-magnetic insulators.

Hence great attention has to be paid to the arrangement of the rectifier cubicle and the transformer, so as to make the connection between their terminals as short as possible. This has been successfully realized in the *Rectiformer* construction, in which the transformer and the rectifier cubicles are installed as near as possible longitudinally. Figure 21.18 shows the schematic arrangement of the rectiformer construction for a 12-pulse system. The secondary terminals of the two rectifier transformers are brought out on either side through the longer side-walls of the tank for ease of connection to the reactifier cubicles.

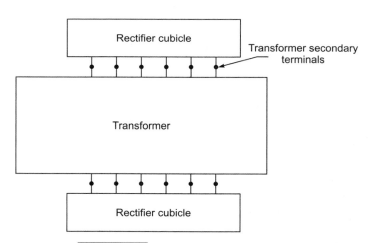

Figure 21.18 *Rectiformer construction.*

In the other type of construction, which is a conventional one, the transformer and the rectifier can be installed at different places and connected together by bus-bars. This would facilitate the installation of the transformer in the outdoor switch-yard outside the building which houses the rectifiers and the electrolytic cells.

21.5.2 Layout

It has been stated earlier that all the associated equipment of the rectifier transformer, like regulating transformers, interphase transformers and transductors are placed in the tank for obvious advantages like smaller space requirement and higher operational reliability. In addition to this the other advantages are:

(a) By placing the regulating transformer in the same tank not only is considerable space saved, but it also reduces the risk of the short-circuit on outgoing terminals of the regulating transformer. It obviates the need for expensive short-circuit protection gear owing to their low impedance voltage, specially when the ratio is small (auto-transformers are not short-circuit-proof on their own).

(b) Materials like steel, cables, oil, copper bus-bars, etc., can be saved by this arrangement. This is particularly the case with the interphase transformer and transductor, since they accommodate heavy copper bus-bars. This also results in lower bus-bar reactance, losses, etc. A typical layout of all the equipment mentioned is shown in Fig. 21.19, which shows the internal arrangement of rectifier transformer tanks for 12-pulse, with the regulating transformer placed perpendicular to the main transformers. The interphase transformers are placed over the top of the main transformers and the transductors in a separate pocket in front on the main transformers.

21.6 Tank Design

The tank dimensions are decided keeping the magnetic clearances in view. The magnetic clearances are very important because the bus-bars carrying heavy currents in the proximity of the structural steel would cause additional stray losses due to the flux linkages. This phenomenon is also called the proximity effect. The additional losses caused by placing a magnetic material close to bus-bar carrying current arise from both hysteresis loss and eddy current effects within the plate. Hysteresis losses are large if the flux density within the plate is high. The material having high permeability and very high resistivity

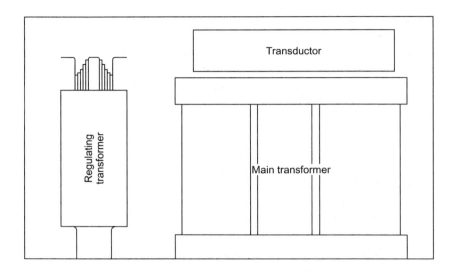

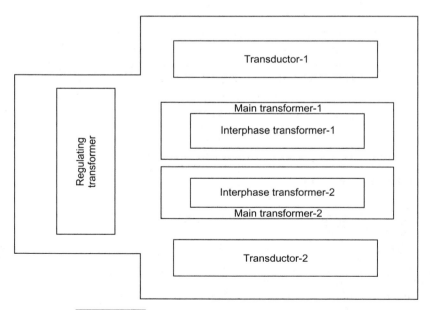

Figure 21.19 *Internal layout of rectifier transformer.*

would increase hysteresis loss, because the flux developed by bus-bar current would concentrate within the low reluctance plates, and because the action of eddy current to counteract the incident flux would be comparatively small in a high resistance material. Eddy current losses depend on the magnetic field strength at the plate and also upon the resistance of the paths available for flow within the plate.

These additional stray-losses can be controlled by the following methods:

(a) By interleaving the bus-bars, so that only the resultant flux links with the tank wall.
(b) By placing the bus-bars in such a way that their width is always perpendicular to the tank wall. It can easily be shown that the loss in the steel plate is far less when the flat surface of the bus-bar is at right angles to the steel plate than when it is parallel.
(c) By increasing the distance between the bus-bars and the tank wall.
(d) The magnetic plate can be shielded with a sheet of good conducting material, such as aluminum, placed so that the magnetic field acts to build up counteracting circulating currents within the conducting sheet. These currents reduce the magnetic field strength near the steel plate considerably.
(e) Non-magnetic steel inserts can be introduced judiciously, or the magnetic plate can entirely be replaced by a non-magnetic steel plate. Non-magnetic steel has low permeability and high resistivity resulting in loss reduction. The strategic locations where an antimagnetic stainless steel plate is provided are where the secondary or interphase transformer terminal bushings are located.

21.7 Testing

The rectifier transformers are designed and manufactured conforming to national or international standards. Due to the complexities of the circuitry of the transformer and its associated equipment, special methods will have to be devised for conducting some of the tests.

Only such tests are described hereunder:

21.7.1 Measurement of Winding Resistance

Since the output leads of the regulating transformer are connected to the hv of the rectifier transformer inside the tank, the rectifier transformer hv winding resistance will add in series to that of the regulating transformer's winding during the winding resistance measurement. For the accurate measurement of the resistance of the individual windings, for the computation of load losses, and further for the measurement of hot winding resistance at the end of a temperature rise test, these two transformers will have to be temporarily isolated. One of the methods is to isolate at the diverter switch if these contacts are accessible externally.

21.7.2 Impedance and Load Loss Measurement

These tests are simultaneously conducted on normal power transformers by short-circuit method. However, in case of a rectifier transformer where the primary and secondary ratings are different (e.g., the rectifier transformer having double star-connection with interphase transformer) and since the impedance is declared on the basis of the primary rating and the load losses are declared on the basis of the load losses in the individual windings at rated currents, this test is conducted separately as below:

(a) For the impedance voltage measurement, all the secondary terminals are shorted and the voltage required for the full rated primary current to flow is applied to the primary.

(b) The load-loss is measured first with one secondary shorted and the other secondary open. The test is repeated with the other secondary shorted and the first one open. During this test, the current in the primary is adjusted to its full rated current. The average of the two measured loss figures gives the total load-loss in the transformer.

Publication 146/1973 –*"Semiconductor convertors"* (also IS : 4540–1968–*Specification for mono-crystalline semi-conductor rectifier assemblies and equipment*), stipulates clearly the test procedure to be followed for load-loss measurement for various circuits that are in practice.

REFERENCES

1. *Copper for Bus-bars*, Copper development association, U.K.
2. Reyner, J.H., *Magnetic Amplifiers*, Rockliff, 2nd ed., 1956.
3. Ryder, J.D., *Engineering Electronics with industrial application and control*, McGraw-Hill, Kogakusha Co., New York, 1957.
4. Schaefer, Johannes, *Rectifier Circuits—Theory and Design*. Wiley & Sons, New York, 1965.
5. *Brown Bovery Review*—March/April 1961.

CHAPTER 22

Convertor Transformers for HVDC Systems

I.C. Tayal
C.M. Sharma
S.C. Bhageria

The transformers used in HVDC systems usually named as the convertor transformers have different design requirements, due to superimposed DC voltages and currents. Convertor transformers for 12 pulse rectification have three windings, one star connected AC winding and two valve windings—one connected in delta and the other in star. The AC winding is connected to AC system, whereas the valve windings are connected to convertors which, in turn, are connected in series to build up the required level of DC voltage. Figure 22.1 shows simplified connection diagram for convertor transformers connected in a typical HVDC system. Due to typical location it has to withstand various abnormal and critical conditions which demand special design and manufacturing features.

22.1 Insulation Design

Insulation system of a convertor transformer has to withstand not only AC voltages and short time over voltages, but also superimposed DC voltages on AC voltages and polarity reversal.

It is noteworthy that while the voltage stress distribution under AC conditions depends on inverse ratio of the dielectric constants of the insulating materials, the DC voltage stress distribution is dependent on the direct ratio of the resistivities.

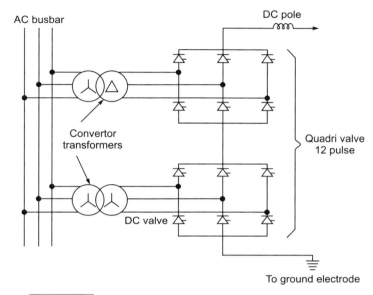

Figure 22.1 *Convertor transformer connected in a typical HVDC system.*

Major insulating materials used in oil barrier insulation system of convertor transformers are oil, paper and precompressed pressboard. The ratio of the dielectric constants for these materials does not vary more than 1 : 2 and is practically independent of external factors. Therefore, the stress distribution under AC is more uniform. Figure 22.2 shows typical AC field distribution in the end insulation.

The stress distribution under polarity reversal is capacitive in the beginning and oil is stressed more than the solid. It will then successively change over to resistive distribution. The ratio of the resistivities of the solid insulating material and oil varies by 2 to 3 power of ten depending on many external factors. Therefore, the solid insulation is stressed more under the DC voltage. Figure 22.3 shows DC field distribution in the end insulation for the ratio of resistivities 1 : 100.

Partial discharges under DC stresses appear in the form of sporadic pulses at random intervals and manifest themselves in different ways:

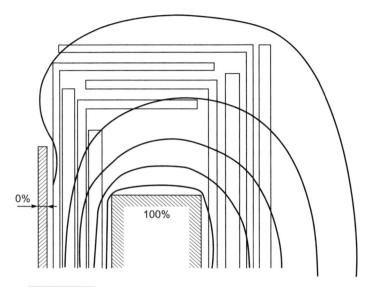

Figure 22.2 *AC field distribution in the end insulation.*

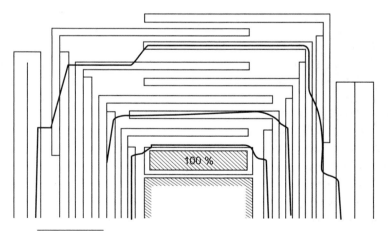

Figure 22.3 *DC field distribution in the end insulation.*

(i) Rapid changes in voltage in oil gap may cause discharges under certain conditions. Generally in well-designed and manufactured insulation systems, low energy level discharges take place which have practically very little effect on insulation.

(ii) Discharges in cellulose insulation generally take place due to imperfections in the insulation. Such discharges slowly disappear as the voltage across the solid insulation decreases. Proper design approach, selection of appropriate materials and careful manufacture ensure elimination or reduction of such discharges as otherwise these tend to reduce the impulse withstand strength.

(iii) Discharges at the oil cellulose interface are also controlled through adequate design and manufacturing practices.

PD measurement is made during a long time DC voltage test and DC polarity reversal test to estimate the extent of these discharges. During the last 10 minutes of the long time DC voltage test, not more than 10 pulses above 2000 pC are allowed and not more than 30 pulses are allowed during the last 29 minutes of the DC polarity reversal test.

The different nature of voltage distribution under AC and DC requires different arrangements of the oil-barrier insulation system than is usually necessary for conventional power transformers. Considerations must be taken for this in designing the windings and insulation system.

22.2 Higher Harmonic Currents

The harmonic content in a convertor transformer is much higher than a conventional AC transformer. There are additional load losses due to the presence of harmonic currents in valve windings. High leakage flux is produced by these harmonics. The stray flux due to these harmonics causes additional stray and eddy losses not only in winding conductors but also in transformer tank and steel structure. These losses may cause local hot spots. Avoidance of such occurrences requires proper selection of winding conductors in critical fields/zones, provision of suitable magnetic shunts, directed oil flow for effective cooling along with additional cooling equipment.

22.3 DC Magnetization

Inaccuracies in valve firing (i.e., firing asymmetry) give rise to DC magnetizing component. Although, with modern firing control, the

DC excitation component is much lower, yet careful selection of suitable flux density is imperative to avoid excessive losses in the core due to presence of DC current.

22.4 DC Bushings

The behavior of an arc under DC and AC conditions is different and, thus, the bushing for DC terminal has to be different. The DC withstand voltage of contaminated insulator is 20% to 30% to that of AC. For satisfactory performance, bushings with higher creepage is required to avoid frequent flashovers. Creepage distance as high as 40 mm/kV is generally specified.

22.5 On Load Tapchanger

Due to the large range of voltage control requirement at both convertor and invertor ends, the tapping range of convertor transformers is generally much wider than conventional AC transformers. Mixed voltage control is also required to meet the voltage requirement. Such requirements call for special tapchanger with higher tapping range and with appropriate make and break capacity.

22.6 Influence of Impedance Variation

Any variation in the specified impedance for a convertor transformer influences the cost of associated equipment in a convertor station. Therefore, the permissible tolerance on impedance of convertor transformer is much lower than that of a conventional transformer. Generally 6% tolerance on the specified impedance and 2–3% variation in measured impedances of different units against a particular design are allowed while in case of conventional transformers acceptable tolerance on the specified impedance is ± 10%.

Closer tolerance on the difference in impedance of AC winding to star-connected and delta-connected valve windings of convertor transformers, reduces the distortion in the DC voltage wave form. Similarly, lesser variation in impedance between two convertor transformers in a convertor station reduces non-characteristic harmonics, thereby reducing cost of AC filters. Closer tolerances on tested impedances of various units permit optimum design of convertor station.

22.7 Connections

22.7.1 3-Phase, 2 Winding (Fig. 22.4)

In this case two designs and one unit of each design is needed for a 12-pulse convertor. One spare unit of each design is recommended.

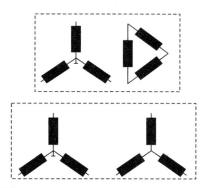

Figure 22.4 *Connection diagram of 3-phase, 2 winding.*

22.7.2 1 Phase, 2 Winding (Fig. 22.5)

In this case 2 designs and 3 units of each design are needed for 12-pulse rectification. One spare unit of each design is recommended.

22.7.3 1 Phase, 3 Winding (Fig. 22.6)

In this case one design with two wound limbs, one limb Y-Y connected (in three-phase bank), the other Y-Δ connected (in three-phase bank) are required. Three units are needed for 12-pulse convertor. One spare unit is recommended.

22.7.4 Extended Delta Connection (Fig. 22.7)

One 3-phase unit or three 1-phase units for a 12-pulse convertor are required. One spare unit is recommended.

532 *Transformers*

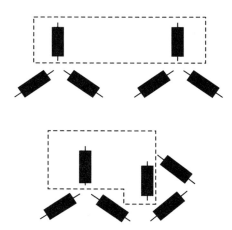

Figure 22.5 *Connection diagram for 1 phase, 2 winding.*

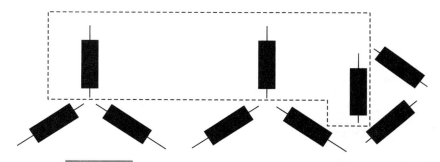

Figure 22.6 *Connection diagram of 1 phase, 3 winding.*

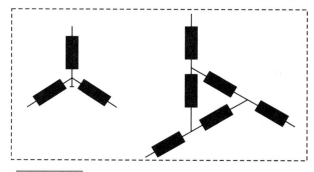

Figure 22.7 *Diagram of extended delta connection.*

22.7.5 3-Phase, 3 Winding (Fig. 22.8)

One 3-phase unit for 12-pulse convertor is required. One spare unit is recommended.

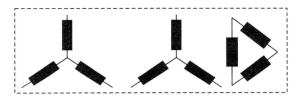

Figure 22.8 *Connection diagram 3-phase, 3 winding.*

22.8 Specifications

The specification of a convertor transformer, due to the presence of DC voltage and current, is different from that of conventional transformer.

The salient features of specification of convertor transformers for National HVDC experimental line project and Rihand-Delhi NTPC Commercial line are given in Tables 22.1 and 22.2 respectively.

22.9 Manufacturing Features

In order to meet onerous demands of a convertor transformer, manufacturing practices have to be adequate to ensure the following:
— Proper selection of materials and sources of materials, particularly insulation items is essential to ensure discharge-free insulation structure, both solid and liquid in AC, DC and polarity reversal conditions.
— Close dimensional tolerances in all windings of transformers are essential to achieve desired impedance and also to ensure minimum deviation between calculated and tested impedance.

> **Table 22.1** Specification of Convertor Transformer for National HVDC

1. kVA Rating		:	40,500 kVA, Single phase
2. Rated Voltage	HV	:	$\dfrac{220 \text{ kV}}{\sqrt{3}}$
	LV	:	$\dfrac{86.6 \text{ kV}}{\sqrt{3}}$
3. HV-LV impedance at principal tap		:	19.3%
4. Connections		:	Star/Star
5. Tappings		:	26 steps, On load
Range		:	-10% to $+22.5\%$
Numbers of steps and variation		:	13 steps, reversing arrangement for $+/-10\%$ HV voltage variation and at $+10\%$ HV voltage 0 to -10% LV voltage variation
6. Type of cooling		:	OFAF/ONAF/ONAN
7. Terminals	HV	:	AC Bushing
	LV	:	DC Bushing
8. Insulation levels			
8.1 Impulse	HV	:	950 kVp Lightning
		:	750 kVp Switching
	LV	:	450 kVp Lightning
			450 kVp Switching
8.2 Induced voltage		:	HV 395 kV, LV 185 kV (fully insulated)
8.3 Long duration DC voltage test associated with partial discharge		:	166 kV for one hour
Short duration DC voltage test		:	220 kV for 2 minutes
Direct voltage polarity reversal test associated with partial discharge		:	88.34 kV

— When several units are required in a convertor station, dimensional accuracies of coils have to be identical to ensure that the difference between impedances of different units is minimum. This calls for highly accurate moulds for windings, utilization of appropriate winding machines with

> **Table 22.2** Specification of Convertor Transformers for Rihand (Rihand–Delhi HVDC Scheme)

1. kVA Rating	:	315,000 kVA, Single phase three winding
2. Rated voltage		
Line Winding (HV)	:	$\dfrac{400 \text{ kV}}{\sqrt{3}}$
Valve Windings (LV)	:	$\dfrac{213 \text{ kV}}{\sqrt{3}}$ star and 213 kV delta
3. HV-LV impedance at principal tap	:	18%
4. Connections	:	Star/Star-Delta
5. Tappings	:	24 steps, On load –12.5% to +17.5%
6. Type of cooling	:	OFAF
7. Terminals	:	HV : AC Bushing
		LV : DC Bushing
8. Insulation levels		
8.1 Impulse HV	:	1300 kVp Lightning
		1080 kVp Switching
LV Star	:	1550 kVp Lightning
		1290 kVp Switching
LV Delta		1175 kVp Lightning
		980 kVp Switching
8.2 Induced voltage test with PD	:	440 kV
8.3 Long duration DC voltage test associated with partial discharge	:	800 kV for one hour
Direct voltage polarity reversal test associated with partial discharge	:	568 kV

    provision for good tensioning arrangement and also adoption of stabilization process of the windings.

— Drying process of the transformer, better quality processed oil, soaking time also govern the quality of convertor trans-former in terms of other electrical parameters like tan-delta value, insulation resistance partial discharges, etc.

— Finally, cleanliness of the working environment, components and accessories used also play their role in the manufacturing process of convertor transformers.

22.10 Tests

In order to simulate the actual operating conditions, the convertor transformers are subjected to both normal AC and DC dielectric tests. Recommendations for various test levels and sequence of tests for valve windings are given in Electra no. 46, Electra no. 157 and IEC 61378-2. The AC windings are tested in the usual manner in accordance with IEC 76/IS: 2026. All other routine type tests are the same as for an AC transformer.

The following tests are performed on the valve winding to assess its ability to withstand various stresses due to DC, AC impulse, polarity reversal, or short time DC over-voltages.

22.10.1 Long Duration DC Voltage Test

This test is performed to demonstrate the ability of the transformer to withstand stresses due to continuous DC voltage. A DC voltage of positive polarity is applied for 60 minutes. The windings not under test are short-circuited and grounded together with the tank. The level of the test voltage (UDC) for one hour is obtained as follows:

$$UDC = 1.5\ [(Z - 0.5)Ud + 0.7\ Uvo]$$

where, Z : The 6-pulse bridge number starting at ground level.
Ud : Nominal direct voltage across one 6-pulse bridge.
Uvo : No load voltage (rms) between the valve winding phase to phase terminals.

During the last 10 minutes of the test, measurement of DC partial discharge is made. The number of discharge pulses exceeding 2000 pico-coulombs should not exceed 10. If the number of pulses exceeds 10, the test is prolonged by another 10 minutes, and if after three ten-minute observation periods the number of pulses exceeds 10 per period, the transformer is rejected.

22.10.2 Polarity Reversal Test Coupled with DC Partial Discharge Test

To demonstrate the ability of the transformer to withstand stresses due to polarity reversal or other rapid DC voltage changes the polarity reversal test coupled with DC partial discharge monitoring are performed.

A DC voltage of negative polarity is applied for at least two hours, at the end of which the polarity is reversed within two minutes, and the reversal voltage is applied for a further 30 minutes. During the last 29 minutes, the number of pulses exceeding 2000 pico-coulombs should not exceeds 30. If the number of pulses exceed 30, the voltage polarity is changed and the whole test is repeated. If the number of pulses measured in the final 29 minutes again exceeds 30, the transformer is deemed to have failed the test. The level of the test voltage (Upr) is obtained as follows:

$$Upr = 1.25\,[(Z - 0.5)\,Ud + 0.35\,Uvo]$$

where Z, Ud, and Uvo are as mentioned earlier under the long duration DC voltage test. During the test, the windings not being tested are short-circuited and grounded together with the transformer tank.

22.10.3 AC Applied Test Along with PD Measurement

This is a good test to develop higher stresses in oil ducts. The recommended test level is 1.5 times the crest operating level for the convertor transformer. The test level would be as follows:

$$\text{Vac applied} = \frac{1.5}{\sqrt{2}}\,[(Z - 0.5)\,Ud + \frac{\sqrt{2}}{\sqrt{3}} \times Uvo]$$

where Z, Ud, and Uvo are as mentioned earlier under the long duration DC voltage test

The AC voltage is applied for one hour along with partial discharge measurement.

22.10.4 Switching Impulse Test

Switching impulse voltage is applied to the valve winding terminals connected together. The other winding terminals are short-circuited and grounded together with the tank. The polarity is negative and the wave shape is kept in accordance with the prevailing standards.

22.10.5 Lightning Impulse Test

Lightning Impulse voltage is applied to each valve winding terminal in turn. The other valve winding terminal and all other winding terminals are grounded together with the tank.

REFERENCES

1. Wahlstrom, Bo., (WG 12.02), Voltage Tests on Transformers and Smoothing Reactors for HVDC Transmission, Electra No. 46, May 1976.
2. Watanabe, M., T. Obata, E. Takahashi, Development of Convertor Transformers and Smoothing Reactors, Hitachi Review, Vol. 34, 1985, No. 5.
3. Mannhein, H., Bold, Static Convertor Transformer for the HVDC Transmission Link Cabora Bassa, Apollo, Brown Boveri Rev. 7, 76.
4. Beletsky, Z.M., I.D. Voevodin, A.K. Lokhanin, Transformer Equipment for DC Transmission Lines, World Electrotechnical Congress, June 1977.
5. A. Lindroth et al. "The relationship between Test and Service Stresses as a function of Resistivity Ratio for HVDC Converter Transformers and Smoothing /Reactors" CIGRE Working Group report WG 12/14.10, Electra No. 157, December 1994.
6. Part 2 : Transformers for HVDC applications, IEC 61378-2, first edition, 2001–02.

CHAPTER 23

Controlled Shunt Reactor

S.C. Bhageria
J.S. Kuntia

Long distance EHV and UHV lines experience fundamental frequency temporary over-voltages under no-load and light-load conditions, and also due to load rejections as per the Ferranti-effect. Such over-voltages can stress the equipment connected to the lines, and need to be controlled. It is a common practice to provide fixed shunt reactors in these lines to compensate the capacitive line charging current, and thereby control the over-voltages under no-load, light-load, and load rejection. These shunt reactors also control the switching over-voltages to some extent.

However, fixed shunt reactors have some major disadvantages. In loaded power lines, these reactors continuously absorb rated reactive power, increase the surge impedance of the line, and thereby reduce the surge impedance loading (SIL) level, that is, the level at which a flat voltage profile along the line can be achieved. These permanently connected fixed shunt reactors also consume active power, which is a continuous loss to the system. The disadvantages associated with permanently connected fixed shunt reactors call for the development of a controlled shunt reactor, that is, a reactor in which reactive power absorption can be varied under changing load conditions.

23.1 Controlled Shunt Reactor (CSR) Principle

The basic principle of the CSR is to control the reactive power by using a thyristor valve, which can provide the necessary speed of switching and control by means of firing angle control. As it is not economical and practical to use the thyristor valves at high transmission voltages like 400 kV, the thyristor valves are connected to the controlled shunt reactor transformer (CSRT) secondary side. A CSRT with primary winding connected to high voltage transmission lines, and secondary winding (control winding) connected to thyristor valves, forms the basis of CSR. Basically, a CSR consists of a high impedance transformer (CSRT) controlled by an antiparallel pair of thyristor valves. The impedance of a CSRT can be controlled by varying the firing angle of the thyristor pair through a controller. The controller is a logical and programmable device, which generates firing pulses based on the input signals. Figure 23.1 explains the underlying principle of the CSR. When the thyristor valve is not conducting the control winding is almost open. In this case, only a very small current (very low MVA) flows through the primary winding because the reactive impedance of a CSR is very high. On the other hand, when the thyristor is fully conductive (i.e., control winding is short-circuited) the inductive impedance of the CSR is nearly 100%. In this case, the current through the primary winding is maximum (nominal MVA) and is approximately equal to the short-circuit current when the secondary of the CSRT is short-circuited.

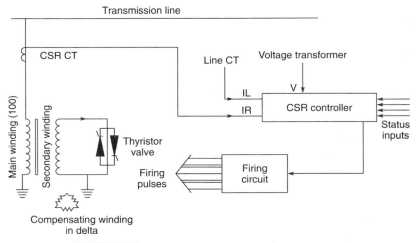

Figure 23.1 *Principle of controlled shunt reactor.*

The CSR generally consists of a controlled shunt reactor transformer with approximately 100% impedance, thyristor valves, a controller, a neutral grounding reactor, the necessary circuit breakers, and other auxiliaries. Figure 23.2 shows the general arrangement of the controlled shunt reactor (CSR) scheme.

BHEL has successfully developed and commissioned a 50 MVAr, 400 kV class controlled shunt reactor (CSR) at the 400 kV Itarsi Substation of Power Grid Corporation of India Ltd.

This type of reactor has been developed for the first time in the country. The CSR is not merely a variable shunt reactor—its advantages are much more than the shunt reactor. The development has been possible due to excellent support provided by the customer M/s. PGCIL. Figure 23.3 shows the installed CSR and Fig. 23.4 shows a panaromic view of the CSR installed at the Itarsi site.

23.2 Controlled Shunt Reactor Transformer (CSRT)

The CSRT is a power transformer with nearly 100% impedance between primary and control winding (secondary winding). In other words, when the control winding is short-circuited with rated voltage applied on the primary winding, the reactive power flow will be similar to the power flow in a shunt reactor of similar rating connected to the same rated voltage.

One end each of the star-connected primary winding of CSRT is directly connected to three phases of line. The neutral end of the primary is grounded through a reactor. The star-connected control windings with grounded neutral are connected to the thyristor valves, and can be controlled independently. It also consists of the compensating winding (tertiary) connected in delta for the compensation of the 3rd and its multiple harmonics.

The CSRT, when its secondary is kept open, draws only the magnetizing current, and consumes negligible reactive power. Another important aspect of the reactor transformer is its magnetization characteristic, which remains practically linear as the main flux path under loaded conditions is mostly through the air. Hence, even under sustained over-voltage conditions, the problem of core saturation does not occur.

The specifications and design of the CSRT depend on the performance specifications of the CSR as a system. The construction of the CSRT is similar to that of a power transformer, which consists of a closed magnetic core, and co-axial windings.

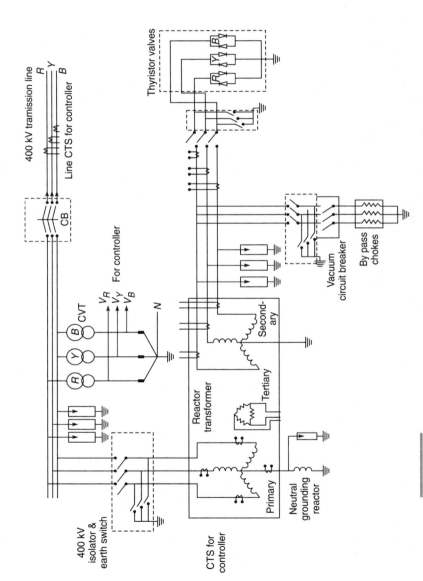

Figure 23.2 General arrangement of controlled shunt reactor (CSR) scheme.

Controlled Shunt Reactor

Figure 23.3 CSR installed at PGCIL itarsi site.

Figure 23.4 Panoramic view of CSR installation at itarsi.

23.3 Special Features of the Controlled Shunt Reactor Transformer

Some of the other important aspects of the design and construction of the CSRT are mentioned below:

(a) Voltage and current rating of winding: The voltage and current rating of the primary winding is decided on the basis of the system voltage and the reactive power requirement. The selection of secondary voltage and current depends on the availability, suitability, and overall economy of thyristor valves. Usually, the compensating winding voltage rating will be the same as that of control winding, and its capacity will be adequate for suppression of harmonics. However, for special application of the CSR viz. compensation of reactive power, the required capacity of compensating winding may be as high as 100% of the CSRT rating.

(b) Winding disposition: The requirement of 100% impedance between primary and control windings can be achieved by a suitable selection of the voltage per turn and the radial gap between these windings. The control winding being the low voltage winding, is generally placed next to the core. The compensating winding is generally placed between the primary and control windings. This not only offers an economical solution, but also has several technical advantages. In this position, the compensating winding is closely coupled with the control winding for very good compensation of harmonics generated in the control winding. On the other hand, the compensating winding is loosely coupled with the primary winding, and helps to reduce the effect of circulating currents on the primary winding. This is particularly important for making the CSR system compatible for single phase auto-reclosure. The interwinding gaps play a very important role in achieving the desired performance and optimum design of CSRT. Figure 23.5 shows the disposition of winding in the CSRT.

(c) Stray loss control: As mentioned earlier the CSRT is a power transformer with approximately 100% impedance between the primary and control windings. The very high leakage impedance and special disposition of the windings make stray losses increase many fold in the case of the CSRT. When the CSR is required to provide full reactive power, the control winding is completely shorted. In this

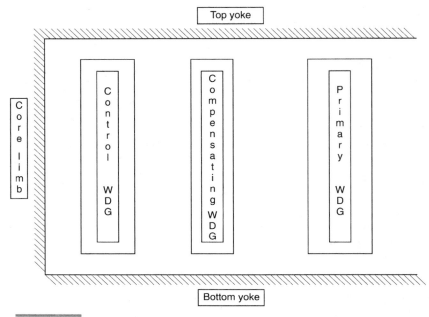

Figure 23.5 *Winding disposition in controlled shunt reactor transformer.*

condition, there is no flux coupling between the primary and control windings. On the other hand, as the primary winding is connected to the nominal voltage, all the flux linking the primary winding is disposed in the radial gap between the primary and control windings. In other words, during the nominal regime of the CSR, there is no flux in the main limb of the CSRT, but is disposed in the interwinding gap of the primary and control windings. This flux is also called the leakage flux, and because of the approximately 100% impedance between the primary and control windings, the leakage flux emanating through the annular space between the windings results in very high stray losses, that is, of the order of 630% of the I^2R loss, without yoke shunts, wall shunts, and non-magnetic clamping structure. Figure 23.6 shows the magnetic field distribution of the newly developed CSRT without yoke shunts and wall shunts. From the field plot, it is evident that in the absence of yoke shunts, there is maximum fringing of flux at the top and bottom. This stray flux is responsible for creating very high losses in the adjacent structure and tank. It may be noted that such high stray losses in the adjacent structures

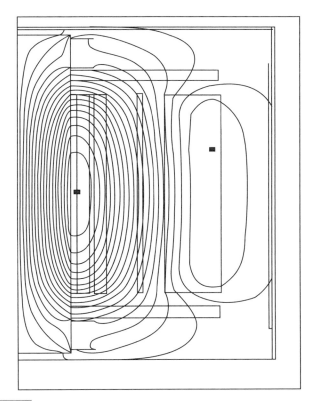

Figure 23.6 *Magnetic field plot without yoke shunts and wall shunts.*

and tank may not allow satisfactory operation of the CSRT, and may lead to temperature rises beyond guaranteed values. The generation of excessive dissolved gasses etc. will alter the operating characteristics of the CSRT, and are not acceptable. Special methods like using yoke shunts and wall shunts, and providing them on the strategic locations with the help of accurate magnetic field analysis were employed to control the stray losses. It is important to note that successful implementation of these methods could reduce the stray losses to about 50% of I^2R losses, compared to stray losses of 630% of I^2R losses without these methods. Figure 23.7 shows the magnetic field plot of the CSR provided with these methods. From the field plot, it is evident that the flux path has become almost parallel to the windings, and fringing at the ends has reduced to a great extent. A very high percentage of flux

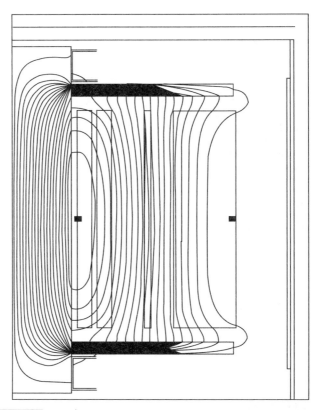

Figure 23.7 *Magnetic field plot with yoke shunts and wall shunts.*

has been collected by yoke shunts and pumped to the core. Similarly, the stray flux near the tank is passing through the wall shunts, and the entry of flux in the tank wall has been reduced to a great extent.

It may be noted that further reduction of stray losses to the extent of approximately 10–15% of I^2R loss is possible by additional use of non-magnetic structures for the end frame and clamp plate.

It is important to mention here that the actual measured losses were approximately in line with the expected losses with the provision of newly developed yoke shunts and wall shunts.

23.4 Controlled Shunt Reactor Much More than a Shunt Reactor

The CSR is not merely a substitute of the shunt reactor—its advantages are much more than that. The major advantages are:

(a) **Fully controllable reactive power**: The CSR provides reactive power support to the line only when it is required, thus reducing the continuous reactive power drawn as in the case of fixed shunt reactors.

(b) **Reduction in dynamic over-voltage limits**: The CSR automatically goes out of circuit during increased line loading conditions, thus eliminating the need to limit the reactive compensation limit to 60% which is the present practice in Indian grids. This feature can be used to provide higher compensation on the lines to limit the power frequency dynamic over-voltages. This shall help in reducing the over-voltage requirements of all the substation equipment, resulting in overall economy.

(c) **Increased power carrying capacity of lines**: The CSR control system automatically takes it out of circuit under full-load condition of the lines. Thus, an increase in the power transmission capacity is achieved. In comparison to the lines compensated with the fixed shunt reactor, the transmission lines provided with CSR can enhance the power transmission capacity by 25–30%.

(d) **Fast response**: The CSR can respond with a short response time of 10 milliseconds for any sudden over-voltage, etc., and can come into the circuit with full reactive power support. This feature eliminates the risk involved in taking out the fixed reactors from the circuit.

(e) **Full compatibility to single phase auto-reclosure**: The fast response time of the CSR makes it suitable to respond to fault conditions like single line to ground and three-phase faults. The control system senses the fault, and brings the reactor to full conduction within a short time, thus providing full compatibility for single-phase auto-reclosure. The need for NGR is similar to that of a fixed shunt reactor.

(f) **Ability to connect directly to EHV level**: The controlled shunt reactor transformer can be directly connected to the EHV level, as against an SVC system, which requires an intermediate transformer in addition to the controllable reactor.

(g) **Minimum harmonics**: The continuously controlled CSR is provided with small filters across the delta connected compensating winding. This is to limit harmonics generated due to variation of the firing angle of the thyristor valves in accordance with the reactive power requirement. The total harmonics generated due to thyristor firing will be less than 2–3% with respect to the nominal capacity of the CSR.

23.5 Conclusion

Thus, the CSR offers various advantages like fully controllable reactive power, reduction in dynamic over-voltages, increased power carrying capacity of lines, fast response, full compatibility to single phase auto-reclosure, economy of size, minimum harmonics, etc.

It is expected that the CSR will provide an economical solution for stable and strong transmission networks. Controlled shunt reactors placed at suitable points in a line can make the line carry surge impedance loading level of power with nearly 1.0 P.U. voltage profile, and theoretically make it possible to make AC power get transmitted over infinite long distances.

REFERENCES

1. S.C. Bhageria, G.N. Alexandrov, S.V.N. Jithin Sunder, and M.M. Bhaway. "Reactor Transformer for Controlled Shunt Reactor Applications." International Conference, Power Transformers 2000 by CBIP.
2. S.V.N. Jithin Sundar, S.C. Bhageria, C.D. Khoday, Amitabh Singhal, A.K. Tripathi, G.N. Alexandrov, M.M. Goswami, I.S. Jha, Subir Sen, V.K. Parashar, "Controlled Shunt Reactor—A member of FACTS family." Eleventh National Power System Conference (NPSC-2000).

CHAPTER 24

Designing and Manufacturing —A Short-Circuit Proof Transformer

T.K. Ganguli
S.K. Gupta

The short-circuit withstand capability of a power transformer is defined as the ability to withstand the full asymmetrical short-circuit currents without impairing its suitability for normal service conditions. Steady increase in unit ratings of transformers and simultaneous growth of short-circuit levels of networks have made the short-circuit withstand capability of the transformer one of the most important aspect of its design.

The short-circuit (SC) test is a very stringent test that involves symmetrical short-circuit currents of the order of 6–7 times the rated current and peak asymmetrical current as high as 15–18 times the rated current. The basic formula for force acting on a current carrying conductor in a magnetic field at any instant is given by:

$F = BIL$ Newtons,

where,

B = Flux density in Tesla
I = Current in Amps
L = Length of conductor in meters

This means that resulting forces shall be tremendously high since they shall increase in square of the current. These forces shall be exerted directly on all current carrying parts of the transformer in a magnetic field, i.e., windings, leads, terminal gear, etc. Short-circuit forces are of pulsating nature since short-circuit current and flux are sinusoidal. Therefore, time variable stresses are applied to the affected structures. These forces are very high during the first asymmetrical peak and come down to the symmetrical *rms* value.

24.1 Forces During Short-Circuit

The forces generated during short-circuit are transferred to all parts mechanically coupled to winding assembly. The overall effect of these forces has been elaborated below:

24.1.1 Radial and Axial Forces on Winding

Due to leakage field distribution during SC, heavy radial and axial forces shall be generated inside windings. This has already been elaborated in Chapter 7. The inner windings under compressive stress have a tendency to fail under buckling as described below.

Buckling Phenomena

This has already been explained in Chapter 7. The buckling strength of a transformer is dependent upon the elasticity of the material, that is, work hardening, conductor radial thickness, base cylinder, etc. Buckling is a very common mode of failure on a transformer subjected to SC test.

24.1.2 Forces on Winding Leads

Winding leads are normally perpendicular to winding conductors in most the cases, as in Fig. 24.1 below.

The leakage flux and current in lead sections shall result in a force perpendicular to the lead length. For inner windings which are under compressive stress, the tendency of winding leads shall be to tighten the screw of the coil by twisting itself towards a smaller diameter winding. This is called the spiralling effect and is one of the common causes of failure in helical and spiral windings. Spiralling has been further explained below.

Spiralling

Spiralling involves a marked deviation of the spacer rows from their original, vertically aligned arrangement. The spacers undergo a progressive displacement, starting from the lower end to the upper end of the winding generally combined with a pronounced relative shift of the last bottom and top turns with respect to turns nearby, the shift

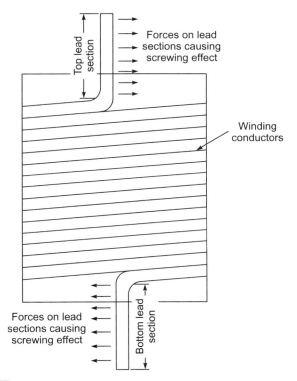

Figure 24.1 *Forces on lead sections during short-circuit spiralling effect.*

being larger for the top turn. This also results in bending of the exit leads and crossovers.

24.1.3 Forces on Base Cylinder and Core

It has already been explained in Chapter 7 that the inside windings, specially the LV winding and inner tap windings, are normally subjected to inside radial stresses. These windings are normally wound over the base cylinder. Although winding is considered as a solid mass, radial stresses are still exerted onto the core through the base cylinder and the insulation arrangement between the core and winding. So, the base cylinder must be adequately designed to withstand short-circuit forces. The radial forces are almost uniform along approximately two-thirds of the winding height. However, it starts decreasing at winding ends due to radial bending of flux lines at ends (see Fig. 7.1).

24.1.4 Forces on Winding Clamping Structure

It is evident that short-circuit results in very high axial forces in the winding, to the order of tens or hundreds of tonnes for large rating power transformers. Normally, the net cumulative axial force on one winding is upwards, while on the other winding it is in the opposite direction, that is, downwards.

A typical axial clamping structure of a large transformer is shown in Fig. 24.2. The axial forces from winding during short-circuit are transmitted to winding end blocks/block washer and top ring, and then to the top yoke or clamping structure. The axial clamping structure, pressing rings, etc., form the backbone of the transformer for withstanding SC forces.

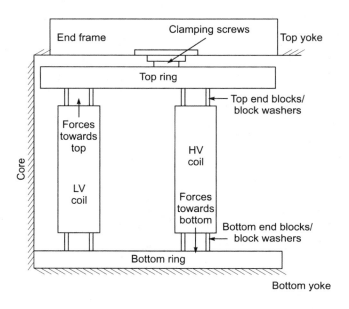

Figure 24.2 *Typical axial clamping structure and typical axial forces pattern during SC test.*

24.1.5 Terminal Gear Items and Their Supports

The lead connections from the winding ends are taken out and connected to form a star/delta inside transformer for a 3-phase transformer and then connected with bushings at the top. This involves a complete terminal gear which is supported by various cleats and supports. During short-circuit leakage flux are produced in the areas adjacent to the windings and approaching towards the tank (refer to leakage field distribution in Chapter 7). Thus, forces are generated in the leads:

(a) Due to the presence of leakage flux of winding. These forces shall be in a direction perpendicular to the length of the lead.
(b) Due to the interaction of two current carrying conductors either in the same direction (for example different parallels of same winding), or in the opposite direction (for example LV and HV leads). The resulting forces in the former case shall be mutually attractive, while in the latter case, it shall be mutually repulsive, as illustrated in Fig. 24.3.

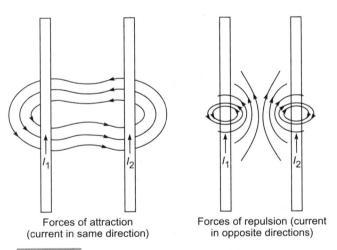

Figure 24.3 *Forces in two current carrying conductors.*

The above forces shall be exerted through these conductors on to their support structure. Low voltage leads of large transformers have very high short-circuit currents and the forces can be quite significant. Tap leads also experience considerable force due to the vicinity of a number of cables.

24.2 Various Considerations to Design a Transformer Suitable for Short-Circuit Duty

The design stage is a very crucial aspect for a short-circuit proof transformer, and if proper care is taken during the design stage, then the short-circuit forces can be controlled. Also, the withstand capability of the various parts can be enhanced to a considerable extent. Areas where proper attention is required are explained below.

24.2.1 Radial Stress Withstand of Winding

Theoretically, radial stress in a winding can be reduced by reducing the current density, or by increasing the impedance. However, there are several practical limitations for a change in the above parameters. It is not always wise to reduce the current density as it is quite uneconomical. Similarly, impedance is also guaranteed as per system requirements and has to be within tolerances.

The radial withstand capability of a winding can be enhanced by increasing the work hardening strength of the winding conductor, by increasing radial thickness of the winding conductor, by increasing the number and dimensions of supporting blocks, etc. However, there are some practical constraints. For instance, very high work hardening means a practical difficulty in manufacture of windings and poor control over manufacturing of conductors. Similarly, increasing the number and size of blocks would reduce the distance between blocks, which may result in inadequate cooling surface or insufficient distance for transposition/crossovers. Therefore, a judicious selection of design of the winding is necessary for a designer. Increasing the radial stress withstand would also improve the buckling strength of the winding.

24.2.2 Axial Forces Withstand of Windings

Axial forces arise out of ampere turn imbalance between primary and secondary windings, and due to the bending of flux lines at winding ends. The ampere turn balance along the height of the winding is therefore one of the most important aspects for reducing the axial force itself, which is the crux of the problem. Ampere turn imbalance occurs due to unequal turns per disc, specially in disc windings or due to unequal duct size distribution along the height of the winding. A good designer always aims at reducing these inbalances by minimizing and adjusting the turns distribution and duct distribution so as to result in minimum possible axial forces.

Another aspect is to enhance the axial withstand capability of the winding itself. This is possible by decreasing the width/thickness ratio of the winding conductor, increasing the number and size of winding blocks, etc.

24.2.3 Top and Bottom Yoke Clearances

Due to design requirements, sometimes the top and bottom yoke clearances are kept unequal. This is specially required if the HV line lead is to be taken out from top. In this case, the only solution is to enhance the axial force withstand capability of the winding.

24.2.4 Location of Winding

Whenever a winding is kept in the main flux due to the design and client's requirement, it shall experience a heavy radial force due to a very high component of axial force in this region. (See Fig. 7.1 of Chapter 7.)

In this arrangement, the radial withstand capability of the winding must be prudently enhanced. The winding ends or sections of winding near winding ends are the most critical area and require an adequate supporting or locking structure to arrest their radial and peripheral movement.

24.2.5 Anchoring and Locking of Leads

Winding ends/leads often become weak points, hence proper anchoring of leads with adjacent sections/turns is essential. Anchoring is very important for tapping winding since a number of leads are taken out from the winding and it often becomes a weak point. A proper locking arrangement to arrest the movement of tapping leads is therefore very essential.

24.2.6 Design of Top Ring

The thickness of the ring is calculated based on the axial short-circuit force, the number and size of clamping screws, the span between two clamping screws, and the number and size of slots for taking out LV and tapping leads. The slot in the top ring should be as small as possible. The material of the top ring, its thickness and size, its method of joining and the material strength should be ensured properly before use. Also, additional pressure pads can be provided if the span between two screws is too high.

24.2.7 Selection and Design of Clamping Screws and Supporting Structure

The capacity and number of clamping screws/pressure pads must be adequate to withstand the maximum axial force expected during short-circuit. The clamping screw dimension, viz. its diameter, length, and basic material is important. The clamping screw is tightened and kept in vertical position to exert uniform pressure to the windings.

Similarly, the complete support structure including the endframe and clamp plate/pin, etc., must be designed to meet the axial force arising during short-circuit.

24.2.8 Terminal Gear and Its Support

All terminal gear leads, especially of LV and tapping coil are to be anchored and supported firmly. The supports must be suitable for the forces arising during the short-circuit test calculated in line with section 24.1. If required, the number of supports per unit length may be increased. The thickness and size of the supports should also be selected based on the short-circuit forces expected.

Spacing between leads is a very important aspect, since force increases in inverse ratio of square of center to center distance between cables. Therefore distance pieces between two cables may be inserted to maintain distance.

24.3 Manufacturing Aspects

Merely a good design cannot make a transformer short-circuit proof. The manufacturing capability is also equally important in all stages of manufacturing, as explained below.

24.3.1 Winding Manufacturing

Proper tightness of winding conductors, proper circular shape of winding without any eccentricity, etc., is to be ensured during manufacturing of winding. Otherwise, even with a good design the transformer might fail under short-circuit. For example the Buckling phenomena as explained in section 24.1 is very complex, and is deeply influenced by the manufacturing process of winding. Poorly wound windings exhibit reduced buckling strength whereas stiff, well-tightened and supported windings provide a significant enhancement in the withstand capability against buckling.

24.3.2 Dimensional Stability

Meeting the dimensional stability of various parts, especially winding is of utmost importance and must be well taken care of. The winding height must be strictly in line with specified dimensions and allowed tolerances.

24.3.3 Proper Curing of Epoxy Bonded Conductors

Proper curing of epoxy bonded conductors is one of the most important aspects since the conductor cannot attain the specified strength if proper curing is not done. The supplier's instructions must be strictly adhered to.

24.3.4 Use of Calibrated Blocks for Windings

Preshrunk and calibrated clacks/blocks should be used to ensure required winding height and for better dimensional stability of the windings. Also, the quality of all base cylinders and end rings must be ensured.

24.3.5 Stabilization of Windings and Coil Assembly

All windings individually and the composite coil assembly are to be properly stabilized as per the manufacturer's standard practice, to ensure proper pressure on the winding assembly.

24.3.6 Crossover/Transposition Locations

In many cases, damage has been observed on the winding due to shorting of conductors at transposition/crossover locations. These areas are weak points, hence they should be critically taken care of to avoid any damage due to sharpness, etc., and additional insulation should be provided in these areas for strengthening.

24.3.7 Tight Lowering of Coils

Tight lowering of coils is very essential in order to achieve a tight assembly of windings. This shall ensure even distribution of force and enhance the strength of the transformer under short-circuit. In case ovality is observed, it should be packed with spacers/cylinders for proper support.

24.4 Quality Aspects

Quality control during each stage of manufacturing is a must in order to manufacture a transformer suitable for short-circuit. Each stage, starting from raw material to various stages like winding, core, assembly, terminal gear, and insulation items should be subjected to strict quality control.

24.5 Conclusion

Unless design, manufacturing, and quality aspects are fully taken care of, it is not possible to manufacture a transformer suitable for short-circuit. The behavior of a large power transformer during short-circuit is quite complex and involves a number of other factors like inner friction in the windings, stability in structural materials, springing and damping within the winding, effects of oil motion, etc. The selection of various fittings, i.e., tapchanger, bushing, etc., should be done so that it is able to withstand the stresses caused during short-circuit.

Thus for designing and manufacturing a short-circuit proof transformer all aspects, viz. design, material selection, selection of its fittings and the manufacturing processes are to be carefully considered and all minor aspects including support structures, etc., are to be verified in detail for adequate strength.

CHAPTER 25

High Voltage Condenser Bushings

R.K. Agrawal
Aseem Dhamija

25.1 Introduction

Bushing is one of the most important components that are fitted to the electrical equipments like the transformer, the switchgear, etc. It is an insulating structure for carrying the HV conductor through an earthed barrier.

The bushing has to

(a) Carry the full load current.
(b) Provide electrical insulation to the conductor for working voltage and for various over-voltages that occur during service.
(c) Provide support against various mechanical forces.

25.2 Classification of Bushings

Bushings are classified according to the following factors:

25.2.1 Application or Utility

(a) *Alternator Bushing*

AC generators require bushings up to 33 kV, but 22 kV is more usual. With modern alternators, current ratings up to 20,000 A are required.

(b) Bushings for Switchgear

In the switchgear, bushings are to carry the conductors through the tank wall, and support the switch contacts.

(c) Transformer Bushings

Transformers require terminal bushings for both primary and secondary windings. In some cases, a high voltage cable is directly connected to the transformer via an oil filled cable box. A bushing then provides the connection between the cable box and transformer winding.

(d) Wall or Roof Bushing

In recent years, many sub-stations for 132 kV and above, in unfavorable situations have been put inside a building. For such applications wall/roof bushings are used.

(e) Loco Bushings

These bushings are used in freight loco and AC EMU transformers for the traction application.

25.2.2 Non-condenser and Condenser Bushings

(a) Non-condenser Bushing

In its simplest form, a bushing would be a cylinder of insulating material, porcelain, glass resin, etc., with the radial clearance and axial clearance to suit the electric strengths. The voltage is not distributed evenly through the material, or along its length. As the rated voltage increases, the dimensions required become so large that this form of bushing is not a practical proposition. The concentration of stress in the insulation and on its surface may give rise to partial discharge. This type of bushing is commonly used as low-voltage bushings for large generator transformers.

(b) Condenser Bushing

In this type, the conducting cylinders are inserted in the insulation to divide the wall thickness into a number of capacitors. In this way, the voltage distribution in the material and along its surface can be controlled.

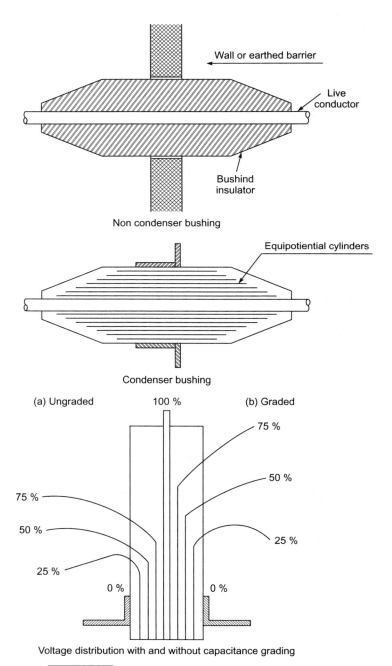

Figure 25.1 *Condenser and non-condenser bushing.*

25.2.3 Insulating Material

The insulating material of bushings is usually paper-based with the following most common types:

(a) *Synthetic Resin Bonded Paper (SRBP)*

In SRBP bushings, one side of the paper is coated with resin which is cylindrically wound under heat and pressure inserting conducting layers at appropriate intervals. However, use of SRBP bushings is limited to low voltages. There is also the danger of thermal instability of insulation produced by the dielectric loss of the resins. The SRBP insulation is essentially a laminate of resin and paper which is prone to cracking. Moreover, paper itself will include air which will cause partial discharges even at low levels of electrical stress.

(b) *Oil Impregnated Paper (OIP)*

OIP insulation is widely used in bushing and instrument transformers up to the highest service voltages. In the manufacturing process, the kraft paper tape or sheet is wound onto the conductor. Aluminum layers are inserted in predetermined positions to build up a stress-controlling condenser insulator. The condenser layer may be closer together, allowing higher radial stress to be used. The bushing is fully assembled before being vacuum impregnated in order to contain the oil.

(c) *Resin Impregnated Paper (RIP)*

RIP bushings are wound in a similar manner as OIP. The raw paper insulation is then kept in a casting tool inside an auto-clave. A strictly controlled process of heat and vacuum is used to dry the paper prior to impregnation with epoxy resin. The resin is cured under a heat and pressure cycle. The resulting insulation is dry, gas tight, and void free.

25.3 Design of Bushing

In its simplest form, a bushing consists of a central conductor embedded in a cylindrical insulation material having a radial thickness enough to withstand the voltage. The design of a bushing depends on the various voltages and over-voltages that it has to come across during service. In order to study the factors which influence the design of a bushing, it is convenient to consider a transformer bushing with one end in the air, and the other in oil. The important factors which affect the design are:

(a) Air-End Clearance

The air-end clearance has to be sufficient to meet the specified over-voltage tests. It is also determined by the creepage distance, and the proportion of it that is protected from the rain. Having determined the air-end length, the air-end dimension of the internal condenser can be determined. It is not necessary to grade 100%. Internal grading of 70% or less will give adequate surface grading for large bushings.

(b) Oil-End Clearance

As internal breakdown unlike air flashover, is more severe, specifications, therefore, demand an internal breakdown with a sufficient margin (about 15%) above the air withstand value. Both power frequency, and impulse voltage withstand tests have been used to specify this characteristic.

(c) Number of Condenser Layers

The number of partial condensers is so chosen that the test voltage of each partial condenser should be between 10 kV to 15 kV. If more foils are introduced, it will cause too many folds and weaken the bushing. Also, will be air introduced in the folds, complicating the manufacture of bushing of high voltage class.

(d) Length of Earth Layer

The length of the earth layer of a bushing is usually determined by the accommodation required for current-transformers, or by mounting considerations, though in some cases it may be allowed to assume its optimum dimension in relation to the radial dimensions. The ratio of Length of first foil ($L1$) and Length of nth foil (Ln) may be taken between 3 to 4. This ratio is denoted by α.

(e) Radial Gradients and Diameters

The radial gradient is limited by the necessity for avoiding damage by discharges at the power-frequency test voltages, whether one minute or instantaneous.

If the ratio of the earth layer diameter to that of the conductor r_n/r_o, is denoted by β, the stresses at the HV end and the earth voltage end will be equal, if the product of α and β is unity.

However, it is not always possible to achieve this value. Hence α and β can vary from 0.8 to 1.2.

If $\alpha. \beta = 1$, then $LnDn = L1. D0$

(f) Equipotential Layer Position

After determining the dimensions of the inner and outer layers of the condenser, the position of the other layers can be calculated. The basis of the design of the condenser bushing is generally equal partial capacitances, which mean equal voltage on them and equal axial spacing between the end of layers.

25.4 Constructional Details and Main Parts of Bushing

25.4.1 Core

The core of bushing consists of a hollow or solid metallic tube over which high grade electrical kraft paper is wound. For condenser cores, conducting layers of metallic foil are introduced at predetermined diameters to make uniform distribution of electrical stress. The winding of the condenser core is done in a dust-free chamber. The core is then processed; this comprises of drying in a high degree of vacuum (0.005 mm), and then impregnating with high quality, filtered and degassed transformer oil.

25.4.2 Porcelain

Bushing for outdoor applications are fitted with hollow porcelain insulators. The OIP bushings are provided with insulators, both at air and oil ends, thus forming an insulating envelope, and the intervening space may be filled with an insulating liquid or another insulating medium.

The function of an insulator is to resist flashover in adverse conditions. This is determined by

   (a) the profile of the dielectric.

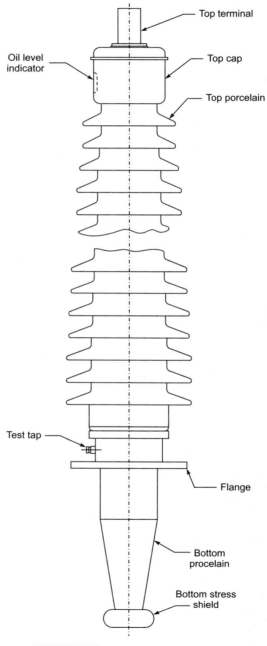

Figure 25.2 *Main parts of bushing.*

(b) the mounting arrangement of the insulator, i.e, vertical, horizontal, or inclined.
(c) the properties of the surface, i.e, hydrophobicity, toughness etc.

For insulators used in the air-end side, a minimum specific creepage distance is required for different pollution levels, as given in Table 25.1.

Table 25.1

Pollution level	Creepage distance
I Light (areas without industries, agricultural or mountain areas)	16 mm/kV
II Medium (areas with industries not producing pollution smoke, low density of houses, areas not exposed to the sea coast)	20 mm/kV
III Heavy (areas with high density of industries, high density of heating plants producing pollution, areas close to the sea)	25 mm/kV
IV Very heavy (areas close to the coast and exposed to sea spray, desert areas, areas exposed to strong winds carrying salt and sand, etc.)	31 mm/kV

Based on the above information, the shapes of porcelain sheds are:
 (a) Plain
 (b) Anti-fog
 (c) Alternate

25.4.3 Top Cap

This is a metallic housing for the spring pack. It serves as an in-built oil conservator to cater for oil expansion, and has an oil level indicator. In many cases, it also serves the purpose of a corona shield.

25.4.4 Mounting Flange

This is used for mounting the bushing on an earth barrier, such as a transformer tank or a wall. It may have the provisions for following:

570 *Transformers*

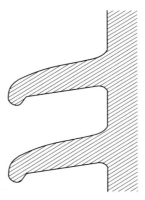

Plain or normal shed

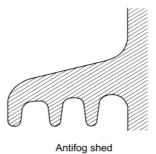

Antifog shed

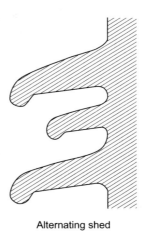

Alternating shed

Figure 25.3 *Shapes of porcelain sheds.*

(a) CT accommodation length
(b) Rating plate giving the rating and identification details of bushing
(c) Test tap
(d) Oil drain plug for sampling of oil
(e) Air release plug

The design of the flange and top cap is such as to minimize the loss due to hysteresis and eddy current effects. When heavy currents are being carried, this loss raises the temperature of the flange and top cap to a noticeable extent. For heavy currents, ordinary cast iron material cannot be used, hence non-magnetic materials such as stainless steel or aluminum are used.

25.4.5 Test Tap

The test tap is provided for measurement of the power factor and capacitance of the bushing during testing and service. The test tap is connected via a tapping lead to the last condenser foil of the core within the bushing. During normal service, this tapping is electrically connected to the mounting flange through a self-grounding arrangement.

25.5 Testing of Bushing

To prove the design and quality of manufacture, bushings are subjected to type tests and routine tests as given in IS 2099 and IEC 137. Where the type tests are done to prove design features of bushing, routine tests check the quality of individual bushing.

25.5.1 Tan Delta and Capacitance Measurement

This test is probably the most universally applied test for all types of condenser bushings. The bushing is set up as in service connected to one arm of the Schering bridge. The voltage is applied in increasing steps, up to the rated voltage. Capacitance and tan delta values are recorded for each voltage. (For bushings, power factor and tan delta values may be regarded as identical.) Tan delta indicates the degree of processing of the condenser core in OIP bushings and also the moisture content.

25.5.2 Dry Power Frequency Voltage Withstand Test

This is the most common routine test used for various classes of electrical equipment. A specified power frequency voltage is applied for one minute. Under wet conditions, the power frequency test is a type test, and is done only for bushings having rating 300 kV and below.

25.5.3 Partial Discharge Test

Partial discharges are the localized electrical discharges within the insulation system, restricted to only one part of the dielectric material, thus only partially bridging the electrodes. They are due to:

(a) voids and cavities present in the solid and liquid dielectrics.
(b) surface discharges that appear at the boundary of different dielectrics.
(c) corona discharges, if strongly non-homogenous fields are present.

The continuous impact of discharges in solid dielectrics form discharge channels called treeing. Every discharge event causes the material to deteriorate by the energy of impact of high electrode ions, causing many types of chemical transformation.

25.5.4 Impulse Voltage Withstand Test

Lightning impulse is a type test, and is applicable for all types of bushings. The bushing is subjected to 15 full wave impulses of positive polarity, followed by 15 full wave impulses of negative polarity of the standard waveform 1.2/50 μs.

For bushings of rated voltage equal to or greater than 300 kV, the switching impulse voltage test is applicable. The impulse tests simulate more closely than the power frequency withstand test, the over-voltages likely to be seen by the bushing in the service.

25.5.5 Thermal Stability Test

The theory of thermal stability states that at an elevated operating temperature, stable thermal equilibrium is assured only when a maximum value of the sustained voltage characteristic of a particular bushing is not exceeded. The magnitude of this voltage serves as a representative measure of the thermal stability. The magnitude depends solely on the quality of the dielectric, its ambient temperature, and the manner in which it is internally cooled. The dimension of the body has no role to play. It has been considered unnecessary to specify thermal stability tests for OIP bushings owing to low dielectric loss. However, it is necessary in large bushings (greater than 300 kV) with high current, to pay attention in the design to the dissipation of the conductor losses which may be several times the dielectric loss.

25.6 Factors Affecting the Performance of Bushing

25.6.1 Dielectric Loss in Bushing

The principal factors governing the dielectric loss in OIP bushings are the quality of oil, and the manner in which the bushing is processed. The quality of dielectric is expressed not so much by the absolute magnitude of the loss as by their constancy with respect to time. The risk of ageing is considerable in OIP bushings. The oil molecule itself does not possess very high resistance, and it is particularly sensitive to high temperature. Low viscosity of oil allows particles of foreign matter and ageing products to migrate to the points at which the field concentration is greatest in an irreversible manner.

25.6.2 Internal Corona

Corona discharges attack the dielectric, and thus weaken the bushing electrically. The corona discharges crack the oil molecule. This produces flammable gases and unsaturated molecular residues, which accelerates the deterioration process. Even low discharge intensities can result in a breakdown after only a short time. Hence, it is essential that corona should not be allowed to persist for long.

25.6.3 Atmospheric Pollution

Failures are also caused due to pollution. When the layer of pollution on an insulator becomes wet, its resistance falls and a leakage current flows over the surface of the insulator. The density of leakage current is high in some parts, usually in the narrowest portions of the insulator. These regions dry more quickly than the rest of the surface, and their resistance increases. This results in the formation of dry bands around the insulator. The voltage on the insulator is virtually impressed across the dry bands, which results in a flash over across the insulator in due course of time.

25.6.4 Temperature Rise of Joints

The current carrying path in a bushing generally has 3–4 joints from the transformer winding to the top terminal. These joints are either soldered, brazed, or bolted. Any looseness will cause a rise in the temperature. If this is not timely checked, the excess heat will pass to the insulation body, ultimately causing damage to the bushing.

25.7 Condition Monitoring at Site

On principle, all solid and liquid insulating materials are exposed to a certain degree of ageing due to various stresses in service. Continuous monitoring at the field provides necessary data of insulation behavior. Safe running of the equipment can be ensured if corrective action is taken in advance, in case of any abnormality. The following can be monitored on site on a regular basis to avoid any mishap:

25.7.1 Monitoring of Dissipation Factor and Capacitance

The tan delta or dissipation factor is considered to be the most dependable parameter in bushing to assess its condition. The increase in value of tan delta as well as capacitance gives very useful information about any insulation deterioration. The measurement at site can be done with a precision Schering bridge at 10 kV by simultaneously balancing the bridge for capacitance and tan delta. The test is conducted in Ungrounded Specimen Test (UST) mode, which eliminates the losses going to the grounded portions of the bushing.

The measured values of tan delta and capacitance can be compared with the factory test results. Though the tan delta value measured at site may not match with the factory results, however, its limiting value for OIP bushings as per standards is 0.007. The value of capacitance may vary from manufacturer to manufacturer, depending upon geometry or construction of the equipment.

Every time the measurements are taken, the value should be compared with the previous readings. If the bushing is more than 5 years of service, an increase in tan delta up to 1% can be allowed. Increase in capacitance value by more than 5% of the original value should always be referred to the manufacturer.

25.7.2 Monitoring of Hot Spots

A thermo-vision camera can be used to monitor the temperature of the bushing, particularly near the top terminal. Abnormally high temperature indicates the following:

(a) Improper engagement of top terminal with pull-through connector.
(b) Looseness of terminal connector.
(c) Improper soldering of cable (lead) with cable adapter.

This may result in overheating or welding of the terminal. In some cases, melting of soldered joints is observed, resulting even in falling of the lead. It is therefore recommended that brazing be resorted to in place of soldering.

25.8 Dos and Don'ts for HV Condenser Bushings

25.8.1 Dos

(a) DO check the packing externally for possible transit damage before unpacking.
(b) DO unpack with care to avoid any direct blow on the bushing or porcelain insulator.
(c) DO store the bushing in a shed or covered with tarpaulin to protect it from moisture and rains.
(d) DO handle the bushing with manila rope slings without any undue force on the porcelain insulator.

(e) DO clean the porcelain insulator thoroughly before taking any measurement or mounting the bushing on the trans-former. Check also for any foreign body adhered to the bottom oil end porcelain.
(f) DO check the oil level by making the bushing vertical.
(g) DO check for leakage of oil from any of the joints. Each bushing is tested with the oil immersed in the oil tank, and so some traces of oil can be found, which actually is not leakage.
(h) DO check the tan delta and capacitance on mounted bushing with the jumper connection removed.
(i) DO maintain a log book of records of periodical checks as mentioned in the supplier's O and M manual.

25.8.2 Don'ts

(a) DO NOT unpack the bushing from the crate unless required to be mounted on the transformer.
(b) DO NOT use metal slings on porcelain and avoid undue jerks while handling.
(c) DO NOT store the bushing outdoors without any protective covering.
(d) DO NOT measure IR value or tan delta value without thoroughly cleaning the porcelain.
(e) DO NOT dismantle or attempt to repair the bushing without prior permission of the manufacturer.
(f) DO NOT fill the oil in the bushing without specific instructions from manufacturer.
(g) DO NOT climb the porcelain to tighten the top terminal. Use an elevator or separate ladder for this purpose.

CHAPTER 26

Computerization—A Tool to Enhance Engineering Productivity

R. Mitra
DGM/TRE

Transformers are custom designed products. Practically every order requires a separate design and a new set of drawings. Though certain assemblies/components have been standardized, these constitute a small part of the total design effort. A typical transformer may consist of as many as 3000 components/sub-assemblies, and the complete Engineering Information may require 500 Drawings/Bill of Material. The effect of this huge design/drawing work is that the issue of Engineering Information takes as much as 25% of the total delivery cycle of the transformer.

All the components/sub-assemblies/assemblies that go into a transformer have to be so designed that they can be assembled properly, any mismatch identified at the time of manufacture is difficult to rectify. In addition, all modifications carry a certain penalty in terms of the cost of the product as well as the time taken to manufacture. Since a large number of drawings have to be made, they are made simultaneously, adding to the chances of mismatch.

Any Engineering Drawing has, basically, two sets of dimensions that are the most important part of the drawing. These are

(a) dimensions which are independent variables, i.e., they are derived from the specification, rating, customers require-ment, etc.
(b) dimensions which are dependent, i.e., they are derived from either the independent dimensions—based on certain formulae, or are derived from certain standard practices or design rules.

Using this logic, it is possible, if the independent dimensions are available, to work out the dependent dimensions, and thus all the dimensions needed to make the drawing can be worked out.

If we take a top down view of the drawing process, we will see that an assembly consists of a number of components. Certain dimensions derived from the assembly are common to a number of components. When the drawings are being made, any mistake in the common dimensions, either for the assembly or the component drawing would create a mismatch. In the ordinary course, all drawings with common dimensions should be checked at the same time to ensure correctness. The fact that all such drawings may not be available, as well as the possibility of human error creates mismatches that have to be rectified later at a cost to the organization.

A new approach has been taken to minimize such errors. Computer programs have been developed, which take the dimensions identified as independent variables as inputs. The logic for calculating the dependent variables is built into the program, and the dependent variables calculated. Based on these dimensions, and the relations between them, a 3-D model of the component/assembly is generated by the program. Finally, the program also generates the 2-D Production Drawings. Since separate programs would cover different assemblies, all the common dimensions are available while making the component drawings. The chances of errors/mismatches are thus reduced. Since all the data required to make the Bill of Material are available, the program generates the Bill of Material also. The advantages of such a system are:

(a) By giving one set of data for an assembly, a number of drawings and the Bill of Material would be generated. Thus the time taken to make the drawings and the Bill of Material is reduced.

(b) Since the common dimensions would be given only once, errors due to mismatch is reduced.

(c) As a 3-D model would be available, the designers can rotate the model and visualize it from different angles and thus make improvements.

(d) Fouling between different components/assemblies, excess clearances would be visible on the computer screen in 3-D, allowing corrections to be made before the drawings are released, or components procured.

(e) The Bill of Material would be generated by the program, and can be stored in a database for use by the planning groups.
(f) Since all calculations are built into the program, there would be commonality of logic. Mistakes due to adopting the wrong formulae by an individual are therefore eliminated.
(g) Any changes in the design philosophy would entail modifications in the program and the need to ensure proper changes in the individual designs would be eliminated. This would ensure better design control.

Based on the above philosophy a **Parametric Drawing Generation System (SYPAD)** has been developed. The system is suitable for the design of transformers following BHEL's design methods. The system is very comprehensive and would generate the drawings and the bill of material for a large variety of transformers with different ratings.

The system has been designed with menu-driven prompts so that a user without any knowledge of 3-D modelling software or database packages can use the system. The only need is that the user knows how to use AutoCAD, which is a common CAD software. The basic modelling software used in the system is M/s Autodesk's Mechanical Desktop. The menu screens for making the system user friendly as well as the calculation etc., has been developed in Visual Basic, while Oracle is the backend database.

As an example of the increase in productivity, let us take the case of the model shown in Fig. 26.1(f). To make the complete assembly seven drawings of the individual components would have to be made. All seven drawings and the Bill of Material can be made through one program by giving 40 independent variables as input. The program calculates, using various formulae, about 160 different dimensions relevant to the seven drawings. One model of the assembly, and seven models of the individual components are generated and the 2-D Production Drawings and Bill of Material are automatically generated. The Billl of Material data is also stored in a database for future use. Normally, about three weeks would have been required to make all the drawings. Using the Parametric Drawing System, the same seven drawings can be made in two days.

580 *Transformers, 2/e*

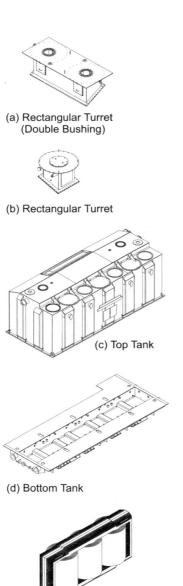
(a) Rectangular Turret (Double Bushing)
(b) Rectangular Turret
(c) Top Tank
(d) Bottom Tank
(e) Limb Core with Coil Assembly

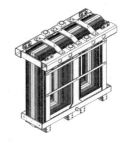
(f) Core and Endframe Assy

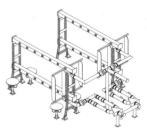
(g) Header Pipe Work

(h) Cooler Pipe Work

(i) Inside Mounting (Top and Bottom Yoke Shunt Assy)

Figure 26.1 *Shows some of the assemblies which have been developed through the system.*

Chapter 27

Condition Monitoring, Residual Life Assessment and Refurbishment of Transformers

C.M. Shrivastava
T.S.R. Murthy

The need for reliable and stable systems is being increasingly felt. Thus, emphasis is now being laid on 100% capacity utilization and the availability of equipment for reliable operation of the system. Fortunately, many tools are now available or underdevelopment, which can be used for condition monitoring of the transformer.

With a view to cover detailed information about DGA and Condition Monitoring, viz residual life assessment, this chapter has been divided into two sections:

Section A: Dissolved Gas Analysis Interpretation.
Section B: RLA and Refurbishment.

SECTION A
Dissolved Gas Analysis Interpretation

Incipient faults in oil filled transformers are usually the result of electrical or thermal stresses in either the transformer oil or insulating materials.

It is known that such excessive stresses produce a mixture of gases characterstics of which give an indication of the type of fault and location associated with the fault.

It is recommended that analysis of dissolved gases in transformer oil by gas chromatograph equipments is made at the time of commissioning and then every six months for transformers of 145 kV class and above.

27.1 Analysis Method

27.1.1 Sampling

Oil in transformers can be sampled through the drain or sampling valve near the bottom of the tank. Special care is to be taken not to introduce air, foreign matter, or dirty oil into the sampling container. For this purpose, first 0.5–1.0 liter of oil from the transformer is to be over-flown through the oil container. The oil sample must not be exposed to air before analysis.

27.1.2 Analysis of Fault Gases

The gases to be analyzed and the criteria for the gases found in the transformer oil are tabulated in Table 27.1.

➤ **Table 27.1** Gases to be Analyzed and Criteria

Sl. No.	Operating condition	Gases to be analyzed
1.	Normal	O_2, N_2, H_2, CO, CO_2, CH_4
2.	Abnormal	H_2, CH_4, C_2H_2, C_2H_4, C_2H_6
3.	Deterioration	CO, CO_2, CH_4

The generation of gases in oil by some typical faults in transformer active part models are shown in Table 27.2.

➤ **Table 27.2** Gas Content in Oil by Faults

Sl. No.	Type of faults	Decomposable gases in transformer oil
1.	Arcing in oil	CH_4, C_2H_4, H_2, (C_2H_6, C_2H_2, C_3H_6, C_3H_8)
2.	Overheating of solid insulating materials	CO, CO_2, (H_2, C_2H_4)
3.	Overheating of oil and paper combination	CH_4, C_2H_4, CO, CO_2, H_2
4.	Arcing of oil and paper combination	H_2, C_2H_2, CO, CO_2, (C_2H_4)

Note. ( ) shows gas contents which appear rarely.

To identify the nature of fault from the DGA results, the ratio technique of IS/IEC is used as per Table 27.3.

> **Table 27.3** DGA Findings and Inspection

Sl. no. Characteristics	Code of range ratios			Findings
	$\dfrac{C_2H_2}{C_2H_4}$	$\dfrac{CH_4}{H_2}$	$\dfrac{C_2H_4}{C_2H_6}$	
1. <0.1	0	1	0	
2. 0.1 to 1	1	0	0	
3. >3	2	2	2	
4. No. Fault	0	0	0	Normal ageing
5. Partial discharges of low energy density	0	1	0	Discharges in gas filled cavities resulting from incomplete impregnation or super saturation or cavities or high humidity.
6. Partial discharges of high energy density	1	1	0	As above, but leading to tracking of perforation of solid insulation.
7. Discharges of low energy	1 → 2	0	1 → 2	Continuous sparking in oil between lead connection of different potential or to floating potential. Breakdown of oil between solid materials.
8. Discharges of high energy	1	0	2	Discharges with power flow through arcing. Breakdown of oil between winding or coils to earth. Selector breaking current.
9. Thermal fault of low temperature <150 °C	0	0	1	General insulated conductor overheating.
10. Thermal fault of low temperature range 150 °C to 300 °C	0	0	2	Local overheating of core due to concentration of flux. Increasing hot spot temperature varying from small hot spots in core, shorting links in core,

(Contd.)

> **Table 27.3** *(Contd.)*

Sl. no.	Characteristics	Code of range ratios			Findings
		$\dfrac{C_2H_2}{C_2H_4}$	$\dfrac{CH_4}{H_2}$	$\dfrac{C_2H_4}{C_2H_6}$	
					overheating of copper due to eddy currents, bad contacts/joints (pyrolitic and carbon formation) up to core and tank circulating current.
11.	Thermal fault of medium temperature range (300°C to 700°C)	0	2	1	-do-
12.	Thermal fault of high temperature range >700°C	0	2	2	-do-

Depending upon the fault gases ratio obtained in the DGA analysis, the intensity of inspection is established. It is recommended that in case of minor change in DGA, results, at least three readings (tested in the same lab) are to be analyzed to arrive at a better conclusion.

27.2 Physical Inspection

Before taking total outage of transformer following preliminary physical inspection, tests should be carried out to establish the reason for the increasing trend of fault gases.

27.2.1 Physical Preliminary Inspection

Sl. No.	Description	Status	
		OK	Not OK
1.	Check for any hot spot developing with the help of thermovision camera.		
2.	Monitor temperature of tank, rim bolts and turret bolts.		

(Contd.)

Condition Monitoring, Residual Life Assessment

Sl. No.	Description	Status	
		OK	Not OK
3.	Conduct PD Measurement by acoustic meter.		
4.	Check core isolation with yoke clamp and tank.		
5.	Check BDV and PPM of transformer oil.		
6.	Check bushing terminals (HV and LV) for any looseness/heating mark, etc.		
7.	Study the temperature log with relation to loading of transformer.		
8.	Check cooling equipment like oil pumps and fans for proper flow of oil and air.		
9.	Check valve circuitry of cooler pipe work and radiator, etc., for proper opening.		
10.	Conduct FRA test if previous benchmark data is available for comparison.		

In the second stage for finding the root cause and doing remedial rectification, the following points should be checked through inspection covers after draining out oil from the transformer.

27.2.2 Core

Sl. No.	Description	Status	
		OK	Not OK
1.	Carry out general inspection inside the tank to see any abnormal heating/hot spot, etc.		
2.	Check for connection and single point earthing of top and bottom yoke shunt.		
3.	Check proper earthing of wall shunt.		
4.	Check for proper connection of earth shield provided on end legs, yoke, etc.		
5.	Core isolation between: (a) Clamp to core (b) Clamp to tank (c) Core to tank (d) Yoke bolt to core (e) Yoke bolt to clamp		
6.	Check isolation between end frame to end frame (400 kV transformers only).		
7.	Check condition of cross beam and end plate area, tapchanger support and cover locking, etc.		

27.2.3 Terminal Gear

Sl. No.	Description	Status OK	Not OK
1.	Check HV/LV line lead connection for heating and burning marks.		
2.	Check bus-bar to bus-bar joint for proper tightening and brazing.		
3.	Check for proper tightness of LV flexible jumpers.		
4.	Check all tapping leads connected on tap-changer for proper tightness and heating impression.		
5.	Check earthing connection of shield provided on turrets.		
6.	Check all barriers for heating and black soothing mark if any.		

27.2.4 Winding

Sl. No.	Description	Status OK	Not OK
1.	Check lead connection after removing paper insulation for proper metalized shielding and brazed joint.		
2.	Check line lead duct for black soothing mark/carbonization.		
3.	Check general condition of barriers provided on winding.		

27.2.5 Bushings

Sl. No.	Description	Status OK	Not OK
1.	Check oil level in bushing conservator.		
2.	Check test tap connection and earthing.		
3.	Carry out physical inspection of oil end of bushing.		
4.	Check proper tightness of terminal provided on top of bushing.		
5	Check any tracking mark on oil end of bushing.		
6.	Check condition of corona shield and fitting.		

27.2.6 Tap Changer

Sl. No.	Description	Status	
		OK	Not OK
1.	Carry out physical inspection of selector terminal connection.		
2.	Check proper tightness of leads connected on selector.		
3.	Check condition of fixed and moving contacts of selector for arcing/carbonization.		
4.	Check for proper alignment of selector assembly.		
5.	Check condition of oil filled in diverter tank.		
6.	Check for any leakages through diverter tank to main tank.		
7.	Check for any damage in diverter switch assembly.		
8.	Check the continuity of diverter transition resistor.		
9.	Check healthiness of copper braid connection.		

27.2.7 Insulating Oil

Sl. No.	Description	Status	
		OK	Not OK
1.	Check BDV and PPM of oil.		
2.	Check tan delta and resistivity of oil at 90°C.		

Note: (a) Internal inspection should be carried out in presence of the manufacturer's representative.

(b) It is possible that some of the checks as listed above may not be accessible without untanking the transformer. All such activity points are to be checked after untanking, and if required the coil assembly is to be removed. All above activities shall be carried out strictly under supervision of the manufacturer's representative only.

SECTION B
RLA and Refurbishment

27.3 Residual Life Assessment (RLA)

Presently, most of the electrical utilities carry out Dissolved Gas Analysis (DGA) of the transformer oil, apart from routine electrical, physical and chemical tests on oil to determine the health of the transformer. However, the DGA gives indication only regarding the incipient fault related to corona, arcing, or overheating that takes place in the transformer. It does not give any conclusive evidence related to the degradation of cellulose paper, and thereby does not give information regarding the health of the transformer winding or remaining life of the transformer. Therefore the Residual Life Assessment (RLA) study is carried out to predict the health of the transformer insulation and remaining life of the transformer.

It is difficult to assess the life of the transformer because of the complex behavior of insulation. The ageing of insulation in transformers is influenced by short term and long term overloads, number and intensity of short-circuits, incidence of lightning, and internal faults.

The ageing behavior is likely to be different for different types of transformers. The life span of the transformer, thus, depends initially on the design and quality of manufacture, and later on service conditions and maintenance standards.

These factors vary considerably and affect the useful span of service life, which therefore needs to be taken into account for residual life assessment. The recent development of various techniques for detecting the incipient fault conditions have improved to some extent the life expectancy of transformers, by resorting to corrective actions in time.

During the natural ageing of transformers, the insulation of winding, which is cellulose paper, deteriorates. Such deterioration can be well assessed by subjecting the cellulose paper to various tests. But the windings of oil filled transformers are generally inaccessible. However, by draining out the oil, an inspection of the outer layer is possible though it is tedious and cumbersome. It is necessary to assess the condition of the cellulose insulation at the hot spot developed within the winding, which is inaccessible. An easy method of finding the integrity of winding insulation is by chemical analysis of the oil in which the winding is immersed.

Cellulose insulation degrades due to heating or electrical breakdown, resulting in the production of furfural derivatives, which dissolve in oil. Hence, the chemical analysis of the transformer oil gives evidence of changes that are taking place in the winding insulation during operation.

The constituents of dissolved gases in oil are utilized in the early detection of incipient fault within the transformer. From the ratios of particular pairs of gases, the fault conditions are diagnosed. It is required to keep a record of gas analysis data from the date of installation, and provide the same for the assessment. In addition to the above, it is recommended to keep conventional test data and the history of transformer in the format provided as per Appendix I.

An analytical technique has been developed to ascertain the remaining life of transformer winding cellulose-based insulation paper. The technique involves extraction of chemical compounds from transformer oil, its quantitive estimation using High Performance Liquid Chromatography (HPLC), conversion to degree of polymerization, and computation of the life. In addition to the above, fault gases are analyzed using expert system software, along with electrical and physiochemical properties of transformer oil monitored from the date of commissioning and service life.

During the life assessment studies, the transformer oil samples drawn are analyzed for all conventional properties in addition to special properties. These inputs are then compared with the data bank available. Depending on the trends observed, recommendations are made about the status of the transformer in terms of residual life and/or need for replacement/reclamation of oil, and frequency of the transformer oil analysis.

27.4 Different Techniques for Life Estimation

Different techniques have been developed to estimate the life of transformer winding insulation. Some of the methods are mere extensions of conventional analytical techniques in terms of data interpretation. The techniques generally followed are summarized below:

27.4.1 Estimation from CO_2 and CO

Dissolved Gas Analysis (DGA) is the most important parameter, which helps in identifying the faults inside the transformer. From the quantities of CO and CO_2, the status of paper insulation is estimated by floating ARV MAP graphs. These ratios are to be supplemented with the degree or polymerization of furfural values for estimation of life.

27.4.2 Estimation from Furfural

Cellulose insulation degrades due to heating or electrical breakdown, resulting in the formation of derivatives of furfurals which are soluble in oil. However, 2-furfural dehyde is the major constituent of the liberated furfural derivatives. By monitoring the levels of 2-furfural dehyde by High Performance Liquid Chromatography (HPLC), and comparing those values with the data bank generated for the transformers of various ratings and age, the remaining life can be assessed. The method has a limitation in terms of solubility of furfural dehyde in transformer oil, viz. its generation, and has to be complemented by other techniques.

Oil Sample Collection

The oil sample is generally taken from the bottom valve of the transformer. Depending on the oil quality and furfural content, it can be decided whether collection of further oil samples for ascertaining the status of transformer are required. Approximately five liters of the oil are to be collected in stainless steel bottles/glass bottles by following relevant standard procedures. Oil samples may be required a minimum three times of over a period of 6–8 months time during the transformer in service, to ascertain the rate of rise of furfural.

Paper Sample Collection

The paper sample must be collected from the transformer after shutdown from lead insulation of HV and LV side after opening the transformer and lowering the oil level. Also, retaping of the paper with fresh paper on the removed portion must be done. The size of paper required for testing of DP is approximately 30 cm^2, weighing around 400 mg to 500 mg.

Compilation and Issue of Test Report

The results of samples collected can be analyzed, indicating the residual life of the transformer.

27.4.3 Estimation from Degree of Polymerization

The Degree of Polymerization (DP) is a direct parameter, which has been established as a linearly varying identity under thermal stress. Hence, a measure of DP of winding insulation paper can give the remaining life. However, such paper samples are inaccessible. DP values of paper samples from the next available location, that is lead insulation, are measured. Hence, the analytical data from this method has to be interpreted along with furfural values measured from the transformer oil.

27.5 Methodology Adopted

The remaining life of the transformer can be estimated by measuring the furfural derivative from oil and the degree of polymerization (DP) of cellulose paper from the lead insulation of windings. The life of the transformer is basically the life of the winding insulation. Normal and abnormal stress on winding results in overheating of the oil or winding insulation paper. Hence, levels of deterioration are monitored by the estimation of furfural from transformer oil, and the degree of polymerization (DP) from lead insulation paper.

The High Performance Liquid Chromatography (HPLC) technique is employed for estimation of furfural content after extraction from the transformer oil sample with a suitable solvent. Since it is the degree of polymerization (DP) which determines the age of the transformer, the furfural content measured is corroborated with the theoretical value of furfural calculated from DP measured for a few winding paper samples of the same transformer. The degree of polymerization is measured for the cellulose paper sample, preferably drawn from all HV and LV leads.

The method involves synthesis of a special solvent, dissolution of paper in solvent, measurement of viscosity, and its conversion to DP. The DP value thus computed is compared with the DP from furfural estimation.

In addition to the above, special tests on the transformer oil samples drawn are conducted for dissolved gas analysis (DGA) and the conventional test for physiochemical and electrical properties as indicated below. The data generated can be logged on to the expert system diagnostic program to ascertain the transformer condition in terms of stresses it has undergone during service.

Final recommendations can then be made, based on the above analysis regarding the residual life and reconditioning/ replacement requirements.

27.6 Conventional and Special Tests on Oil Samples

27.6.1 Physical Tests

  (a) Specific gravity ASTMD 1298–85 (IS 1448-P-16) 1967
  (b) Refractive index ASTMD 117–89
  (c) Interfacial tension ASTMD 971–91 (IS 6104–1971)
  (d) Kinematic viscosity – IS 1448P–25 1960
  (e) Flash point IS 1448P–10 1971

27.6.2 Chemical Tests

  (a) Total acidity ASTMD 974–92
  (b) Sludge content ASTMD 1698–84
  (c) Moisture content ASTMD 1533–88

27.6.3 Electrical Tests

  (a) Tan delta ASTMD 924–92
  (b) Resistivity
  (c) BDV ASTMD 1816–84 a

27.6.4 Special Tests

  (a) Rotary Bomb Oxidation test RBOT ASTMD-2212
  (b) Elemental analysis
  (c) Dissolved gas analysis (DGA) ASTMD 361–93

(d) DGA Interpretation – IEC 60599/IS 10593/IEEE Std.62–1995, ROGERS RATIO METHOD
(e) Furfural content – IEC 61198
(f) Degree of polymerization (DP) test—to be done on availability of paper sample as per IEC 60450 and calculation of furfural content in oil from DP value

27.7 Refurbishment

The transformer is to be withdrawn from service as per planned shutdown. For refurbishment, the core and coil assembly is taken out of the tank. The time required for refurbishment (servicing/reinstalling and commissioning) of the transformer can add up to to about four weeks.

27.7.1 Activities Involved

- Low voltage tests, i.e.
    (i) Winding resistance.
    (ii) Magnetizing current.
    (iii) Magnetic balance.
    (iv) Ratio.
    (v) Insulation resistance.
    (vi) BDV of oil.
- Draining of complete oil.
- Removal of bushings, turrets and pipe work.
- Shifting of transformer to service bay nearby where servicing is planned.
- Removal of tank cover.
- Lifting of core and coil assembly.
- Cleaning and removal of deposit, dirt and sludge.
- Checking of yoke bolt and core insulation.
- Tightening of terminal gear and core.
- Servicing of tapchanger and replacement of defective components.
- Retanking of core and coil assembly in tank and refitting of cover.
- Vacuum pulling and dry-out.

594   *Transformers*

- Rewrite as Vacuum pulling and dry-outs.
- Shifting of tank to plinth.
- Assembly of turrets and bushings and other accessories.
- Oil filling, pressure testing for ensuring that there is no oil leakage, and final dry-out by hot oil circulation.
- Final painting of transformer and accessories.
- Pre-commissioning checks.

27.7.2 Facilities Required

- Service bay with mobile crane or mobile covered shed, fabricated with scaffolding pipes and GI sheets, tarpaulin, rails, rollers, etc.
- Steel tray made of sheet.
- N_2 gas of 99.9% purity and dry –50°C dew point.
- Industrial concealed heaters.
- Vacuum pumps sufficient to maintain vacuum in the tank up to 0.5 torr.
- Oil filter machine, high vacuum 4500/6000 lit/hr.
- Oil storage tank to accommodate the total quantity of transformer oil.
- Mobile crane suitable to lift the core and coil assembly from tank.
- Platform/ladders.
- Hand tools, electrician tools.
- Testing equipment for dielectric test, water content, ratio, resistance of winding, megger, multimeters, etc.
- Skilled and unskilled manpower.

27.7.3 Materials Required for Opening and Overhauling the Transformer

- One set of complete erection gasket.
- All manufacturing gaskets related to tank assembly.
- Any other item found defective to be replaced.

27.7.4 Advantages of Refurbishment

- Pre-determination of incipient pre-matured fault (if any).
- Improvement in cooling due to removal of deposit, dirt and sludge.

- Ensuring tightness of all bolted joints.
- Reduction in noise and vibration after tightness of core clamp and bolts.
- Re-conditioning of insulating oil.

Many transformers designed and manufactured 15 to 20 years back may not be able to perform satisfactorily because of the auxiliaries, fittings, and terminal equipments with old technologies. To improve their performance, it is possible to replace these vital fittings with fittings of the latest technology. Important fittings which can be changed during refurbishment are:

- Bushings
- Tapchangers
- Buchholz relay
- Oil conservator with Atmoseal
- Oil temperature indicators
- Winding temperature indicators
- *Removal of tertiary winding:* Tertiary winding is one of the weak parts in transformers. Many transformers have failed due to tertiary. As per latest CBIP recommendations, it is not mandatory to provide tertiary winding for transformers up to 100 MVA. This will not only help increase the reliability of the transformer, but will also provide extra space for uprating of transformer capacity.
- *Redesign windings*: The windings are a vital part of the transformer. With the advent of latest type of windings, like interleaved disc and partially interleaved disc windings, improved quality of insulating materials and transformer oil, and computer aided design techniques for analysis of the transformer under transient and power frequency conditions, it is now possible to optimize the winding design and provide 10% to 50% more capacity in the same window as available for old transformers.

27.7.5 Methodology Proposed for Refurbishment

- Use the existing core, tank and fittings. These are reusable without having an adverse effect on the performance under uprated condition. These cover about 50%–60% of the cost of the transformers.

- Redesign the windings with the available computer-aided tools/technology so that it is able to meet the latest IEC requirement withstand capability. Modify the clamping arrangement to increase the short-circuit power frequency conditions strength.
- On the existing tank, provide extra cooling arrangement to meet the dissipation of extra losses due to increased capacity, and meet the oil and winding temperature rises within limit. Considering the high return of scrapped copper winding, the total cost of uprating is about 40–45% of the cost of new transformers of the same rating.

27.8 Conclusion

Due to paucity of funds, it is not practical and economical to replace old units with new ones. In such cases, refurbishing/retrofitting is an economical and viable alternative. The main advantage of refurbishment on site is that there is no need for rail or road transport. Consequently, the transportation cost and risks are eliminated by RLA and refurbishment. The transformer life is extended, and the transformer made suitable for providing satisfactory services/performance. Uprating of the transformer along with refurbishment helps the user in meeting system demands uninterruptedly.

Appendix I

Format for Data Collection History of Transformer

1. Type of transformer :
2. Rating :
3. Date of manufacture :
4. Sl. No. and make :
5. Date of commissioning and location :
6. Insulation details :
   (a) Weight of oil
   (b) Weight of paper
7. Type of winding used (tick) : Spiral/Cont.Disc.
8. Type of cooling :
9. Make of oil used :
10. Quantity of oil used :
    (a) In tank
    (b) In conservator
11. Date and number of topping ups done :
12. Quantity of oil used for each topping and last date of topping :
13. Details of breakdowns (minor) :
14. Details of breakdowns (major) :
15. Temperature data :
    (a) Oil temp.
    (b) Winding temp.
16. DGA and other data
    H_2, CO, CO_2, CH_4, C_2H_6, C_2H_4, C_2H_2
    Oil parameters
    Breakdown voltage (kV) :
    Moisture (PPM) :
    Resistivity X 10^{12} (ohm.cm) :
    Tan delta (90°C) :
    Total acidity (mg/g) :
    IFT at 27 °C (N/m) :
    Flash point (°C) :
    Furfural content :
17. Remarks for any follow-up :

Signature:
Name:
Designation:

Date:

CHAPTER 28

Transformers: An Overview

S.N. Roy
P.T. Deo

In the previous chapters, various technological aspects of transformers, starting from core design to final testing and commissioning at site is explained in detail. Though a transformer is a simple and static electrical equipment, its outage can cause interruption to all spheres of daily life.

Specification is the basic document on the basis of which the design is finalized, and final product is realized. To begin with, the user of the machine has to study all site condition requirements and spell out the critical parameters. The basic data requirement is defined in various chapters and has to be strictly followed. On the basis of this basic requirement, the design of a transformer is conceptualized. In this era of computers, most of the design is completed by feeding the input data. Since the programs are already in use, the logics behind these programs are not realized by the new engineers. It is necessary that a fresh engineer must study the formulae in detail, and try to understand the logic which has been used in the programs.

Next is the incoming material for which different manufacturers are already available in the market. The need for vendor development is true today from a quality point of view as well as prompt delivery. In an organization, methods and tools are available to locate a vendor and then evaluate the work from a technical competence and quality point of view. The raw material has to be arranged in a way that the different assemblies and sub-assemblies are completed for realizing a final product. The major activities which constitute formation of a transformer are core, winding, tank, tapchanger, bushing, cooling equipment, etc. So, for all such sub-groups, one has to look for raw material, working platform/machines, skilled manpower, etc. Each activity has to be critically planned and monitored at every stage.

Producing a quality transformer is the need of the day and the factors which can give reliable service can be summarized as below:

(a) The specification should cover the system requirement and any other non-standard duty-performance that the trans-former has to see.

(b) The designers should make use of all combinations of duty cycle requirements and put these into a computer program to generate the most economical design. The facility available with the manufacturer for producing core, winding, assembly, and processing must be taken into account. Otherwise, the finalized design may not be possible to handle in the shop floor.

(c) The selection of components like bushing, tapchanger, protective devices, etc., must take care of the over-voltage, transient overloading and other such conditions.

(d) The raw material and component sourcing must be done from limited vendors which meet the specification requirement in totality. To meet the short delivery of transformers, the green-channel vendor development and supply may be considered. This means that the vendor is capable of producing in accordance with the job specifications and quality assurance procedures, and hence quality control checks at every stage can be withdrawn.

(e) A well-trained workforce is necessary to understand the complex drawings, and produce the equipment. A slight deviation unnoticed shall make the transformer susceptible to failure at the test bed or in service prematurely.

(f) Cleanliness in the shop floor particularly insulation, winding, assembly, etc., must be maintained. Dirt particles, which can also be conductive, are extremely harmful for transformers, which are normally high voltage equipment.

(g) The insulation items used in transformer winding, core, assembly, etc., are mostly hygroscopic in nature. Processing of raw material before starting manufacture of insulation items can give a very stable dimension to these items. The items used directly within the winding play an important role for dimensional stability. Any difference in the dimension of

windings other than as specified by the designer, can produce extremely high amounts of short-circuit forces.
(h) System should prevail in every sphere of design, production, testing, etc. Every design rule, operation instruction, etc., must be documented, and made available at the work place.
(i) There should be a system for feedback from the shop floor, testing, and the customer, and records maintained in a well-formatted data bank format. Such information helps in re-looking into the design for any corrective measures and improvements from a service reliability point of view.

For transformers with higher ratings with respect to MVA and kV, even a minor deviation if not detected at the processing stage may create problems during final assembly and testing. The process can be either design, raw material, manufacturing, assembly, or testing. It is imperative to study all these processes into the minutest detail. We call such study of the processes as "Process Mapping." Under this head, each activity is critically examined. The areas can be defined as below:

(a) Specification and Design.
(b) Manufacturing.
   (i) Core
   (ii) Winding
   (iii) Assembly
   (iv) Processing
   (v) Testing and dispatch
(c) Site assembly, pre-commissioning.
(d) Loading to the capacity and feedback.

The specification which defines the basic requirement of the equipment and site condition is the tool to start a design. Knowing the capitalization rates of no-load loss, load-loss, and auxiliary loss, the initial design work starts. One is supposed to know the transport limitation with respect to dimensions and the weight which can be handled. A compromise is made with respect to the material cost and loss-capitalized cost, and the design is finalized. A low loss design shall need more weight of material, i.e., core, copper, oil, etc. But the transport restrictions imposed may cause a re-look into the design and make a final feasible design from all aspects. One can look into the strengthening of the small road bridges, etc. from transport point of view. A thorough route survey is undertaken in such a case.

The manufacturing process for the sub-groups as defined in (b) can be looked into in detail. The data bank and feedback of the shop floor for a minimum of 24 months becomes a tool for taking up the corrective measures. The non-conformities are tabulated, and the critical failures identified. A group can be formed with design, production, technology, and quality-control, to review these non-conformities. The reason is identified and the corrections are put into the new technology/process specification. For success, a process-owner can be identified. He can take the lead role in corrective measures and final implementation.

After final testing at works, the site assembly and commissioning becomes quite important from an operation point of view. Only qualified technicians should be allowed to handle the transformer at the site. Most of the transformer manufacturers have their field checkpoints. During assembly, all such checks should be verified and recorded in a format. Any deviation must be critically examined, and referred to the manufacturer for final acceptance or corrective action. The check-list jointly signed by the supplier and the owner shall form as the base document. This can be maintained as a history-card. After final commissioning, loading to the required capacity is done. In case of any interruption or maloperation of the equipment, the same must be immediately referred to the manufacturer. These incidents can be recorded in the history-card, which shall give the overall view of this vital equipment.

For easy access to the manufacturer and to seek urgent attention, it is now possible to establish a centralized contact station (website), and all the users can be connected directly on line. Even the users at remote areas can make use of the latest advances of information technology to get prompt services of the manufacturer.

The transformer industry is now more than 100 years old. We at BHEL have produced more than 3700 transformers, contributing approximately 230000 MVA to the power system in India and abroad. Producing these units is a craftsman's job since every stage of manufacture is manually accomplished. As such, it needs a highly skilled, trained, and motivated workforce to achieve the desired quality. In the present day scenario, it is possible to identify the technical institutions that can be involved with the training in different areas of manufacturing, i.e., core, insulation, winding and assembly, etc. This will help in generating qualified and trained manpower for the industry.

Finally, it is always possible to simplify the processes of the manufacturing stages rather than training our people for the complex processes generated by us. It is necessary to think in the direction of making our activities simpler. Maybe we can adopt the modular concept of design standardization and manufacture of the various sub-assemblies so that mistakes are minimized, and we produce a reliable piece of equipment for years to come.

CHAPTER 29

Solved Examples

1. A 60 MVA, 11/132 kV three-phase, 50 Hz, generator transformer has guaranteed parameters as, load loss 150 kW at full load at 75°C, no load loss 30 kW and percentage reactance 9% at rated load. Calculate

   (a) percentage regulation at

   (i) full load and unity power factor
   (ii) 3/4 load and 0.85 lagging p.f.

   (b) Efficiency at

   (i) full load and unity p.f.
   (ii) 3/4 load and 0.85 lagging p.f.

   (c) Maximum efficiency and load at which it occurs.

Solution

(a) (i) $r = (150 \times 100)/(60 \times 1000) = 0.25\%$, $x = 9\%$, $\cos \theta = 1.0$, $\sin \theta = 0.0$

Percentage regulation = $r \cos \theta + x \sin \theta + (x \cos \theta - r \sin \theta)^2/200$

$= 0.25 \times 1 + 0.0 + (9 \times 1 - 0.25 \times 0.0)^2/200$

$= 0.655\%$

(ii) $r = \{150 \times (0.75)^2 \times 100\}/(60 \times 0.75 \times 1000) = 0.1875\%$, $x = 6.75\%$

$\cos \theta = 0.85$, $\sin \theta = 0.53$

Percentage regulation = $0.1875 \times 0.85 + 6.75 \times 0.53 + (6.75 \times 0.85 - 0.1875 \times .53)^2/200$

$= 3.9\%$

(b) (i) Percentage efficiency = (output × 100)/(output + Cu loss + iron loss)

$= (60 \times 1000 \times 100)/(60 \times 1000 + 150 + 30)$

$= 99.70\%$

(ii) Percentage efficiency = $(60 \times 1000 \times 0.75 \times 0.85 \times 100)/\{60 \times 1000 \times 0.75 \times 0.85 + 150 \times (0.75)^2 + 30\}$

= 99.70%

(c) Maximum efficiency at unity p.f. at load = F.L. × √(iron loss)/(F.L. Cu loss)

= 60 × √(30/150) = 26.83 MVA

Maximum efficiency occurs when Cu loss = Iron loss hence, max. efficiency = (26.83 × 1000 × 100)/(26.83 × 1000 + 30 + 30)

= 99.78%

2. A 100 MVA, 220/66 kV, Y/Y, three-phase, 50 Hz transformer has iron loss 54 kW. The maximum efficiency occurs at 60% of full load. Find the efficiency of transformer at:

(a) full load and 0.8 lagging p.f.
(b) 3/4 load and unity p.f.

Solution

Cu loss at 60% of full load = 54 kW
Cu loss at full load = 54 × (1/0.6)² = 150 kW

(a) Percentage efficiency = (100 × 1000 × 0.8 × 100)/(100 × 1000 × 0.8 + 54 + 150)

= 99.75%

(b) Percentage efficiency = (100 × 1000 × 0.75 × 1.0 × 100)/{100 × 1000 × 0.75 × 1.0 + 54 + 150 × (0.75)²}

= 99.81%

3. A 33/11 kV, 50 Hz, single-phase transformer has core loss of 10 kW and exciting current of 0.9 amp, when its HV side is energized at rated voltage. Calculate the two components of the exciting current.

Solution

Exciting current Io = 0.9 A, supply voltage $V1$ = 3300 V, Core loss Pc = 10 kW

Hence core loss component $Iw = Pc/V1$

= (10 × 1000)/33000

= 0.303 A

and magnetizing component $I\mu = \sqrt{(I^2o - I^2w)}$

= √(0.9² − 0.303²)

= 0.847 A

Solved Examples

4. Two transformers of same voltage ratio, rated at 315 MVA each are connected in parallel to supply a load of 700 MVA at 0.8 p.f. lagging. The per phase resistance and per phase reactance of the first transformer are 2% and 11% respec-tively, and of the second transformer are 1.5% and 12% respectively. Calculate the load shared by each transformer.

Solution
Percentage impedance of first transformer $Z_1 = (2 + j11)\% = 11.18 \angle 79.69°$
Percentage impedance of second transformer $Z_2 = (1.5 + j12)\% = 12.09 \angle 82.87°$
$Z_1 + Z_2 = (3.5 + j23)\% = 23.26 \angle 81.35°$
Load $S = 700 \angle -\cos^{-1} 0.8 = 700 \angle -36.9°$
Load shared by first transformer $S_1 = S \times Z_2/(Z_1 + Z_2)$
$= 700 \angle -36.9° \times 12.09 \angle 82.87°/23.26 \angle 81.35°$
$= 363.84 \angle -35.38°$
$= 363.84$ MVA at 0.81 p.f. lagging
Load shared by second transformer $S_2 = S \times Z_1/(Z_1 + Z_2)$
$= 700 \angle -36.9° \times 11.18 \angle 79.69°/23.26 \angle 81.35°$
$= 336.45 \angle -38.56°$
$= 336.45$ MVA at 0.78 p.f. lagging

5. Calculate the percentage reactance of a 50 MVA, 33/220 kV, 50 Hz, 3-phase star/star connected transformer having the following manufacturing information: Volts/turn-90, height of HV and LV coil-2100 mm, radial depth of HV and LV coils-80 mm and 50 mm respectively, mean length of HV and LV turns-3.2 m and 2.2 m respectively and the radial gap between the HV and LV coils-90 mm.

Solution
The percentage reactance $X = 2\pi f \times \mu_0 \times L_{mt} (AT)/(L_c \times E_t) \times \{a + (b_1 + b_2)/3\} \times 100$
AT: current in HV $= (50 \times 10^3)/(\sqrt{3} \times 220) = 131.2$ A
turns in HV coil $= (220 \times 10^3/\sqrt{3})/90 = 1411$
hence AT $= 131.2 \times 1411 = 185123$,
$L_{mt} = (3.2 + 2.2)/2 = 2.7$ m
$X = 2 \times \pi \times 50 \times 4\pi \times 10^{-7} \times 2.7 \times 185123/(2.1 \times 90) \times \{0.09 + (0.08 + 0.05)/3\} \times 100$
$= 13.92\%$

6. Calculate the air core reactance of an HV winding of 40 MVA, 21/11.5 kV, 3-phase, 50 Hz, delta/star connected transformer having the following manufacturing details. HV coil: number of turns-260, inside diameter of coil-940 mm, radial thickness of coil-90 mm, and height of coil-1960 mm.

Solution

Air core reactance is calculated by the formula $X_a = 2\pi f L_{sat}$

where $L_{sat} = N^2 \times D_{wm} \times K$

where N = no. of turns in the coil

D_{wm} = inside diameter of coil

Factor $K = 9.9 \times 10^{-7}/(0.45 + \alpha + \rho + 0.5 \times \rho)$

where α = height of coil/inside diameter of coil

ρ = radial thickness of coil/inside diameter of coil

hence,

$f = 50$, $N = 260$, $D_{wm} = 940$, $\alpha = 1960/940 = 2.085$,

$\rho = 90/940 = 0.096$

$K = 3.70 \times 10^{-7}$

So, $X_a = 2 \times \pi \times 50 \times 260^2 \times 0.94 \times 3.7 \times 10^{-7}$

$= 7.40$ ohm

Percentage air core reactance = $(7.4 \times 100)/\{(21 \times 10^3)/634.92\}$

$= 22.37\%$

7. Show the thermal ability to withstand short-circuit of dura-tion 3 seconds of oil-immersed 50 MVA, 21/11.5 kV, 3-phase, delta/star connected transformer having 12% impedance. The Cu conductor cross-section area of HV and LV coils are 350 mm² and 1050 mm² respectively. The allowed temperature rise of the windings at rated condition is 55° C over ambient temperature of 50° C. The short-circuit apparent power of the HV system is 9500 MVA and of the LV system is 500 MVA.

Solution

Calculation method for thermal ability to withstand short-circuit of transformer is given in Cl. 4.1 of IEC 60076–5 (2000).

The symmetrical short-circuit current $I = U/\{\sqrt{3} \times (Z_t + Z_s)\}$ in kA

Where $Z_s = U_s^2/S$ in ohms

U_s = rated voltage of the system in kV

S = short-circuit apparent power of the system in MVA

U = rated voltage U_r of the winding under consideration in kV

$Z_t = (z_t \times U_r^2)/(100 \times S_r)$ in ohms
z_t = short-circuit impedance of transformer in percentage
S_r = rated power of transformer in MVA

The average temperature θ_1 attained by winding after short-circuit shall be calculated by the formula $\theta_1 = \theta_0 + \{2 \times (\theta_0 + 235)\}/[\{106000/(J^2 \times t)\} - 1]$ for Cu winding.

where θ_0 = initial winding temperature in °C
J = short-circuit current density in Amp/mm^2
t = short-circuit duration in seconds

HV winding

$Z_s = 21^2/9500 = 0.046$ ohms
$Z_t = (12 \times 21^2)/(100 \times 50) = 1.06$ ohms
$I = 21/\{\sqrt{3} \times (0.046 + 1.06)\} = 10.96$ kA line current
   $= 6.3$ kA phase current
$J = 6.3 \times 1000/350 = 18$ Amp/mm^2
$\theta_0 = 50°$ C $+ 55°$C $= 105°$C
$\theta_1 = 105 + \{2 \times (105 + 235)\}/[\{106000/(18^2 \times 3)\} - 1]$
   $= 111.3°$C (Allowed 250°C)

LV winding

$Z_s = 11.5^2/500 = 0.2645$ ohms
$Z_t = (12 \times 11.5^2)/(100 \times 50) = 0.3174$ ohms
$I = 11.5/\{\sqrt{3} \times (0.2645 + 0.3174)\} = 11.4$ kA line current
$J = 11.4 \times 1000/1050 = 10.9$ Amp/mm^2
$\theta_0 = 50°$C $+ 55°$C $= 105°$C
$\theta_1 = 105 + \{2 \times (105 + 235)\}/[\{106000/(10.9^2 \times 3)\} - 1]$
   $= 107.3°$C (Allowed 250°C)

Index

Acrylonitrile 67
Ambient, weighted 382, 384, 387, 389
Amorphous steel 84
Ampere turn
   lancing 205
   diagram 207, 208
Analysis of gaseous incipient fault 450
Analytical method 159
Angle ring 56
Arc suppression reactor 483
Askarel, see Oil
Auto transformer 23
Auxiliary limb 88, 91
Axial flux 177, 156, 160

Bell-shaped tank 272
Breather 260, 265
Buchholz relay 251, 253
Bushing
   D.C. 530
   and cable sealing box 257
   insulating material 565

Cable sealing box 257
Capacitance 118
   distribution 118
CBIP 371
CEA 370
Chopped wave impulse 117
Clamp plate 92

Coil
   assembly 169
   treatment 68
Cold rolled steel 86, 87
Combined fault scheme 463
Commissioning of transformer 447
Compressibility 57
Condenser
   bushing 563
   grading 564
Conservator diaphragm type 264
Construction
   features of traction transformers 489
   of tap changers 148
Continuous transposed conductor 59, 62
Control of onload tapchangers 151
Controlled proof stress copper 84
Conventional drying 278
Converter
   transformer connections 531–533
   transformer test 536–538
Cooling 177
   classification 177
Copper silver alloy 84
Copper weight 210
Core
   fault 463
   insulation 292
   less shunt reactor 476

Index

saturation 87
steps 199
types 200
Correction factor 414–416
Cross grain direction 76
Crystallite 68

Design features of traction
 transformers 489
Dispatch 439
Dielectric
 dissipation
  loss 129
 test 297
Differential protection 462
Diffusion coefficient 278
Dimensions of coil 204
Directed oil flow 125, 126, 127,
 178, 179
Dirichlet boundaries 133
Dissolved gas 280
 analysis 450
Driving force 195
Drying
 by stream line filter 447
 of transformer 446
Duplex reactor 467

Earth fault 459
Earthing transformer 483
Eddy current loss 21
Efficiency 22
Electric field poltting methods 130, 131
Electrolytic grade copper 62
Electromagnetic
 forces in winding 154, 158
Electromotive force 7
Electrostatic
 field plot 134
 shield 120
Energy dissipation in reactor 406

Epstein square 100
Equipotentials 52
Equivalent circuit 16–20
Explosion vent 257

Failure electric circuit 166
Faraday's law 6
Fault
 core 463
 delection 267
 earth 459
 interturn 461
 short circuit 456
 winding 461
Feeder reactor 466
Filtration 280
Finite element method 236
Finite difference method 131
Flange joints 272
Flash point 45
Flash test core insulation 292
Flashover 307
Flexible copper conductor 66
Forces
 electromgnetic 154–161
 mechanical 218
Frictional head loss 192, 194
Full wave impulse 117

Gap volume in reactor 481
Gas sealing conservator 260
Generator line reactor 466
Grain oriented steel 68, 87
Grapped core shunt reactor 477

Handling 440
Harmonics of no-load current 304
Heat carrier 281
Heat transfer in winding 125
Hi-B steel 69
High tensile strength steel 81
High voltage disturbances 457

Index

Hot spot 386
   temperature 92
Hygroscopic 274, 279
Hysteresis loss 21

Ideal transformer 7, 10
Image method 212
Impedance 205
Impregnated cellulose 84
Impulse
   calculation 218
   chopped wave 306
   full wave 306
   generation 309
   recording 315
   test 305
   circuit 309
   testing 305–320
   voltage measurement 305
   voltage wave shape 306, 309
   wave front 309
   wave shape control 309
Induced EMF 6
Induced over-voltage test 298
Inductance 118, 121
   distribution 118
Initial voltage distribution 118
Installation
   of bushing 443
   of transformer 441
Insulation
   ageing 379
   ageing law 381
   deterioration 378
   major 128
   minor 127
   pressboard 51, 128
   resistance test 297
Interfacial tension 43
Interleaved
   joint 100

   winding 113
Interphase transformer 513
Interpretation of oscillogram 318
Interturn failure 461
Interturn fault 461
Iron
   loss 210
   test 249
   weight 210

Ladder network 122
Lamination preparation 101
Lamination slitting 102
Laplace equation 131
Leakage flux 212
Leakage reactance 15
Lightning impulse test 306
Lightning surge 116
Line trap 470
List of standards 372–376
Load curve 385
Load loss 22
Loss measurement 345

Magnetic
   leakage 14
   yoke shunt 214
Magnetizability 68
Magnetization D.C. 529
Magnetizing
   current 7
   inrush current 463
Magnetostriction 68, 76, 78, 81
   effect on noise 76
Maintenance 449
   of transformer 449
Major insulation 128
Measurement
   of impedance 295
   of insulation resistance 297
   of load loss 295

of no-load current 296
of no-load loss 296
of winding resistance 294
Mechanical force 218
Metal limiter 67
Metallic glass 84
Method of potential connection 147
Minor insulation 127
Mitered joint 100
Moisture content 44
Montsinger relation 381
Motor starting reactor 468
Multiwinding transformer 122

Neaumann boundaries 133
Need for standardization 370
Neutral
    couplers 483
    earthing reactor 482
Neutralization value 45
Nitrile butadiene 66
Nitrile rubber bonded cork 66
Noise 76, 304
    vibration in reactor 479
No-load
    current 10
    loss 21
Non-magnetic inserts 217
Numerical method 159

Off circuit tap switch 138, 139
Oil 34, 42, 47
    base stock 42
    filling 445
    level indicator 256
    maintenance 279, 445
    preservation 258
On load tapchanger 141
Operating condition, abnormal 404
Operation control of tapchanger 151

Oscillograph recurrent surge 308
Over-voltage test 342–345
Overload effect 406
Oxidation stability 46

P.D.
    direct current 527
    direct current measurement 529
    measurement 321
    detection 329
Parallel operation 27
Partial discharge 321
Partially interleaved winding 114
Peltier effect 265
Permeability air 48
Phase sequence 31
Pitch of transposition 63
Polarity 30, 31, 292
    test 292, 537
Polymerization degree 47
Poly-vinyl acetal based enamel 62
Porcelain bushing 562
Pour point 45
Power frequency test 297, 298
Preliminary test 292
Pressure relief valve 255
Protection 456
    differential 462
    earth fault 459
    internal fault 462
    winding fault 461
Proximity effect 510
PTFE 66

Quality factor 30

Rabin's method 212
Radial
    flux 212
    force 155, 159
Ranga-Kutta method 120

Ratio test 292
Reactance 205–210
Reactor
  for capacitor bank 469
  magnetically shielded 475
Reconditioning of bushing 450
Rectifier circuit 499
Recurrent surge oscillography 308
Reed type switch 253
Regulating transformer 503
Regulation 21
Reinforcement of insulation 120
Resistivity 45, 51
Restricted e/f protection 461
Ripple 497
Rotary transposition 112
Roth method 163, 212
RSO technique 308

Sandwich coil 208
Sealing 66
Separate source withstand test 338
Series reactor 463
Series reactor general character 471
Series reactor oil immersed coreless 472
SF6 insulated 84
Shielded layer winding 115
Short circuit 78, 456
  fault 456
  forces 154, 168, 456
  inter-turn 461
  temperature rise test 300
Shunt reactor 473
Shunt reactor general character 475
Skin effect 509
Sludge and sediment 46
Smoothing reactor 469

Solvent 281
Specification transformer 368
Specification convertor transformer 535, 536
Stacking factor 87, 92
Standard specification of transformer 368
Stiffeners 230
Stiffeners matrix 240
Stored energy 121
Strain energy 237
Stray loss 22, 211, 213, 216
Stray loss control 214
Stress
  in pipes 192
  relief of lamination 102
  sensitivity 78
Structural steel 81
Superconducting material 84
Surge
  divertor 457
  voltage behavior 116
  voltage in winding 116
Switching impulse 306, 307, 313
Synchronizing reactor 468

Tank
  stress 248
  tests 244
  type 227
Tapchanger type 202
Tapchanging circuits 143, 144
Temperature
  alarm 253
  gradient 219
  indicator 253
  rise limits 300
  variation 384, 395
Terminal gear 136
  assembly 271
Tertiary winding 25, 26

Tests
    impulse voltage 345
    induced overvoltage 298
    core insulation 292
    impedance 295
    insulation resistance 297
    load loss 295
    no-load current 296
    no-load loss 296
    phase rotation
        polarity 292
    ratio 294
    separate source voltage 297
    winding resistance 294
    of reactor 335
Thermal
    conductivity 125
    head 180
    image 254
Thermosyphon filter 263
Third harmonic measurement 340
Three winding transformer 25
Tie line reactor 467
Top oil rise limit 300
Traction transformer 493
Transductors 517
Transformation
    ratio 9, 23
Transformer
    auxiliaries 251
    life 378
    utility factor 506
Transient over-voltage 116
Transposed conductor 62
Transposition 112, 217
Tuning, filter reactor 482

Types of traction transformers 486

Utilization factor 93

Vacuum
    dehydration 280
    impregnation 280
Vector group 32
Veneer 58
Vibration measurement 339
Viscosity 44
Voltage regulation, see Regulation

Water cooled transformer 178, 182
Weighted ambient 384, 389, 392
    calculation 386, 397
Winding
    assembly 269
    continuous disc 110, 111, 268
    crossover 113
    direction of coil 110
    fault 461
    gradient 302
    helical 110, 111, 268
    insulation 204
    insulation resistance test 297
    polarity 292
    preparation 268
    resistance test 294
    spiral 110
    temperature rise 301
    treatment 279
    types 201

Young's modulus 174

Zero phase sequence impedance 303